Prentice Hall Advanced Reference Series

Physical and Life Series

BINKLEY *The Pineal: Endocrine and Nonendocrine Function*
CAROZZI *Carbonate Rock Depositional Models: A Microfacies Approach*
CHAFETZ *Nutrition and Neurotransmitters: The Nutrient Bases of Behavior*
EISEN *Mathematical Methods and Models in the Biological Sciences: Linear and One-Dimensional Theory*
EISEN *Mathematical Methods and Models in the Biological Sciences: Nonlinear and Multidimensional Theory*
FRASER *Stratigraphic Evolution of Clastic Depositional Sequences*
HAZELWOOD *The Endocrine Pancreas*
JEGER, ED. *Spatial Components of Plant Disease Epidemics*
MCLENNAN *Introduction to Nonequilibrium Statistical Mechanics*
MEFFE AND SNELSON *Ecology and Evolution of Livebearing Fishes (POECILIIDAE)*
PLISCHKE AND BERGERSEN *Equilibrium Statistical Physics*
SORBJAN *Structure of the Atmospheric Boundary Layer*
VALENZUELA AND MYERS *Adsorption Equilibrium Data Handbook*
WARREN *Evaporite Sedimentology: Importance in Hydrocarbon Accumulation*

STRUCTURE OF THE ATMOSPHERIC BOUNDARY LAYER

Zbigniew Sorbjan

Associate Professor of Atmospheric Sciences
The University of Wisconsin, Milwaukee
Adjunct at the Warsaw Polytechnic University, Poland

PRENTICE HALL, Englewood Cliffs, New Jersey 07632

Library of Congress Cataloging-in-Publication Data

Sorbjan, Zbigniew.
Structure of the atmospheric boundary layer / Zbigniew Sorbjan.
p. cm.
Bibliography: p.
Includes index.
ISBN 0-13-853557-4
1. Boundary layer (Meteorology) 2. Boundary layer. I. Title.
QC880.4.B65S67 1989
551.5--dc19 88-27420
CIP

Prentice Hall Advanced Reference Series

Editorial/production supervision and
interior design: Merrill Peterson
Cover design: Karen Stephens
Manufacturing buyer: Mary Ann Gloriande

Cover art composed from sodar images
obtained by R.L. Coulter, Argonne National
Laboratory

A Division of Simon & Schuster
Englewood Cliffs, New Jersey 07632

Printed in the United States of America

10 9 8 7 6 5 4 3 2 1

ISBN 0-13-853557-4

Prentice-Hall International (UK) Limited, *London*
Prentice-Hall of Australia Pty. Limited, *Sydney*
Prentice-Hall Canada Inc., *Toronto*
Prentice-Hall Hispanoamericana, S.A., *Mexico*
Prentice-Hall of India Private Limited, *New Delhi*
Prentice-Hall of Japan, Inc., *Tokyo*
Simon & Schuster Asia Pte. Ltd., *Singapore*
Editora Prentice-Hall do Brasil, Ltda., *Rio de Janeiro*

CONTENTS

To my mother

PREFACE

This book discusses processes shaping the atmospheric boundary layer and methods of their mathematical modeling. It provides a theoretical and empirical background on the flow, turbulence, and diffusion in the lower portion of the atmosphere. The book is based on a series of lectures I presented at the Warsaw Polytechnic University, Poland and at the University of Wisconsin–Milwaukee.

The book addresses two groups of readers. First, it is designed as a textbook for senior undergraduate and graduate students in meteorology. The text is self-contained; however, a number of topics require previous introductory courses in mathematics, fluid mechanics, and atmospheric science. Some portions of this book go beyond the material recommended for a one semester course. This material addresses scientists in boundary layer meteorology and other relative fields who are interested in more detailed information about physical processes in the lower atmosphere.

The selection of material is always a difficult matter for a writer. The material included in this book is conceived by the author to be most important. However, many topics had to be excluded simply for the sake of internal harmony. Generally, the book focuses on the vertical structure of the atmospheric boundary layer. Phenomena connected with horizontal inhomogeneities or sea and land breeze winds, mountain and valley circulations, development of the internal boundary layers due to changes of surface roughness are excluded.

The book contains eight chapters. The first chapter begins with a brief description of the atmospheric boundary layer and ends with a few historical notes. Chapter 2 deals with the derivation and discussion of basic equations. Chapter 3 considers the structure of atmospheric turbulence and the mathematical methods of its description. A framework of the similarity theory is used in Chapter 4 to present results of atmospheric boundary layer measurements. A microstructure of the turbulent flow is discussed in terms of spectral analysis in Chapter 5. Chapter 6 deals with closure schemes of the Reynolds equations and numerical modeling of the atmospheric boundary layer. Chapter 7 outlines the basics of turbulent diffusion in the atmospheric boundary layer. Finally, Chapter 8 describes both the instruments and techniques used to detect basic characteristics of atmospheric turbulence.

Readers should benefit greatly from solving the problems presented at the end of each chapter. The problems are designed to illustrate and, in some cases, to extend the material presented in the text.

An Instructor's Manual is available to instructors using *Structure of the Atmospheric Boundary Layer*. It contains numerical solutions to 79 problems in the text and includes many figures, computer programs (written in BASIC), comments, and additional references. Solutions provided in this manual often include extra calculations and programs that can be used by instructors to develop more problems if necessary.

I would like to thank all of those who helped me prepare this book. I am indebted to Dr. J. C. Kaimal of the National Oceanographic and Atmospheric Agency (NOAA) in Boulder, Colorado; Dr. J. M. Schneider of the University of Oklahoma in Norman; and to Dr. D. H. Lenschow of the National Center for Atmospheric Research (NCAR) in Boulder, Colorado, for reading the manuscript and offering many helpful suggestions for improvement. I am also grateful to Dr. J. C. Kaimal, Dr. E. W. Eloranta, and G. D. Vassiliou of the University of Wisconsin in Madison; Dr. R. L. Coulter of the Argonne National Laboratory; and Dr. J. Wilson (NCAR) for providing photographs and materials; and to D. E. Oseen (University of Wisconsin–Milwaukee) for editorial help. Very special appreciation is addressed to Annabelle, Grace, and Charles Sherba for proofreading and editing the manuscript.

Zbigniew Sorbjan

Milwaukee, Wisconsin

LIST OF SYMBOLS

Vectors are denoted by bold symbols

A	similarity function
a	similarity function
$\mathbf{a}$	acceleration
B	similarity function
B_λ	filter function
b	similarity function
$\mathscr{B}$	spatial spectrum
C	similarity function (for temperature)
C	concentration
C_*	concentration scale
C	cospectrum
C_g	geostrophic bulk coefficient
c_p, c_v	specific heat at constant pressure and constant volume
C_p, C_v	heat capacities at constant pressure and constant volume
C_T	structure parameter for temperature
C_V	structure parameter for wind velocity
D	parameter, $D = \Delta / z_i$
D	similarity function (for humidity)
D_i	turbulent mass flux in x_i-direction
E	latent heat flux
E	turbulent kinetic energy
E	electric voltage
$\mathscr{F}$	Fourier transform
f	Coriolis parameter
f	nondimensional frequency $f = nz/U$
f_m	nondimensional peak frequency
G	heat flux to the ground
G	magnitude of the geostrophic wind
g	gravity acceleration
H	height of the atmospheric boundary layer
H	entropy
H_e	effective height of the source
$\mathbf{H}$, H_i	turbulent heat flux and its i-component
h	height of the stable ABL
$\mathscr{H}_{11}$, $\mathscr{H}_{22}$	longitudinal and lateral spectra
$\mathscr{H}$	spatial one-dimensional energy spectrum

Symbol	Meaning
I	intensity of radiation
i	$i = \sqrt{-1}$
i	electrical current
K	dimensionless eddy diffusivity
k_h	eddy diffusivity for the heat flux
k_m	eddy diffusivity for the momentum flux
k	wavenumber
L	Monin-Obukhov length
l	mixing length
L_E	Ekman's length scale
$\mathscr{L}$	integral length scale
M_x, M_y	dimensionless baroclinicity parameters
m	mass
N	Brunt-Våisålå frequency
N	destruction rate of the temperature variance
$N_{q\theta}$	destruction rate of the temperature-humidity covariance
N_q	destruction rate of the humidity variance
n	frequency in Hz
p	pressure
Pr	Prandtl number
Q	heat
Q	emission strength
Q	quadrature spectrum
Q_0	surface heat flux
q^2	half of the turbulent kinetic energy
q	specific humidity
R	individual gas constant, $R = 287$ J/kg/K
R	correlation function
R	dimensionless heat flux
R	ratio of heat fluxes on the top and the bottom of the ABL
Ra	Rayleigh number
Re	Reynolds number
R_f	flux Richardson number
Ri	gradient Richardson number
R_{ij}	Reynolds stress tensor
R_n	net radiation flux
Ro	Rossby number
R_u	universal gas constant
S	spectral density
S, S_*	bulk stability parameters
S_{ij}	rate of strain
T	absolute temperature
T_L	Lagrangian time scale
T_E	Eulerian time scale
T_*	temperature scale
$\mathbf{T}_*$	temperature scale in the stable outer ABL
T_v	virtual temperature
t	time
t_f	free convection temperature scale
$\mathscr{T}$	integral time scale
U	x component of the mean wind vector
$\mathbb{U}$	internal energy
U_1, U_2, U_3	components of mean velocity vector
$\mathbf{U}_*$	velocity scale in the outer stable ABL
$\mathbf{U}_b \mathbf{U}_t$	velocity scales in the convective ABL
u_f	free convection velocity scale
U_g	x component of the geostrophic wind
u_*	friction velocity
u_1, u_2, u_3	components of the velocity vector
u''	x component of the ageostrophic wind
V	y component of the mean wind vector
$\mathbf{V}$	mean wind vector

$\mathbb{V}$	volume
V_g	y component of the geostrophic wind
$\mathbf{v}$	wind velocity vector
v''	y component of the ageostrophic wind
W	mean vertical velocity
W_e	entrainment velocity
w_*	convective velocity scale
w_h	velocity on the top of the boundary layer
X	dimensionless component of the Reynolds stress
x	space variable
x_1, x_2, x_3	space variables of the coordinate system
Y	dimensionless component of the Reynolds stress
y	space variable
$\mathbf{Z}$	dimensionless height, $\mathbf{Z} = z/H$
Z	dimensionless height, $Z = z/L_E$
Z_T	dimensionless height of the stable ABL
z	height
z_i	inversion layer
z_0	roughness parameter
α	extinction coefficient
α	angle
α_0	surface cross isobar angle
α_θ	ratio of heat and momentum eddy diffusivities, reciprocal of the Prandtl number
β	buoyancy parameter, $\beta = g/T_0$
Γ	coherence
γ	temperature lapse rate
γ_a	adiabatic lapse rate
Δ	depth of the interfacial layer on the top of the mixed layer
$\Delta\Theta$	temperature jump on the top of the mixed layer
$\Delta U,\ \Delta V$	wind velocity jump on the top of the mixed layer
$\delta(x)$	Dirac's delta
δ_{ij}	Kronecker's delta
ε	dissipation rate
ε	error
ε_{ij}	dissipation rate of the velocity covariance $\overline{u_i'u_j'}$
ε_{ijk}	alternating unit tensor
ζ	dimensionless height $\zeta = z/L$
η	Taylor's microscale
$\eta_x\eta_y$	dimensionless components of the thermal wind
ϑ	reciprocal of the reference potential temperature, $1/\Theta_0$
Θ	potential temperature
Θ_m	mean mixed layer temperature
Θ_*	convective temperature scale
Θ_v	virtual potential temperature
Θ'	temperature fluctuation
Θ''	temperature deficit
κ	von Karman constant
Λ	Monin-Obukhov length for the stable outer ABL
λ	wavelength
λ_m	peak wavelength
μ	Kolmogorov's microscale
μ_h	stability parameter, $\mu_h = h/L$
μ_i	stability parameter, $\mu_i = z_i/L$

μ_0	stability parameter, $\mu_0 = L_E/L$
ν	coefficient of viscosity
ν_c	molecular diffusivity
ν_Θ	thermal diffusivity
π	$\pi = 3.1415\ldots$
ρ	air density
Σ	surface
σ	variance
σ_x, σ_y	dispersion coefficients
τ	time scale
τ	kinetic Reynolds stress
τ_{ij}	stress tensor component
$\tau_{1/2}$	half-life time
$\varPhi$	temporal spectrum
Φ	similarity function
ϕ	surface layer similarity function
Ψ	cross spectrum
ψ_E	three-dimensional energy spectrum
ψ	slope angle
Ω	Earth's angular velocity
ω	frequency, $\omega = 2\pi/\mathrm{T}\ (s^{-1})$
ω_j	component of the angular velocity vector
d / dt	total derivative
$''$	mesoscale perturbation from the reference state
$'$	turbulent fluctuation
$(\ )_0$	reference state valve
$\overline{(\ )}$	averaged valve
$\langle\ \rangle$	averaged valve
ABL	atmospheric boundary layer

CHAPTER ONE

INTRODUCTION

1.1 GENERAL REMARKS

The earth's atmosphere, like a giant heat engine, transforms available energy into the movement of huge masses of air. Practically all "fuel" for this engine is supplied by the sun. The contribution of all other sources (e.g., the earth's interior) is smaller than 0.02%.

The earth's atmosphere is semitransparent to incoming solar radiation. The atmosphere obtains only about 20% of its energy strictly by absorption. About 30% is reflected or scattered to space. The rest of the energy passes through the atmosphere and is absorbed by the surface. Later this energy is transferred back, primarily to the lowest kilometer of the atmosphere. This lowest portion of the atmosphere, which intensively exchanges heat as well as mass (water, pollutants) and momentum with the earth's surface, is called the *atmospheric boundary layer* (ABL). The atmospheric boundary layer has great practical and scientific importance. All human and biological activities take place in this layer. Mass and energy transfer within the boundary layer regulates a broad variety of processes in the entire atmosphere.

The flow in the atmospheric boundary layer has a *turbulent* character. *Turbulent motion* is understood to be an irregular condition of flow in which the various quantities show random variation in time and space. Turbulence is an essential part of the mechanism which disperses air pollutants resulting from anthropogenic activities. Turbulence is also crucial for the efficiency of many natural processes, such as evaporation of water, dissipation of fog, and dispersion of plant seeds. It also contributes to the structural fatigue of man-made constructions.

Turbulent flow is often compared with *laminar motion*, which is smooth, regular, and predictable movement of the individual particles of a fluid.

Laminar flow occurs in the lowest, several-millimeter-thick layer of the atmosphere called the *laminar sublayer*.

Flow in the atmospheric boundary layer is controlled by the diurnal cycle of the surface energy budget. The energy balance at the surface is expressed as

$$R_n + G + H + E = 0 \tag{1.1}$$

where R_n is the flux of net radiation (global solar radiation received by the surface plus atmospheric radiation minus terrestrial radiation), G is the vertical heat flux into the soil, and H and E are the sensible (conduction) and latent (resulting from water phase changes) heat fluxes to the atmosphere.

The diurnal changes in the energy balance components are shown in Figure 1.1. In the figure, a quantity has a positive value when energy is transferred away from the interface and a negative value in the case of transfer toward the interface. Net radiation flux R_n is negative during the day, reaching minimum values at local solar noon. At night, R_n is positive, illustrating loss of energy by terrestrial radiation. R_n is zero just before sunset and just after sunrise. At night, the term E can become negative if dew forms.

During the day, energy gained by the surface is transferred to the atmosphere, to the soil, and is also used in the evaporation processes. This transfer of heat from the ground surface to the air directly above can generate

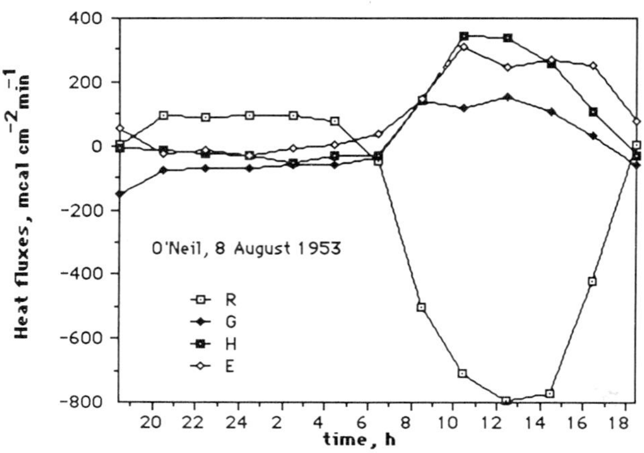

Figure 1.1 Diurnal distribution of heat fluxes at the Earth's surface, observed during the 1953 O'Neill experiment. R is the flux of net radiation on the surface, G is the heat flux to the ground, H is the sensible heat flux, and E is the latent heat flux from the surface to the atmosphere.

convection. Convection redistributes heat throughout the atmospheric boundary layer.

The influence of the surface sensible heat flux decreases with height. As a result, the temperature amplitude also decreases with height, becoming zero on the top of the atmospheric boundary layer (Figure 1.2).

The sensible heat flux influences the vertical distribution of temperature in the boundary layer. Resulting temperature stratification can be classified into three distinct categories: *stable*, *unstable*, and *neutral*. In the case of *stable stratification*, the temperature of the atmosphere is distributed such that any fluid particle moving vertically is impeded by Archimedean force. This suppresses turbulence and can lead to laminar flow. Stable cases usually occur at night.

In *unstable stratification*, the parcel of air moving vertically is accelerated by Archimedean force. This mechanism increases turbulence. Unstable cases usually occur during daylight hours. In the *neutral case*, observed during transitions between day and night, the buoyancy forces vanish.

The effects of thermal stability described above are illustrated in Figure 1.3. The figure shows diurnal changes of the vertical velocity of air measured 10m above the ground. During the day convection-driven turbulence is intense and, as a consequence, the amplitude of the vertical velocity is large. The amplitude decreases after sunset and remains low until sunrise.

A typical vertical distribution of meteorological parameters in a diurnal cycle is shown in Figures 1.4 to 1.6. After sunrise on a clear day, the surface

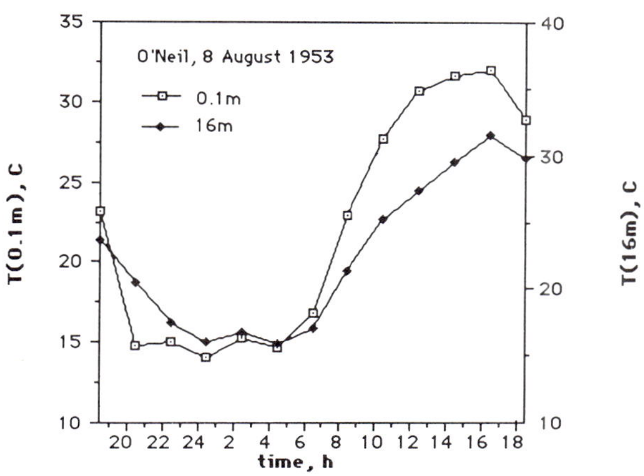

Figure 1.2 Diurnal distribution of temperature at two levels, observed during the 1953 O'Neill experiment.

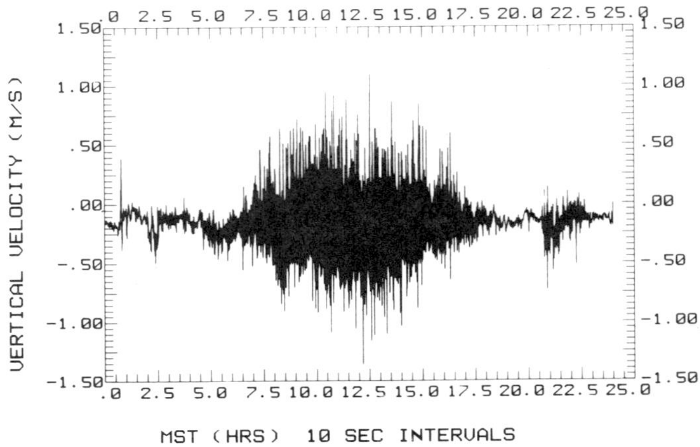

Figure 1.3 Vertical velocity of air on a level of 10 m plotted in 10-second intervals observed at the BAO Tower in Boulder, Colorado, on 9/26/1986. (Courtesy of A. A. Tsonis.)

becomes warmer than the air above and the sensible heat is transferred from the ground to the air, causing intense convective mixing within the boundary layer. As a consequence, in its upper part, called the *mixed layer*, meteorological characteristics such as wind velocity, wind direction, and *potential temperature* are almost uniform with height (Figure 1.4). Vigorous turbulence, driven primarily by surface heating, also uniformly mixes pollutants in the vertical. The top of the mixed layer is marked by a temperature inversion which suppresses the turbulence and buffers the free atmosphere above. Warmer air from this layer is *entrained* into the mixed layer, generating gravity waves on the border of these two layers. The mixed layer reaches a depth ranging from a few hundred to a few thousand meters by late afternoon. Due to the entrainment from the capping-inversion layer, the mixed layer can deepen at a rate of a few centimeters per second.

In the late afternoon, just before sunset, radiative cooling starts. The ground surface is at nearly the same temperature as the air above it. Consequently, the sensible heat flux in the atmosphere vanishes (Figure 1.5). This state of neutral stability is short-lived.

As the sun sets, the radiative energy input at the surface stops and completely shuts off the energy supply to the convective motions. The resulting heat flux from the atmosphere toward the surface causes the gradual cooling of the air aloft and a decay of the ABL turbulence. The surface temperature inversion layer increases and, at the same time, the height of the base of the upper inversion decreases. Eventually, the upper and surface inversions join

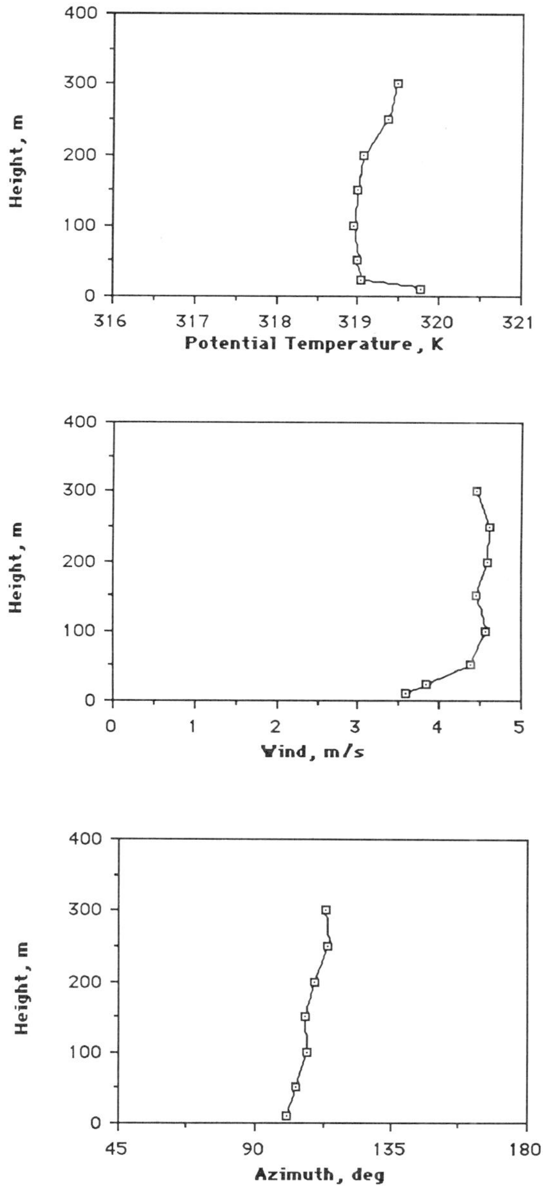

Figure 1.4 Typical distribution of potential temperature, wind velocity, and direction in the convective boundary layer. Data collected at the BAO Tower in Boulder, Colorado, on 9/9/1983 at 13:40.

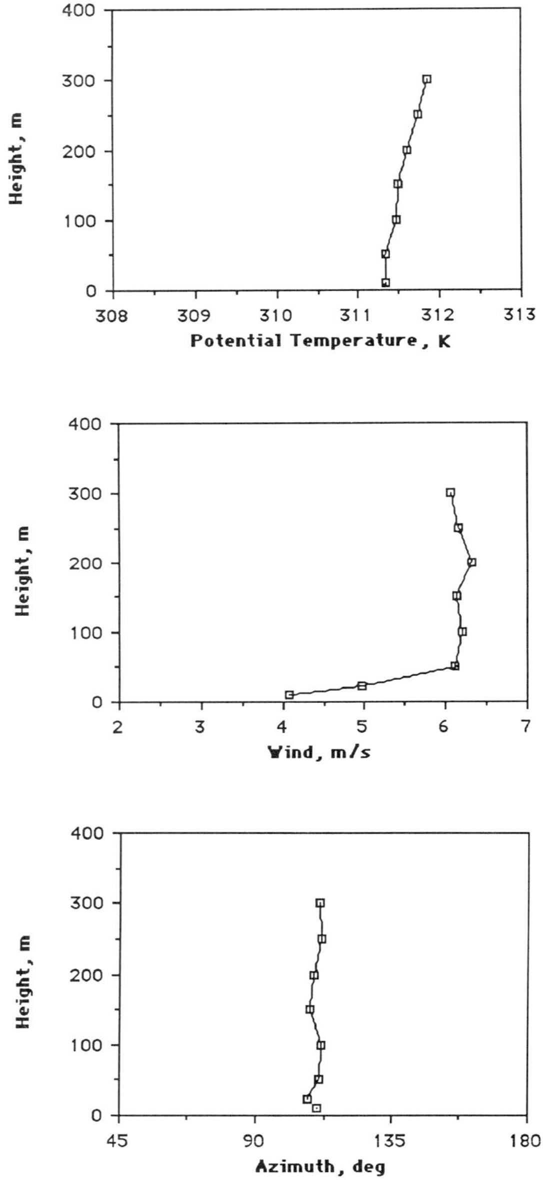

Figure 1.5 Typical distribution of potential temperature, wind velocity, and direction in the ABL during day–night transition. Data collected at BAO Tower in Boulder, Colorado, on 9/12/1983 at 17:00.

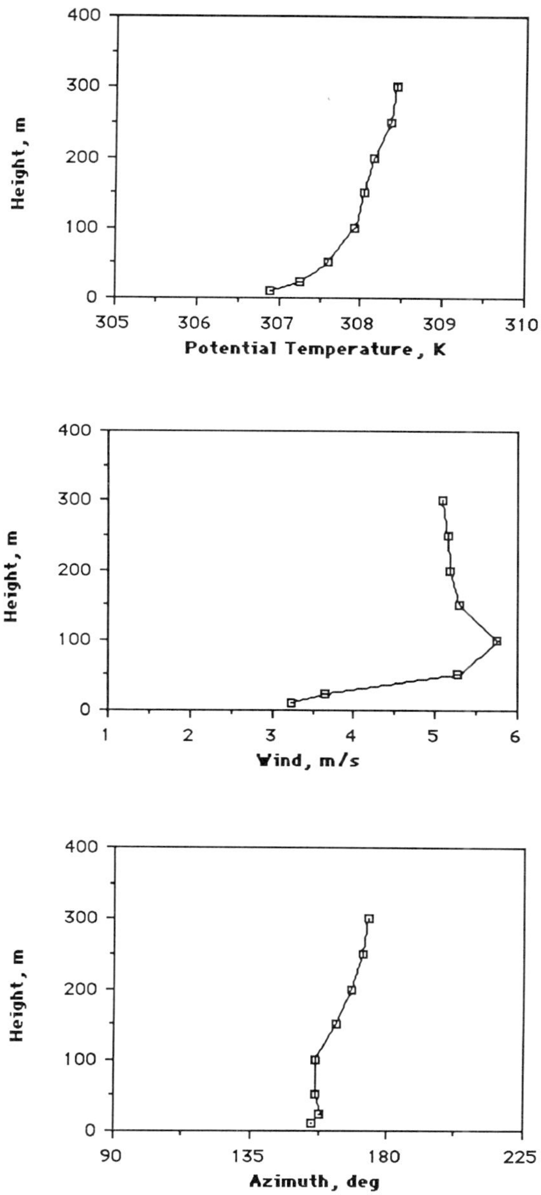

Figure 1.6 Typical distribution of potential temperature, wind velocity, and direction in the stable boundary layer. Data collected at the BAO Tower in Boulder, Colorado, on 9/10/1983 at 18:20.

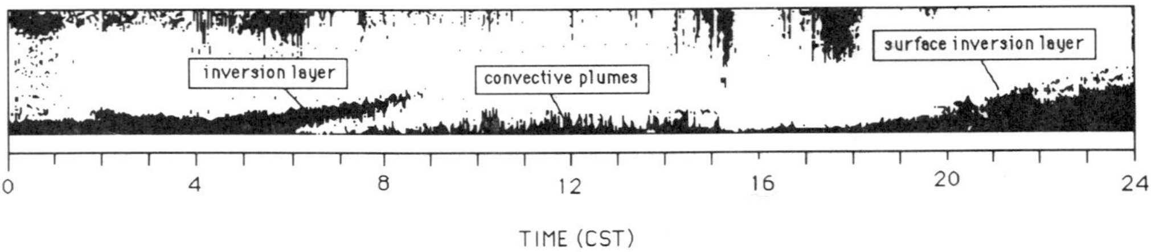

Figure 1.7 Sodar image of the diurnal changes in the ABL.

together (Figure 1.6). Sometimes turbulence ceases completely, causing a laminar flow, which can break into *gravity waves*. The wind velocity often reaches its maximum (the so-called *low-level jet stream*) in the first few hundred meters (Figure 1.6).

The typical diurnal cycle of the ABL can also be depicted on the picture obtained from the *sodar* sounding in Figure 1.7. The picture shows a surface inversion layer (dark) developed at night. As the surface layer warms up after sunrise, the surface inversion layer formed during the previous night vanishes. During the day, convective plumes form. In the late afternoon, the turbulence

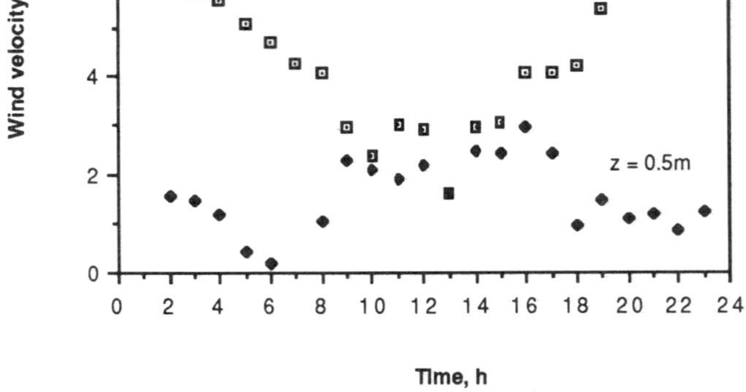

Figure 1.8 Diurnal variation of wind velocity on two levels (z = 0.5 m and z = 50 m) observed during day 33 of the Wangara experiment, Australia.

in the mixed layer becomes weaker as the surface heat flux is reduced to zero and changes sign shortly before sunset. At the same time, a shallow surface inversion layer evolves as the downward heat flux increases with time, due to the radiative cooling of the surface.

The diurnal changes of temperature are accompanied by changes in wind velocity and wind direction. Above the surface layer the winds are observed to reach maximum speeds at night and minimum speeds during the day. But near the surface, the opposite behavior of wind velocity is observed (Figure 1.8). The surface (at $z = 0.5$ m) wind speed increases after sunrise, peaks in the early afternoon, and decreases near sunset. Above the surface layer ($z = 50$m), the diurnal wave is nearly 12 hours out of phase.

The atmospheric boundary layer is also influenced by other factors. Some of them are connected with irregularities in the earth's surface (mountains, trees), the shape of the surface (slope), and its roughness. Also, clouds which often develop as a result of processes within the atmospheric boundary layer modify its regime.

1.2 HISTORICAL NOTES

The concepts of *turbulence* and *boundary layers* were originated in the field of fluid mechanics. Turbulent flows were first investigated at the end of the nineteenth century by O. Reynolds. Reynolds reduced the problem of solving full, time-dependent *Navier–Stokes equations* for random variables to an easier task of finding the mean quantities, which are deterministic.

Boundary layers were first investigated at the beginning of the twentieth century by Prandtl, who proved that the flow about the solid body can be divided into two regions: a thin layer in the neighborhood of the body (boundary layer), where friction plays an essential role, and the remaining region, where friction can be neglected. The first geophysical application of the boundary layer theory belongs to W. V. Ekman, who in 1905 presented his famous solution of simplified equations of motion, known as the *Ekman spiral*, describing a flow in the ocean.

Since its beginning, the atmospheric boundary layer theory has concentrated on two areas: the first, devoted to studies of a fine-scale turbulence, and the second, directed toward research on mean properties of turbulent flows. Theoretical and numerical milestones include *K-theory* modeling in the 1950s and 1960s, the *Monin–Obukhov similarity theory* (1954), and the *Rossby-number similarity theory* in the 1960s. The K-theory failed to describe many features of the boundary layer and in the 1970s the *higher-order closure* approach was developed. Simultaneously, numerical experiments with *large-eddy simulation* models were pioneered by Deardorff in 1972. New directions in research on turbulence include the theory of *strange attractors* and the theory of *fractals*.

In addition to theoretical accomplishments, since the early 1950s, improvements in probing techniques have been made. The *sonic anemometer* was perfected to provide accurate, fast-response measurements of the three-dimensional velocity field. Fine-scale measurements of wind velocity components and temperature were made possible with *hot-* and *cold-wire sensors*. These and other new devices, including remote sensors such as *radars*, *lidars*, and *sodars*, have provided a strong data base on the structure of the ABL. As a result, numerous experimental studies performed in the 1950s, 1960s, and 1970s have provided substantial knowledge of the atmospheric boundary layer (the Great Plains 1953 Turbulence Field Program, the Wangara 1967 Experiment, the Kansas 1968 Field Program, the Minnesota 1973 Experiment).

The growth made in understanding the ABL structure is evident in monographs written over the last 30 years. The first was Sutton's *Micrometeorology* (1953). A few years later Priestley published his *Turbulent Transport in the Lower Atmosphere* (1959). These two monographs provided the foundation for the theoretical and experimental work for the next decade. In the 1960s and early 1970s new interest in ABL was established and resulted in a rapid growth in the data base. Publication of Lumley and Panofsky's textbook, *The Structure of Atmospheric Turbulence*, in 1964 and of the *Workshop on Micrometeorology* (ed. Haugen), later in 1973, strengthened the theoretical foundations of the atmospheric turbulence and presented the most complete discussion of the observed structure of the ABL. Another milestone was Monin and Yaglom's *Statistical Fluid Mechanics* (1965), which appeared in English translation in 1971. In 1972, Tennekes and Lumley published *A First Course in Turbulence*. Two Soviet textbooks worth mentioning are Zilitinkevich's *Dynamics of the Atmospheric Boundary Layer* (1970) and Laichtman's *Physics of the Atmospheric Boundary Layer* (1970). The latest developments in ABL turbulence and diffusion have been summarized in *Engineering Meteorology* (1982), edited by Plate; *Atmospheric Turbulence and Air Pollution Modelling* (1983), edited by Nieuwstadt and van Dop; *Atmospheric Turbulence* (1984), by Panofsky and Dutton; and *An Introduction to Boundary Layer Meteorology* (1988), by Stull.

PROBLEMS

1.1. Given the radiosonde observations of the temperature in the atmospheric boundary layer at 12:00 local time (day 33 of the Wangara experiment; see Section 8.6.2):

Height (m)	Pressure (mb)	Temperature (°C)
0	1021	12.6
50	1015	11.2
100	1009	10.4
	1003	10.0
200	997	9.5
	991	9.1
300	985	8.6
	979	7.9
400	973	7.3
	967	6.9
500	961	6.6
	955	6.2
600	949	5.8
	944	5.3
700	938	4.7
	932	4.1
800	926	3.5
	921	3.1
900	915	2.7
	909	2.3
1000	904	1.9
1100	893	1.3
1200	882	1.4
	871	2.4
1400	860	2.5
	849	2.5
1600	839	2.4
	829	1.2
1800	818	0.5

Plot the vertical distribution of the potential temperature [see (2.39)]. Find the height of the ABL (the upper inversion height). Show on the plot the four distinct layers of the daytime ABL: the surface layer, mixed layer, interface entrainment layer, and inversion layer.

CHAPTER TWO

EQUATIONS

Discussion of the structure of the atmospheric boundary layer has to begin with a consideration of the basic equations. The set of governing equations describing the atmospheric flow consists of three differential equations expressing conservation of mass (the continuity equation), conservation of momentum (the Navier-Stokes equations), and conservation of energy (the first law of thermodynamics). In addition, a fourth equation must be added in order to close the system. This is the equation of state for an ideal gas.

2.1 THE CONTINUITY EQUATION

The equation that results from applying the conservation of a mass law to a fluid flow is called the *continuity equation.* This equation expresses the balance of mass in a volume $\mathbb{V}$. If the change of mass is due only to the outflow through the closed surface Σ, bounding the volume $\mathbb{V}$, then

$$\underbrace{-\frac{\partial}{\partial t}\iiint_{\mathbb{V}} \rho \, d\mathbb{V}}_{\substack{\text{decrease of mass} \\ \text{in volume } \mathbb{V}}} = \underbrace{\iint_{\Sigma} (\rho \mathbf{v}) \cdot \mathbf{d\Sigma}}_{\substack{\text{outflow through} \\ \text{the surface } \Sigma}} \tag{2.1}$$

where ρ is the air density, $\mathbf{v}$ is the wind velocity vector, t the time, and the vector $\mathbf{d\Sigma}$ is an outward-drawn vector whose length is equal to the surface element $d\Sigma$ (Figure 2.1).

Assuming that volume $\mathbb{V}$ does not change with time and applying the *Gauss theorem* (see the Appendix), we obtain, from (2.1),

$$\iiint_{\mathbb{V}} \left[\frac{\partial}{\partial t}\rho + \operatorname{div}(\rho \mathbf{v}) \right] d\mathbb{V} = 0 \tag{2.2}$$

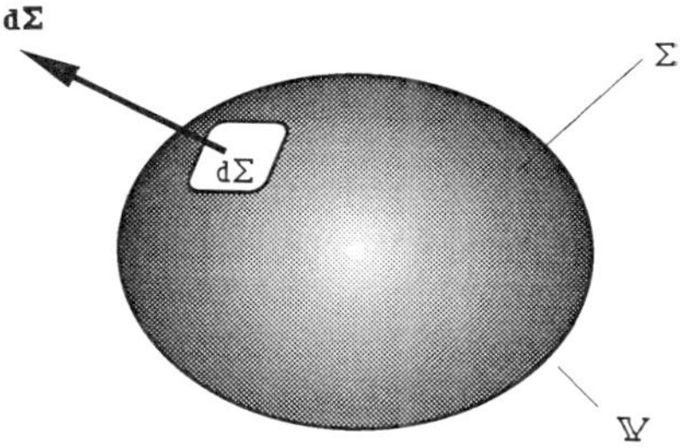

Figure 2.1 Volume $\mathbb{V}$ bounded by surface Σ.

As (2.2) must be satisfied for any volume $\mathbb{V}$, the integrand must vanish everywhere. This gives a differential relation between the density and velocity fields:

$$\frac{\partial}{\partial t}\rho + \operatorname{div}(\rho \mathbf{v}) = 0 \tag{2.3}$$

A fluid is said to be *incompressible* if the changes of density are considered to be negligible. In this case (2.3) takes the simple form

$$\operatorname{div} \mathbf{v} = 0 \tag{2.4}$$

Various substances, including water vapor, carbon dioxide, and sulfur dioxide, are transported through the ABL. Their concentration C can be described by an equation analogous to (2.3):

$$\frac{\partial}{\partial t}C + \operatorname{div}(C\mathbf{v}) = S \tag{2.5}$$

where S is the sum of the source and sink functions.

The continuity equation is often written slightly different than the (2.3) form. The new form is obtained by using an identity: $\operatorname{grad}(\rho \mathbf{v}) = \mathbf{v} \cdot \operatorname{grad} \rho + \rho \operatorname{div} \mathbf{v}$, which yields

$$\underbrace{\frac{\partial}{\partial t}\rho + \mathbf{v} \cdot \operatorname{grad} \rho}_{\mathbf{d}\rho/\mathbf{d}t} = -\rho \operatorname{div} \mathbf{v} \tag{2.6}$$

In (2.6) the terms on the left-hand side are denoted by $\mathbf{d}\rho/\mathbf{d}t$. The operator $\mathbf{d}/\mathbf{d}t$ is called the *total derivative.* The first term of the total derivative describes the local time changes and the second one describes spatial (advective) changes of the quantity. The total derivatives can be obtained by using the chain rule and recognizing ρ to be a function of space distances x, y, z and time t.

2.2 THE NAVIER–STOKES EQUATIONS

Equations of motion are the form of Newton's second law for conservation of momentum (formulated by Isaac Newton in 1687 in his monumental work *Principia Mathematica*). They are customarily called the *Navier–Stokes equations,* named after the two scientists who derived them independently (Navier's derivation was presented in 1847).

Let us assume that the *total momentum* changes in a volume $\mathbb{V}$ are a result of forces acting on the surface bounding this volume (surface forces) and forces acting on the whole volume (volume forces). For the i component of momentum vector, we obtain

$$\iiint_{\mathbb{V}} \rho \frac{\mathbf{d}}{\mathbf{dt}} u_i \, d\mathbb{V} = \iint_{\Sigma} \mathbf{T}_i \cdot d\Sigma + \iiint_{\mathbb{V}} \rho G_i \, d\mathbb{V} \tag{2.7}$$

where u_i is a component of vector $\mathbf{v}$, G_i a component of a body force $\mathbf{G}$ (force per unit of mass), and $\mathbf{T}_i$ a surface force in the i direction. Body forces are to be regarded as given external forces, but surface forces depend on the fluid velocity field.

Applying the Gauss theorem to (2.7) yields

$$\rho \frac{\mathbf{d}}{\mathbf{d}t} u_i = \operatorname{div} \mathbf{T}_i + \rho G_i \tag{2.8}$$

The ensemble of all vectors $\mathbf{T}_i$ for all possible orientations of the surface describes the state of *stress* in the body. In three-dimensional space, the vectors $\mathbf{T}_1$, $\mathbf{T}_2$, and $\mathbf{T}_3$ define a *stress tensor* (Figure 2.2):

$$\Pi = \begin{pmatrix} \mathbf{T}_1 \\ \mathbf{T}_2 \\ \mathbf{T}_3 \end{pmatrix} = \begin{pmatrix} \tau_{11} & \tau_{12} & \tau_{13} \\ \tau_{21} & \tau_{22} & \tau_{23} \\ \tau_{31} & \tau_{32} & \tau_{33} \end{pmatrix} \tag{2.9}$$

where τ_{ii} are normal components and τ_{ij} are parallel components (shearing stress).

It can be shown that the stress tensor and the corresponding matrix are symmetric, $\tau_{ij} = \tau_{ji}$. If the fluid is "frictionless," all shearing stresses τ_{ij} vanish and only normal stresses τ_{ii} remain in the equation. Moreover, they are equal and their negative is defined as the pressure:

$$\tau_{11} = \tau_{22} = \tau_{33} = -p \tag{2.10}$$

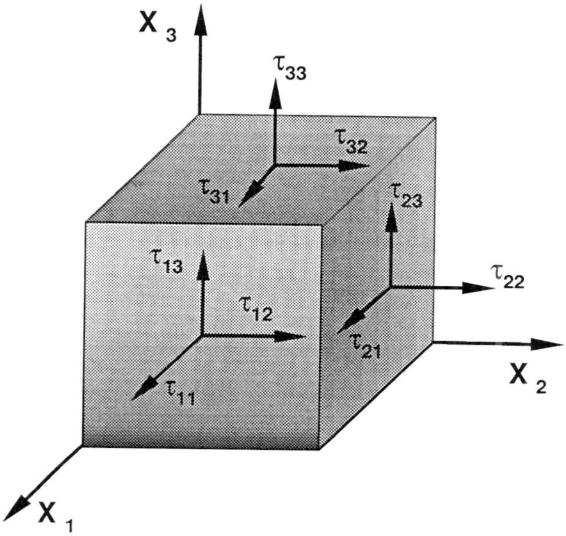

Figure 2.2 Orientation of components of the stress tensor.

Equation (2.10) is equivalent to

$$\frac{\tau_{11} + \tau_{22} + \tau_{33}}{3} = -p \tag{2.11}$$

Since the left-hand side of (2.11) is one-third of the trace of matrix Π, it follows that pressure is invariant with respect to transformations of the system coordinates.

A hypothesis is usually made that the relation between τ_{ij} and *rate of strain* S_{ij}, defined as

$$S_{ij} = \frac{1}{2}\left(\frac{\partial u_i}{\partial x_j} + \frac{\partial u_j}{\partial x_i}\right) \tag{2.12}$$

is linear and of the form (Newton's hypothesis):

$$\tau_{ij} = a\delta_{ij} + bS_{ij} \tag{2.13}$$

where a and b are constants. For $\tau_{\alpha\alpha} = -3p$ and $b = 2\nu'$ we obtain:

$$a = -p - \tfrac{2}{3}\nu' S_{\alpha\alpha} \tag{2.14}$$

where ν' is the molecular viscosity coefficient and δ_{ij} is the Kronecker delta, defined as $\delta_{ij} = 1$ for $i = j$ and $\delta_{ij} = 0$ for $i \neq j$. In (2.14) the so-called *Einstein notation* is used:

$$S_{\alpha\alpha} = S_{11} + S_{22} + S_{33} = \frac{\partial u_1}{\partial x_1} + \frac{\partial u_2}{\partial x_2} + \frac{\partial u_3}{\partial x_3} = \frac{\partial u_\alpha}{\partial x_\alpha} = \operatorname{div} \mathbf{v} \tag{2.15}$$

which means that if any index in an expression is repeated, it indicates a summation with respect to this index.

Using (2.13) and (2.14), the stress tensor may be written in the form

$$\tau_{ij} = -p\delta_{ij} + \nu'\left[\left(\frac{\partial u_i}{\partial x_j} + \frac{\partial u_j}{\partial x_i}\right) - \frac{2}{3}\delta_{ij}\frac{\partial u_\alpha}{\partial x_\alpha}\right] \tag{2.16}$$

(Hereafter the Greek-letter notation will be used to emphasize that there is a summation.) The stress tensor is frequently separated in the following manner:

$$\tau_{ij} = -p\delta_{ij} + \sigma_{ij} \tag{2.17}$$

where σ_{ij} represents the viscous stress tensor:

$$\sigma_{ij} = \nu'\left[\left(\frac{\partial u_i}{\partial x_j} + \frac{\partial u_j}{\partial x_i}\right) - \frac{2}{3}\delta_{ij}\frac{\partial u_\alpha}{\partial x_\alpha}\right] \tag{2.18}$$

Now using (2.17) and (2.18), the Navier–Stokes equations (2.8) can be rewritten in the following form:

$$\rho\frac{\mathbf{d}u_i}{\mathbf{d}t} = -\frac{\partial p}{\partial x_i} + \nu'\frac{\partial^2 u_i}{\partial x_\alpha^2} + \frac{1}{3}\nu'\frac{\partial^2 u_\alpha}{\partial x_i\,\partial x_\alpha} + \rho G_i \tag{2.19}$$

For an incompressible fluid, $\partial u_\alpha/\partial x_\alpha = 0$ and (2.19) simplifies to

$$\left(\frac{\mathbf{d}u_i}{\mathbf{d}t}\right)_n = -\frac{1}{\rho}\frac{\partial p}{\partial x_i} + \nu\frac{\partial^2 u_i}{\partial x_\alpha^2} + G_i \tag{2.20}$$

where $\nu = \nu'/\rho$ and the subscript n refers to the nonaccelerating (inertial or fixed) coordinate system that has been used until now.

Because the motion of the atmosphere is relative to the rotating earth, it is more convenient to refer to the coordinate system fixed to the earth. Since the earth is rotating with constant angular velocity $\mathbf{\Omega}$, the velocity $\mathbf{v}$ of a parcel of air can be written as the sum of the velocity relative to the earth $\mathbf{V}$ and the velocity resulting from rotation:

$$\mathbf{v} = \mathbf{V} + \mathbf{\Omega} \times \mathbf{R} \tag{2.21}$$

where $\mathbf{R}$ represents the position vector of the parcel of air as measured from the origin of the earth's center. The time differential operator can similarly be described by the sum of a derivative relative to the earth's surface and changes resulting from the rotation rate of the earth:

$$\left(\frac{\mathbf{d}}{\mathbf{d}t}\right)_n = \frac{\mathbf{d}}{\mathbf{d}t} + \mathbf{\Omega} \times \tag{2.22}$$

Substituting (2.21) into (2.22) yields

$$\left(\frac{\mathbf{dv}}{\mathbf{d}t}\right)_n = \left(\frac{\mathbf{d}}{\mathbf{d}t} + \boldsymbol{\Omega} \times\right)(\mathbf{V} + \boldsymbol{\Omega} \times \mathbf{R})$$

$$= \underbrace{\frac{\mathbf{dV}}{\mathbf{d}t}}_{\substack{\text{acceleration measured}\\ \text{on the earth}}} + \underbrace{2\boldsymbol{\Omega} \times \mathbf{V}}_{\substack{\text{Coriolis}\\ \text{acceleration}}} + \underbrace{\boldsymbol{\Omega} \times (\boldsymbol{\Omega} \times \mathbf{R})}_{\substack{\text{centripetal}\\ \text{acceleration}}} \tag{2.23}$$

The Coriolis acceleration operates only when a parcel is in motion. The centripetal acceleration acts on a parcel permanently.

In the case of the air parcel of the atmosphere, force **G** in (2.20) is the gravitational force directed toward the center of the earth with a magnitude proportional to the mass of the earth, mass of the parcel, and inversely proportional to the square of the distance of a parcel from the center of the earth.

It is convenient to define an effective gravity acceleration **g** such that

$$\mathbf{g} = \mathbf{G} - \boldsymbol{\Omega} \times (\boldsymbol{\Omega} \times \mathbf{R}) \tag{2.24}$$

Since the second term on the right-hand side is much smaller than the first, $\mathbf{g} \approx \mathbf{G}$.

In a noninertial reference frame rotating with the earth, the system of the equation of motion will read

$$\frac{\mathbf{d}u_i}{\mathbf{d}t} = -\frac{1}{\rho}\frac{\partial p}{\partial x_i} + \nu\frac{\partial^2 u_i}{\partial x_\alpha^2} + g\delta_{i3} - 2\epsilon_{i\alpha\beta}\omega_\alpha u_\beta \tag{2.25}$$

where u_i are the components of the wind vector **V** (the same notation is retained). It is assumed in (2.25) that the third axis is vertical and that the vector **g** has components $(0, 0, g) = g\delta_{i3}$. Components of the Coriolis acceleration vector **C** are expressed in the form

$$C_i = 2(\boldsymbol{\Omega} \times \mathbf{V})_i = 2\epsilon_{i\alpha\beta}\omega_\alpha u_\beta \tag{2.26}$$

where ω_j is a component of angular velocity vector $\boldsymbol{\Omega}$ and ϵ_{ijk} is the alternating unit tensor, $\epsilon_{ijk} = 1$ for even permutations of i, j, and k (ijk = 123, 231, 312); $\epsilon_{ijk} = -1$ for odd permutations of i, j, and k (ijk = 321, 213, 132), and $\epsilon_{ijk} = 0$ otherwise.

If the x_1 axis is oriented to the east, x_2 to the north, and x_3 is vertical (Figure 2.3), then $\omega_1 = 0$, $\omega_2 = \Omega \cos\phi$ and $\omega_3 = \Omega \sin\phi$, where ϕ is the latitude and Ω is the angular velocity. In this coordinate system the components of the Coriolis force are

$$\begin{aligned} C_1 &= -2\Omega \sin\phi u_2 + 2\Omega \cos\phi u_3 \\ C_2 &= \Omega \sin\phi u_2 \\ C_3 &= -2\Omega \cos\phi u_1 \end{aligned} \tag{2.27}$$

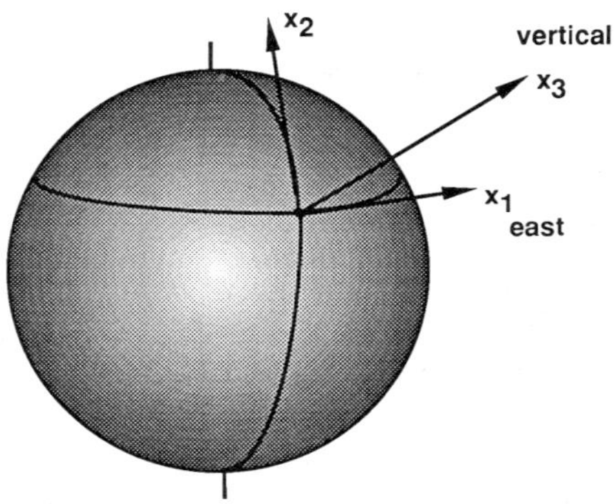

Figure 2.3 Coordinate system.

2.3 THE EQUATION OF STATE

The system of four differential equations consisting of the continuity equation and the Navier–Stokes equations contains five unknowns (ρ, p, u, v, and w). It is obvious that two additional equations are required to close the system. These two equations can be obtained by determining the relationship that exists between the thermodynamic variables.

First, the relation known as an *equation of state* will be discussed. An example of an equation of state is the *perfect gas equation of state.* "*Perfect gas*" is understood to be the limiting behavior of any gas at temperatures well above its *critical point* and at vanishingly low density. Such a gas obeys *Boyle's law* ($p\mathbb{V}$ = const) at any temperature. With the temperature defined on the *perfect gas scale,* the equation of state becomes

$$p\mathbb{V} = nR_uT \tag{2.28}$$

where $\mathbb{V}$ is the volume, R_u the *universal gas constant* ($R_u = 8314.3$ J kmol^{-1} K^{-1}), n the number of moles ($n = m/\mu_0$, where m is mass and μ_0 is a mass of 1 mol), and T the *absolute temperature.* The *absolute thermodynamic temperature scale* is proportional to the pressure of a perfect gas when held at a constant volume, or proportional to the volume of a perfect gas held at a constant pressure. The term *absolute* indicates no negative temperature and the word *thermodynamic* means that the temperature scale is independent of a thermometric substance.

Equation (2.28) can be rewritten in the form

$$p = \rho RT \tag{2.29}$$

where $R = R_u/\mu_0$ is the *gas constant.* For a standard condition and for dry air, $R = 287$ m^2 s^{-2} K^{-1}.

When moisture is present in the atmosphere, (2.29) has a slightly different form. The content of water vapor can be characterized by a *specific humidity* $q = \rho_v/\rho$—the ratio of the water vapor density ρ_v to the density of the moist air $\rho = (\rho_v + \rho_d)$ (ρ_d is the density of dry air). It should be remembered that water vapor is lighter than moist air and moist air is much lighter than dry air: $\rho_v \ll \rho < \rho_d$.

The atmospheric pressure p is a sum of partial pressures of water vapor e and dry air p_d (*Dalton's law*). Assuming that air and water vapor are ideal gases, the following is obtained:

$$p = p_d + e = \rho_d RT + \rho_v R_v T = \rho RT\left(\frac{\rho_d}{\rho} + \frac{1.61\rho_v}{\rho}\right) = \rho RT(1 + 0.61q)$$

or

$$p = \rho RT_v \tag{2.30}$$

where $T_v = T(1 + 0.61q)$ is a *virtual temperature*—a temperature that a sample of dry air at given pressure has to achieve in order to have the same density as a parcel with specific humidity q at the same pressure; and also, $R_v = 1.61R$. Hence the form of the equation of state is identical for dry and moist air, and what differs in these two cases is the calculation of the temperature.

In the saturation case, when both vapor and liquid water are present, $\rho = (\rho_v + \rho_L + \rho_d)$, where the subscript L denotes liquid water, and $T_\iota = T(1 + 0.61q - q_L)$, where $q_L = \rho_L/\rho$.

2.4 THE FIRST LAW OF THERMODYNAMICS

The second equation relating thermodynamic variables is the *first law of thermodynamics,* derived in 1850 by Rudolf Clausius. The law asserts that the system possesses an internal energy content $\mathbb{U}$, which is a function of the variables necessary to specify the *state.* The internal energy includes energy of all types of internal motion and the energy of the interaction of all particles making up the system (atoms, molecules, ions, etc.). For a simple one-component system, the state is fully defined by specifying any two of the three parameters: pressure p, volume $\mathbb{V}$, and temperature T_v. If the system undergoes a small change in which it remains close to *thermal equilibrium* and the quantity of heat energy entering through the walls is dQ while the volume increases by $\mathbf{d}\mathbb{V}$ against the pressure p, then the change in internal energy $\mathbf{d}\mathbb{U}$ is

$$\mathbf{d}\mathbb{U} = dQ - p\,\mathbf{d}\mathbb{V} \tag{2.31}$$

In other words, the system exchanges energy with its surroundings only by taking in heat dQ or by performing external work $p\,\mathbf{d}\mathbb{V}$. Internal energy increases when heat is delivered to the system (i.e., $dQ > 0$) and decreases when the system expands, performing work against external pressure (i.e., $\mathbf{d}\mathbb{V} > 0$). The quantities $\mathbf{d}\mathbb{U}$ and $\mathbf{d}\mathbb{V}$ are exact differentials of functions depending on variables defining the state, while dQ is not the exact differential. This comes about because $p\,\mathbf{d}\mathbb{V}$ is not an exact differential, since its integral between two states A and B depends on the path of intermediate states from A to B. If the change occurs at a constant volume, no external work is done. All of the added heat increases the internal energy.

The *heat capacity at constant volume* C_v is defined as $C_v = (dQ/dT_v)_v$ and is given by

$$C_v = \left(\frac{\partial \mathbb{U}}{\partial T_v}\right)_v \tag{2.32}$$

Equation (2.31) can be written in another form:

$$dQ = \mathbf{d}H - \mathbb{V}\,\mathbf{d}p \tag{2.33}$$

where H is a new function of the state of the system, called its *enthalpy,* defined as

$$H = \mathbb{U} + p\mathbb{V} \tag{2.34}$$

In a change occurring at constant pressure, all the heat added to the system increases the enthalpy. The *heat capacity at constant pressure* C_p is defined as $C_p = (dQ/dT_v)_p$ and is therefore given by

$$C_p = \left(\frac{\partial H}{\partial T_v}\right)_p \tag{2.35}$$

An important attribute of the perfect gas is that its internal energy $\mathbb{U}$ is independent of its volume and is a function of T_v alone.

$$\mathbb{U} = C_v T_v \tag{2.36}$$

For a perfect gas the following relation exists:

$$H = C_p T_v \tag{2.37}$$

For the *specific heat,* defined as the heat capacity of a unit of mass, there is:

$$c_p - c_v = R \tag{2.38}$$

where c_p is the specific heat at constant pressure and c_v is the heat capacity at constant pressure. For the dry air $c_p = 1007\ \text{m}^2\ \text{s}^{-2}\ \text{K}^{-1}$ and $c_v = 720\ \text{m}^2\ \text{s}^{-2}\ \text{K}^{-1}$.

An *adiabatic process* is one in which no heat is taken in or given out; thus $dQ = 0$. From (2.33) and (2.37) we find that in an adiabatic process, $T_{v1}/T_{v2} = (p_1/p_2)^{R/c_p}$ (*Poisson's equation*). The *virtual potential temperature* Θ is defined from Poisson's equation as the virtual temperature achieved by a parcel displaced from a pressure level p to the 1000-mbar level:

$$\Theta = T_v\left(\frac{1000}{p}\right)^{R/c_p} \tag{2.39}$$

If in (2.39) the virtual temperature is replaced by the temperature, the quantity defined is called the *potential temperature.* For a dry adiabatic process the potential temperature is invariant and the vertical change in temperature $dT/dz = -g/c_p = -\gamma_a$, where $\gamma_a \approx 1°\text{C}/100$ m is the dry *adiabatic lapse rate.* From (2.39) it also follows that $d\Theta/dz = \Theta/T(dT/dz + \gamma_a)$. If $d\Theta/dz > 0$, thermal stratification is called *stable*; if $d\Theta/dz < 0$ it is called *unstable.* Finally, when $d\Theta/dz = 0$, the case is called *neutral.*

From the definition of the potential temperature, it follows that

$$\frac{\mathbf{d}\Theta}{\Theta} = \frac{\mathbf{d}T_v}{T_v} - \frac{R}{c_p}\frac{\mathbf{d}p}{p} \tag{2.40}$$

Dividing (2.33) by the time increment dt and taking (2.37) into consideration, we obtain

$$\frac{\mathbf{d}T_v}{\mathbf{d}t} - \frac{1}{\rho c_p}\frac{\mathbf{d}p}{\mathbf{d}t} = S_\theta \tag{2.41}$$

where the term

$$S_\theta = \frac{1}{c_p}\frac{dQ}{dt} \tag{2.42}$$

represents sources and sinks of heat. With the help of (2.40), we finally have

$$\frac{\mathbf{d}\Theta}{\mathbf{d}t} = \frac{\Theta}{T_v}S_\theta \tag{2.43}$$

The contributors to S_θ include the sum of the following processes:

S_θ = [+ condensation − evaporation] + [+ freezing − melting] + [+ deposition − sublimation]

+ [+ radiative flux convergence − radiative flux divergence] + [dissipation of kinetic energy by molecular motion]

− [dissipation of heat by molecular diffusion]

2.5 SIMPLIFICATION OF THE BASIC EQUATIONS

It is possible to reduce a number of variables in the governing equations by eliminating one—density ρ. For this purpose each of the thermodynamic variables will be split into two parts: the synoptic scale reference variables, subscripted 0, and the mesoscale perturbation, denoted by a double prime:

$$\begin{aligned} p &= p_0 + p'' \\ \rho &= \rho_0 + \rho'' \\ T &= T_0 + T'' \\ \Theta &= \Theta_0 + \Theta'' \end{aligned} \tag{2.44}$$

It will be assumed that the thermodynamic state of the atmosphere does not depart too much from a reference state that is *hydrostatic* (balance of pressure gradient and gravity forces) and *adiabatic*. The assumptions above lead to the *anelastic approximation.*

The reference variables satisfy

$$\begin{aligned} -\frac{1}{\rho_0}\frac{\partial p_0}{\partial x_2} &= fU_{g1} \\ \frac{1}{\rho_0}\frac{\partial p_0}{\partial x_1} &= fU_{g2} \\ \frac{1}{\rho_0}\frac{\partial p_0}{\partial x_3} &= -g \\ p_0 &= \rho_0 R T_0 \end{aligned} \tag{2.45}$$

where U_{g1}, U_{g2} are components of the *geostrophic wind* and $f = 2\Omega \sin \phi$ is the *Coriolis parameter.* It is also assumed that all perturbations are much smaller than reference values.

Introducing (2.44) and (2.45) into the equation of state yields

$$\begin{aligned} p'' &= \rho_0 R T'' + \rho'' R T_0 + \rho'' R T'' \\ &\approx \rho_0 R T'' + \rho'' R T_0 \end{aligned}$$

or

$$\frac{p''}{p_0} = \frac{T''}{T_0} + \frac{\rho''}{\rho_0} \tag{2.46}$$

If the vertical scale of the movements is small compared to the scale height of the atmosphere, an assumption is made that $p''/p_0 \approx 0$. From this it follows that

$$\frac{T''}{T_0} = -\frac{\rho''}{\rho_0} \tag{2.47}$$

A simple experiment can be performed to show the validity of (2.47) in the atmosphere. An empty glass can be warmed to the temperature of the boiling water (by filling it with boiling water for a few minutes) and then, inverted, placed into a pot of cold water.

The air in the glass will decrease its volume during the cooling process and water from the pot will fill the space vacated by the air. From the equation of state it can be found that pressure changes during the process are

$$\frac{\Delta p}{p} = \frac{\Delta \rho}{\rho} + \frac{\Delta T}{T} = -\frac{\Delta \mathbb{V}}{\mathbb{V}} + \frac{\Delta T}{T} \quad \text{or} \quad \frac{\Delta h}{h_a} = -\frac{\Delta h}{h_g} + \frac{\Delta T}{T}$$

where Δh is the change of water level in the glass, h_a the atmospheric pressure expressed as a height of water column, h_g the height of the glass, and ΔT the change of the temperature of air in the glass. For values obtained during an experiment, $\Delta h = 1$ cm, $h_a = 1{,}000$ cm, $h_g = 3.5$ cm, $\Delta T = 80$ K, and $T = 280$ K, it is observed that the relative pressure changes ($\Delta p/p = \Delta h/h_a = 1/1000 = 0.001$) and are much smaller than relative density changes ($-\Delta\rho/\rho = \Delta h/h_g = 1/3.5 \approx 0.3$) and relative temperature changes ($\Delta T/T = 80/280 \approx 0.3$). This yields $\Delta\rho/\rho \approx -\Delta T/T$.

From the definition of the potential temperature (2.39) and from the equation of state (2.29), it follows that

$$\begin{aligned} \frac{\Theta''}{\Theta_0} &= \frac{T''}{T_0} - \frac{R}{c_p}\frac{p''}{p_0} \\ &\approx \frac{T''}{T_0} \end{aligned} \tag{2.48}$$

since $p''/p_0 \approx 0$ and $R/c_p \approx 0.286$.

Finally, the equation of state for perturbations takes the form

$$\frac{\Theta''}{\Theta_0} = -\frac{\rho''}{\rho_0} \tag{2.49}$$

The pressure gradient term and the gravity term of the Navier–Stokes equation (2.25) can be rewritten in the form

$$\begin{aligned} \frac{1}{\rho}\frac{\partial p}{\partial x_i} + g\delta_{i3} &= \frac{1}{\rho}\left(\frac{\partial p}{\partial x_i} + \rho g\delta_{i3}\right) \\ &= \frac{1}{\rho_0 + \rho''}\left[\frac{\partial p_0}{\partial x_i} + \frac{\partial p''}{\partial x_i} + (\rho_0 + \rho'')g\delta_{i3}\right] \\ &\cong \left(fU_{g\alpha}\epsilon_{i\alpha 3} + \frac{1}{\rho_0}\frac{\partial p''}{\partial x_i} + \frac{\rho''}{\rho_0}g\delta_{i3}\right) \\ &= \left(fU_{g\alpha}\epsilon_{i\alpha 3} + \frac{1}{\rho_0}\frac{\partial p''}{\partial x_i} - \frac{\Theta''}{\Theta_0}g\delta_{i3}\right) \end{aligned} \tag{2.50}$$

Notice that the density fluctuations were neglected in (2.50) except when they were multiplied by gravity. This simplification is called the *Boussinesq approximation.* The Boussinesq approximation can be used to describe most of the atmospheric flows, with the vertical scale less than about 1 km. Its advantage is in incompressibility and suppression of the acoustic waves, which simplifies the numerical simulation.

Using the results above and (2.27), the Navier–Stokes equations (2.25) can be expressed as

$$\frac{\mathbf{d}}{\mathbf{d}t}u_i = f(u_\alpha - U_{g\alpha})\epsilon_{i\alpha 3} + g\frac{\Theta''}{\Theta_0}\delta_{i3} - \frac{1}{\rho_0}\frac{\partial p''}{\partial x_i} + \nu\frac{\partial^2 u_i}{\partial x_\alpha^2} \tag{2.51}$$

In (2.51) an additional simplification was made, neglecting the term with the vertical velocity in the Coriolis force term for $i = 1$, and the Coriolis term for $i = 3$.

2.6 AVERAGING PROCEDURES

It is generally accepted that flow in the atmosphere can be described by the Navier–Stokes equations. Analytical solution of these equations is yet unknown. It is possible to solve them only numerically. However, because of a wide range of time and space scales of turbulent motion, the number of required grid points and time steps puts the practical computation of turbulent flows outside the capability of present computers.

This numerical inconvenience can be overcome by adopting the approach of Reynolds, who in 1895 proposed decomposing any flow variable in the form of a *mean quantity* (capital letters) and a *fluctuation* about the average (letters with primes):

$$\begin{aligned} u_i &= U_i + u_i' \\ p'' &= P + p' \\ \Theta'' &= \Theta + \theta' \\ c &= C + c' \\ &\text{etc.} \end{aligned} \tag{2.52}$$

The method proposed by Reynolds allows obtaining the mean quantities by a numerical solution of the averaged Navier–Stokes equations on the grid with convenient spacing.

The averaging procedure can be defined in different ways, as a *time average, space average,* or *ensemble average.* In conventional averaging procedures the mean quantities are ensemble averages (Figure 2.4). It is assumed that atmospheric flows are members of an *ensemble* whose individual realizations obey the Navier–Stokes equations. Ensemble averages have the following

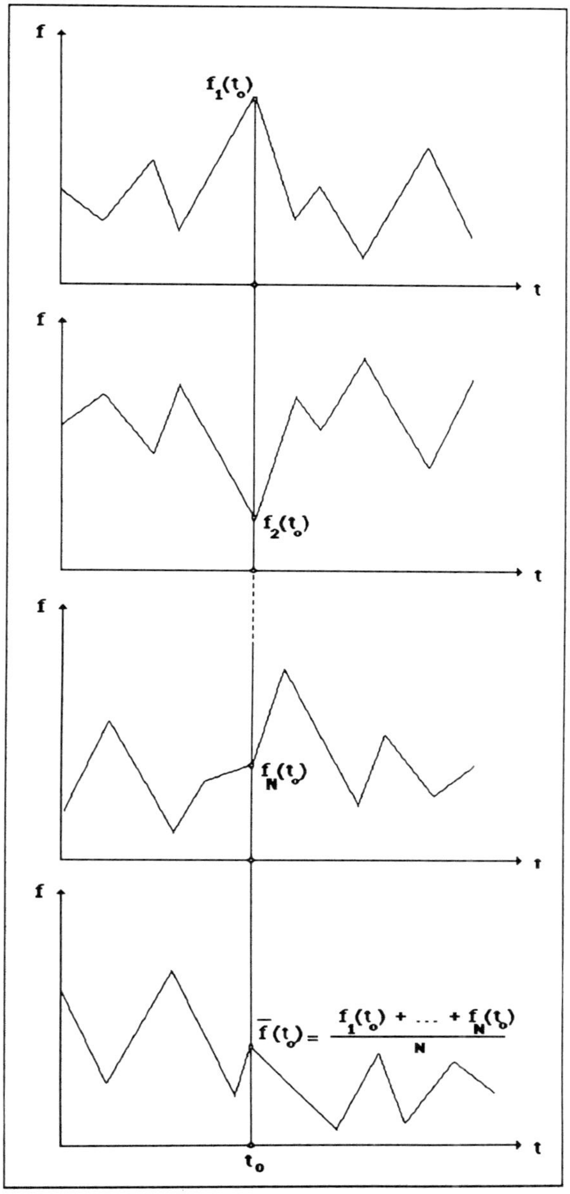

Figure 2.4 Illustration of ensemble averaging. The mean function is obtained for each moment of time by averaging random variables $f_1(t), f_2(t), f_3(t), \ldots$.

properties:

$$
\begin{aligned}
\overline{u_i}(t) &= \lim_{N\to\infty} \frac{u_i(t;1) + \cdots + u_i(t;N)}{N} \\
\overline{u_i} &= U_i \qquad \overline{u_i'} = 0,\ u_i = U_i + u_i' \\
\overline{U_iU_j} &= U_iU_j \qquad \overline{u_i'U_j} = \overline{U_iu_j'} = 0 \\
\overline{u_i + u_j} &= U_i + U_j \\
\overline{\frac{\partial u_i}{\partial t}} &= \frac{\partial U_i}{\partial t} \qquad \overline{\frac{\partial u_i}{\partial x_j}} = \frac{\partial U_i}{\partial x_j}
\end{aligned}
\tag{2.53}
$$

where $u_i(t;\ j)$ is the jth member of an ensemble of i components of the wind velocity, and the bar denotes the ensemble averaging operator. From (2.53) it follows that an ensemble average of a primed quantity is zero and that doubled averaging produces the same average. In addition, if a function is continuous and differentiable, its ensemble average is also continuous and differentiable.

The average of the nonlinear term u_iu_j has the form

$$\overline{u_iu_j} = U_iU_j + \underbrace{\overline{u_i'u_j'}}_{\substack{\text{Reynolds}\\ \text{stress}}} \tag{2.54}$$

Thus ensemble averaging of the velocity product u_iu_j adds new terms in the equations of motion, called *Reynolds stress* terms. Analogically, the average of the nonlinear temperature term has the form

$$\overline{\Theta''u_j} = \Theta U_j + \underbrace{\overline{\theta'u_j'}}_{\text{turbulent flux}} \tag{2.55}$$

Ensemble averaging of the product $\Theta''u_j$ adds the so-called *kinematic heat flux* components. Averaging (2.51), (2.4), and (2.43) yields the so-called *Reynolds form* of the equations:

$$
\begin{aligned}
&\frac{\partial U_i}{\partial t} + \frac{\partial U_iU_\alpha}{\partial x_\alpha} = f(U_\alpha - U_{g\alpha})\epsilon_{i\alpha 3} + g\frac{\Theta}{\Theta_0}\delta_{i3} - \frac{1}{\rho_0}\frac{\partial P}{\partial x_i} + \nu\frac{\partial^2 U_i}{\partial x_\alpha^2} - \frac{\partial \overline{u_i'u_\alpha'}}{\partial x_\alpha} \\
&\frac{\partial U_\alpha}{\partial x_\alpha} = 0 \\
&\frac{\partial \Theta}{\partial t} + \frac{\partial U_\alpha\Theta}{\partial x_\alpha} = \nu_\theta\frac{\partial^2\Theta}{\partial x_\alpha^2} - \frac{\partial\overline{u_\alpha'\theta'}}{\partial x_\alpha} + S_\theta \\
&\frac{\partial C}{\partial t} + \frac{\partial U_\alpha C}{\partial x_\alpha} = \nu_c\frac{\partial^2 C}{\partial x_\alpha^2} - \frac{\partial\overline{u_\alpha'c'}}{\partial x_\alpha} + S_c
\end{aligned}
\tag{2.56}
$$

In the potential temperature equation, the first term on the right-hand side describes the molecular diffusion (which was set apart from the sink/source term S_θ), and ν_θ is the thermal conductivity of air. The mean concentration equations are derived in a way similar to the potential temperature equation. In this equation S_c is a sink/source term and ν_c is the molecular diffusivity. The order of magnitude for both ν_θ and ν_c is 10^{-1} cm^2/s.

The main difference between the averaged (often called the Reynolds) set of equations and the primary one is the presence of second-order moments of the type $\overline{u_i'u_j'}$ or $\overline{u_i'\theta'}$, which are a consequence of the nonlinearity of the basic equations.

In practice, it is more convenient to use time or volume averaging instead of ensemble averaging. A simple time (or volume) average over a given interval of time (or in a given volume) is a constant. Applying this type of averaging operator to a given function will produce an averaged function that is discontinuous. To overcome this difficulty, a moving average can be considered. Generally, this type of averaging can be expressed in the following way:

$$\bar{u}(\mathbf{x}) = \int_{-\infty}^{\infty} u(\zeta) B_\lambda(\mathbf{x} - \zeta)\, d\zeta \tag{2.57}$$

where $\mathbf{x}$ is time or length, B_λ a filter function, and λ a wavelength characterizing the averaging operator. B_λ is supposed to be a symmetric (even) function of its argument, with an integral equal to 1. The simplest example of such a function is shown in Figure 2.5. For this function (2.57) can be modified as

$$\bar{u}(\mathbf{x}) = \frac{1}{2\lambda} \int_{\mathbf{x}-\lambda}^{\mathbf{x}+\lambda} u(\zeta) B_\lambda(\mathbf{x} - \zeta)\, d\zeta \tag{2.58}$$

and in the interval $(x - \lambda, x + \lambda)$, the function $B_\lambda = 1$.

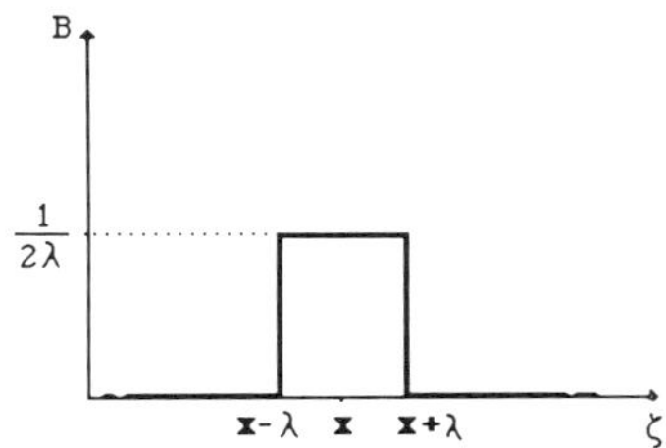

Figure 2.5 Filter function.

To discuss the properties of (2.58), we assume that

$$u = A \exp(ik\mathbf{x}) \tag{2.59}$$

where $k = 2\pi/L$ and L is a wavelength of the mode being considered. Substituting (2.59) in (2.58) gives

$$\bar{u}(\mathbf{x}) = Fu \tag{2.60}$$

where

$$F = \frac{\sin(2\pi\lambda/L)}{2\pi\lambda/L} \tag{2.61}$$

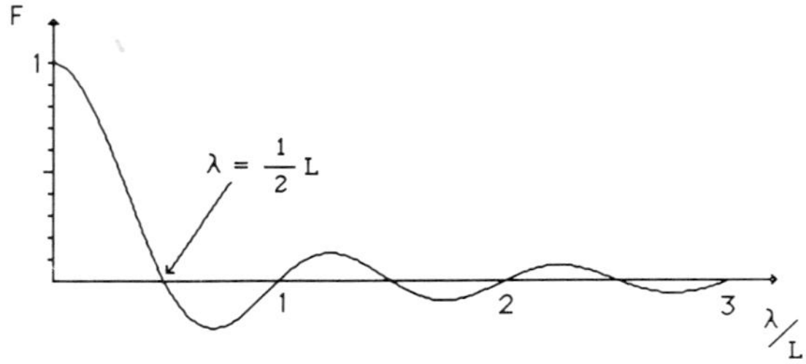

Figure 2.6 Function F given by (2.61).

The function F is displayed in Figure 2.6. The figure indicates that operation (2.58) filters practically all modes for which $L < 2\lambda$ [i.e., small (high-frequency) eddies (B_λ is a low-pass filter)].

It can be noticed from (2.59) and (2.60) that

$$\frac{\partial \bar{u}}{\partial x} = \overline{\frac{\partial u}{\partial x}} \tag{2.62}$$

From the definition it follows that

$$u' = u - \bar{u} = u\left(1 - \frac{\sin k\lambda}{k\lambda}\right) \tag{2.63}$$

and

$$\overline{u'} = u\frac{\sin k\lambda}{k\lambda}\left(1 - \frac{\sin k\lambda}{k\lambda}\right) \tag{2.64}$$

The result above indicates that time or space moving averaging has different properties than ensemble averaging and that

$$\overline{u'} \neq 0 \tag{2.65}$$

Applying the space-averaging operation to equations of motion, it is found that

$$\begin{aligned}
&\frac{\partial \bar{u}_i}{\partial t} + \frac{\partial \overline{u_i u_\alpha}}{\partial x_\alpha} = f(\bar{u}_\alpha - U_{g\alpha})\epsilon_{i\alpha 3} + g\frac{\overline{\Theta''}}{\Theta_0}\delta_{i3} - \frac{1}{\rho_0}\frac{\partial \overline{P''}}{\partial x_i} + \nu\frac{\partial^2 \bar{u}_i}{\partial x_\alpha^2}\\
&\frac{\partial \bar{u}_\alpha}{\partial x_\alpha} = 0\\
&\frac{\partial \overline{\Theta}}{\partial t} + \frac{\partial \overline{u_\alpha \Theta}}{\partial x_\alpha} = \nu_\theta \frac{\partial^2 \overline{\Theta}}{\partial x_\alpha^2} + \overline{S}_\theta\\
&\frac{\partial \overline{C}}{\partial t} + \frac{\partial \overline{U_\alpha C}}{\partial x_\alpha} = \nu_c \frac{\partial^2 \overline{C}}{\partial x_\alpha^2} + \overline{S}_c
\end{aligned} \tag{2.66}$$

Assuming that

$$R_{ij} = \overline{u_i u_j} - \bar{u}_i \bar{u}_j$$
$$H_i = \overline{u_i \Theta} - \bar{u}_i \overline{\Theta} \qquad (2.67)$$
$$D_i = \overline{u_i C} - \bar{u}_i \overline{C}$$

we obtain from (2.66):

$$\frac{\partial \bar{u}_i}{\partial t} + \frac{\partial \bar{u}_i \bar{u}_\alpha}{\partial x_\alpha} = f(\bar{u}_\alpha - U_{g\alpha})\epsilon_{i\alpha 3} + g\frac{\overline{\Theta''}}{\Theta_0}\delta_{i3} - \frac{1}{\rho_0}\frac{\partial \overline{P''}}{\partial x_i} + \nu\frac{\partial^2 \bar{u}_i}{\partial x_\alpha^2} - \frac{\partial R_{i\alpha}}{\partial x_\alpha}$$

$$\frac{\partial \bar{u}_\alpha}{\partial x_\alpha} = 0 \qquad (2.68)$$

$$\frac{\partial \overline{\Theta}}{\partial t} + \frac{\partial \bar{u}_\alpha \overline{\Theta}}{\partial x_\alpha} = \nu_\theta \frac{\partial^2 \overline{\Theta}}{\partial x_\alpha^2} - \frac{\partial H_\alpha}{\partial x_\alpha} + \overline{S}_\theta$$

$$\frac{\partial \overline{C}}{\partial t} + \frac{\partial \bar{u}_\alpha \overline{C}}{\partial x_\alpha} = \nu_c \frac{\partial^2 \overline{C}}{\partial x_\alpha^2} - \frac{\partial D_\alpha}{\partial x_\alpha} + \overline{S}_c$$

Equations (2.68) are identical in form to (2.56), obtained by applying ensemble averaging to the equation of motion. Neither set (2.56) nor (2.68) is closed, so they require parameterization of the Reynolds terms. Methods of closing governing equations in the Reynolds form are discussed in Chapter 6.

2.7 FURTHER SIMPLIFICATIONS

The most common simplification of the equation of motion assumes *horizontal homogeneity.* This assumption implies that the structure of the atmosphere does not change in the horizontal plane and allows the omission of horizontal gradients. From the equation of continuity, we also obtain that if the vertical velocity on the surface is zero, it vanishes in the whole boundary layer. As a result, we obtain

$$\frac{\partial U_i}{\partial t} - f(U_\alpha - U_{g\alpha})\epsilon_{i\alpha 3} = -\frac{\partial \overline{u_i' u_3'}}{\partial x_3} \qquad i = 1,2 \qquad (2.69)$$

$$\frac{\partial \Theta}{\partial t} = -\frac{\partial \overline{\theta' u_3'}}{\partial x_3} \qquad (2.70)$$

Denoting the left-hand sides of (2.69) and (2.70) by $\mathscr{T}_i$ and $\mathscr{T}_h$ yields

$$\mathscr{T}_i = -\frac{\partial \overline{u_i' u_3'}}{\partial x_3} \qquad i = 1, 2 \tag{2.71}$$

$$\mathscr{T}_h = -\frac{\partial \overline{\theta' u_3'}}{\partial x_3} \tag{2.72}$$

Integrating (2.71) and (2.72), we can obtain

$$\int_0^z \mathscr{T}_i \, dz = \overline{u_i' u_3'}\Big|_0 - \overline{u_i' u_3'}\Big|_z \qquad i = 1, 2 \tag{2.73}$$

$$\int_0^z \mathscr{T}_h \, dz = \overline{\theta' u_3'}\Big|_0 - \overline{\theta' u_3'}\Big|_z \tag{2.74}$$

The upper limit of integration can be chosen such that integrals of $\mathscr{T}_i$ and $\mathscr{T}_h$ are small (e.g., equal to 10% of the surface fluxes). This defines a layer, called the *surface layer,* in which

$$\begin{aligned} \overline{u_i' u_3'}\Big|_0 &\approx \overline{u_i' u_3'}\Big|_z \qquad i = 1, 2 \\ \overline{\theta' u_3'}\Big|_0 &\approx \overline{\theta' u_3'}\Big|_z \end{aligned} \tag{2.75}$$

From this it can be seen that in the surface layer the Reynolds stress and turbulent heat flux are approximately constant with height. The surface layer occupies the lowest 10% of the whole ABL, which is about 10^0 to 10^2 m.

The coordinate system can be rotated so that the x-axis of a new system is parallel to the Reynolds stress vector. In a new coordinate system,

$$\begin{aligned} -\overline{u'w'}\Big|_0 &= \text{const} = u_*^2 \\ -\overline{u'w'}\Big|_0 &= 0 \\ -\overline{\theta' w'}\Big|_0 &= \text{const} = u_* T_* \end{aligned} \tag{2.76}$$

where a new notation has been introduced:

$$\begin{aligned} x_1 &= x \qquad & U_1 &= U \qquad & u_1' &= u' \\ x_2 &= y \qquad & U_2 &= V \qquad & u_2' &= v' \\ x_3 &= z \qquad & U_3 &= W \qquad & u_3' &= w' \end{aligned} \tag{2.77}$$

The quantity u_* introduced above has the dimension of velocity and for that reason is called the *friction velocity.* The friction velocity is a characteristic scale of wind velocity in the surface layer. The quantity T_* has the dimension of temperature and is considered as a temperature scale in the surface layer. Scales u_* and T_* are treated more fully in Chapter 4.

PROBLEMS

2.1. What is the difference in tangent velocities of two points on the Earth's surface, the first point is on the equator and the second one on the tropic of cancer?

2.2. Find the lateral deflection of an object due to Coriolis acceleration. The object moves eastward at constant velocity $v = 20$ m/s. Air resistance and surface friction are to be neglected. The latitude is: **(a)** 45°N; **(b)** 0°N; **(c)** 45°S; **(d)** 90°N.

2.3. Find the divergence of the geostrophic wind.

2.4. Using (2.45) and assuming that $f = 10^{-4}$ s^{-1}, $U_g = V_g = 10$ m/s, $\rho = 1.1$ kg/m^3, $R = 287$ J kg^{-1} K^{-1}, pressure at point A $(0, 0, 0)$ $p_A = 1000$ mbar, find the temperature and pressure at point B (10 km, 10 km, 1 km).

2.5. Prove that for ensemble averaging,

$$\overline{u_i'} = 0$$

2.6. Find the potential temperature at point B, $\Theta(B)$, assuming that the air parcel moves from point A to point B, $\mathbf{d}\Theta/\mathbf{d}t = f_t$, $\Theta(A) = 273$ K, and

(a) $f_t = 0$

(b) $f_t = 1$ K/h, $U = 1$ m/s, $V = W = 0$, and the distance between A and B is 1 km. Assume that on the way from A to B, temperature does not change with time.

2.7. Prove that:

(a) $\gamma_a = g/c_p$

(b) $d\Theta/dz \approx dT/dz + \gamma_a$

(c) Θ = constant, for adiabatic processes

2.8. Define:

(a) Time average

(b) Volume average

(c) Ensemble average

2.9. Assuming that stress is proportional to the wind gradient,

$$\overline{v'w'} = -k_m \frac{dV}{dz}$$

where k_m (called the eddy viscosity) is a positive function of z, prove that the wind direction in the surface layer does not change with height.

2.10. Find the value of $\epsilon_{1\alpha 3}\Omega_\alpha U_3$.

2.11. What is the value of the geostrophic wind on the equator?

2.12. Given are instantaneous measurements of the potential temperature and vertical velocity:

Time, (s)	1	2	3	4	5	6	7	8	9	10
w (m/s)	0.5	0.7	0.3	0.4	0.1	−0.2	−0.4	0.7	0.3	−0.2
$\boldsymbol{\Theta}$ (K)	281.0	281.5	280.9	280.7	281.2	281.4	281.9	281.5	281.7	281.1

Find $\overline{w'^2}$, $\overline{\theta'^2}$, and $\overline{w'\theta'}$.

2.13. Assuming that

$$\overline{w'\theta'} = 1 - \frac{z}{z_i} \qquad \text{K} \cdot \text{m/s}$$

and $z_i = 500$ m, determine how the potential temperature will change in the next hour.

2.14. Assuming wind constant with height and equal to 10 m/s, find the horizontal gradient of the mean potential temperature for the same heat flux as in Problem 2.13. Assume also that the temperature does not change with time.

2.15. What are the density fluctuations corresponding to temperature fluctuations of 0.1 K? Assume that $\Theta = 300$ K and $\rho = 1.1$ kg/m^3.

2.16. What is a difference between p' and p'' in (2.52)?

CHAPTER THREE

STRUCTURE OF TURBULENCE

3.1 TRANSITION AND INSTABILITIES

The answer to the question of how turbulence arises is not easy; even the issue has a century-long history. The transition between laminar and turbulent flow is considered the most complex problem in fluid mechanics.

Theoretical investigations of transition are based on the assumption that laminar flows are affected by certain small disturbances. If the disturbances decay with time, the main flow is considered *stable*; on the other hand, if the disturbances increase with time, the flow is considered *unstable* and there exists the possibility of a transition to a turbulent pattern.

The theory of the stability of laminar flows decomposes the motion into a mean flow (whose stability constitutes the subject of the investigation) and into a disturbance superimposed on it. For the neutrally stratified regime the mean flow is described by its Cartesian velocity components, U, V, W, and its pressure, P. The corresponding quantities for nonsteady disturbance can be denoted by u', v', w', and p', respectively. Usually, additional assumptions on the mean flow are also made [e.g., that $U = U(z)$, $V = W = 0$, and $P = P(x, z)$]. The decomposition and additional assumptions allow linearization of the governing equations and their solution.

Linearized equations for u', v', w', and p' can be solved by introducing a stream function ψ for the perturbation flow field. For example, the stream function representing a single oscillation of the disturbance, in the two-dimensional case, has the form

$$\psi(x, z, t) = \phi(z)e^{i(\alpha x - \beta t)} \tag{3.1}$$

Components of the perturbation velocity can be expressed as

$$u' = \frac{\partial \psi}{\partial z} = \phi'(z) e^{i(\alpha x - \beta t)}$$

$$v' = -\frac{\partial \psi}{\partial x} = -i\alpha\phi(z) e^{i(\alpha x - \beta t)} \tag{3.2}$$

where $\phi = \phi_r + i\phi_i$ is the complex amplitude. Physical meaning is attached only to the real part of the stream function. The quantity β is also complex, $\beta = \beta_r + i\beta_i$, where β_r is the circular frequency of the partial oscillation, whereas β_i determines the degree of damping or amplification. If $\beta_i < 0$, the disturbances are damped and the laminar flow is stable. When $\beta_i > 0$ the laminar flow is unstable.

Substituting (3.2) into linearized equations of motion, after eliminating pressure, yields an ordinary, higher-order, differential equation for the amplitude $\phi(z)$, commonly referred to as the *Orr–Sommerfeld equation.* The issue of stability reduces to an eigenvalue problem of the Orr–Sommerfeld equation (Hinze, 1959).

In 1916, Lord Rayleigh successfully compiled several important general theorems concerning the stability of laminar velocity profiles. Derived by him, the *point-of-inflection criterion* asserts that velocity profiles which possess a point of inflection are unstable. It can be proven that the existence of a point of inflection constitutes a *necessary condition* for the occurrence of instability.

Turbulent flows can also be unstable. Instability of the turbulent flow in the atmospheric boundary layer leads to initialization of secondary motions often called *coherent structures.* Their development is also investigated in terms of the Orr–Sommerfeld equation (e.g., Drazin and Reid, 1981).

Figure 3.1 shows flows initiated in the atmosphere boundary layer by inflection-point instability and observed on many satellite pictures as horizontal roll vortices. The presence of horizontal roll vortices is marked by cloud streets that form in the regions of upward-moving air. Studies by Brown (1970, 1972, 1974, 1980) and LeMone (1973) show that the angle between the roll direction and the free stream vary from about 30° in the stable case to about 5° for the convective case. Kuettner (1959, 1971) found the following typical values for rolls: length: 20 to 500 km, spacing: 2 to 8 km, height: 0.8 to 2 km, width/height ratio: 2 to 4 : 1. The structure of such rolls is very difficult to determine from tower measurements since their axes are parallel to the wind, their crosswind propagation velocity is small, and also because the spacing between rolls is several kilometers.

The intensity of turbulent motion is closely connected with the *static stability* of the atmosphere and its thermal stratification. To demonstrate this in a most elementary way, consider balance of the gravity and the pressure gradient forces:

$$-\frac{1}{\rho}\frac{\partial p}{\partial z} = g \tag{3.3}$$

where ρ is the density of the air. It can be assumed that one parcel of air changes its density from the old value ρ to the new value $\mathrm{\rho}$. As a result of this change the parcel starts moving vertically with acceleration a:

$$-\frac{1}{\mathrm{\rho}}\frac{\partial p}{\partial z} = g + a \tag{3.4}$$

Substituting (3.3) into (3.4) and employing the equation of state (2.29) yields

$$\begin{aligned} a &= -g\frac{\mathrm{\rho} - \rho}{\mathrm{\rho}} = -g\frac{T - \mathbf{T}}{T} \\ &\approx -\frac{g}{T_0}\left(\frac{dT}{dz} - \frac{d\mathbf{T}}{dz}\right)z = -\frac{g}{T_0}\frac{d\Theta}{dz}z \end{aligned} \tag{3.5}$$

where T and $\mathbf{T}$ are the temperatures of the atmosphere and the parcel, respectively; T_0 is the average temperature in the lower atmosphere; Θ is the potential (or equivalent-potential) temperature; and z is the height above the earth's surface. Equation (3.5) can be transformed into the form

$$\frac{d^2z}{dt^2} = -N^2z \tag{3.6}$$

where N is the Brunt–Våisålå frequency, defined as

$$N^2 = \frac{g}{T_0}\frac{d\Theta}{dz} \tag{3.7}$$

The general solution of the differential equation (3.6) is

$$z = Ae^{iNt} + Be^{-iNt} \tag{3.8}$$

If $N^2 > 0$, the stratification is stable and (3.8) describes a vertical oscillation with angular frequency N. Period $2\pi/N$ is typically a few minutes long. These oscillations can be interpreted as *gravity waves*, since their nature is strongly affected by the action of gravity.

If $N^2 < 0$, the stratification is unstable. In this case it is convenient to define $\mathbf{N} = iN$, where $\mathbf{N}$ is real. Then

$$z = Ae^{\mathbf{N}t} + Be^{-\mathbf{N}t} \tag{3.9}$$

describes displacement which increases exponentially with time (at least initially) which can lead to turbulent motion. Usually, the exponential increase is limited later by nonlinear or dissipative effects.

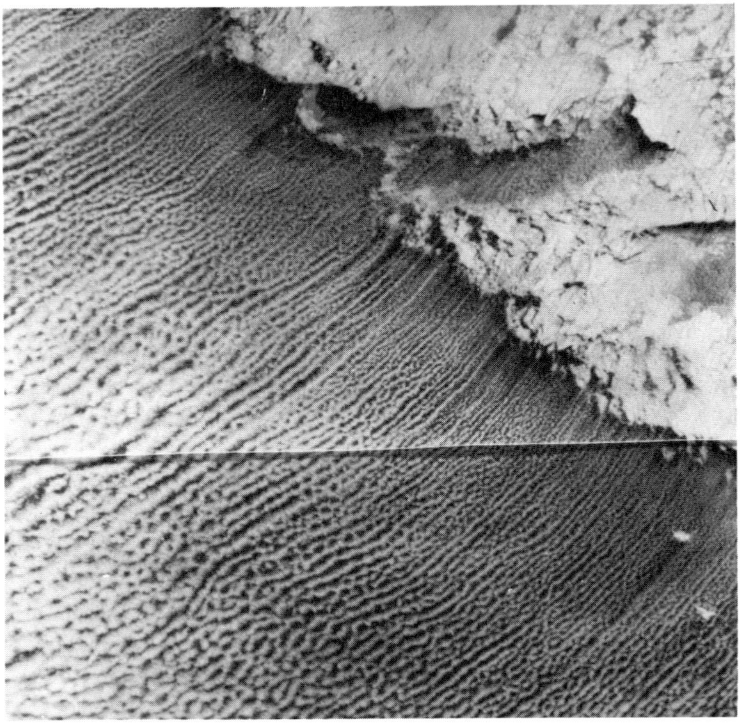

Figure 3.1 Cold air flowing off an ice pack and over the Bering Sea, February 22, 1983. (NOAA Satellite photo.)

Important criteria for the existence of turbulence can be derived by considering the energy balance of the flow. Let us assume that the horizontally homogeneous flow is affected by a disturbance characterized by length scale l, velocity scale v_l, and temperature scale Θ_l. The rate at which the kinetic energy of the motion is reduced by the kinetic energy of the disturbance in the unit of time is

$$R_k \sim \frac{v_l^2}{\tau_l} \sim \frac{v_l^3}{l} \tag{3.10}$$

The kinetic energy of the disturbance is later employed as work against the viscosity force,

$$R_v \sim \nu \frac{v_l^2}{l^2} \tag{3.11}$$

and also against the buoyancy force,

$$R_b \sim \frac{g}{\Theta_o}\Theta_l v_l \tag{3.12}$$

The disturbance persists only if $R_k > R_v + R_b$.

If the role of the buoyancy force is small and $R_v \gg R_b$, then $R_k > R_v$, which is equivalent to

$$\frac{l v_l}{\nu} = \mathrm{Re}_l > 1 \tag{3.13}$$

where Re_l is the Reynolds number for the disturbance. In contrast, if the role of the buoyancy force is large and $R_b \gg R_v$, then $R_k > R_b$, which is equivalent to

$$\frac{g}{\Theta_o}\frac{\Theta_l l}{v_l^2} = \mathrm{Ri}_l < 1 \tag{3.14}$$

where Ri_l is the Richardson number for the disturbance.

In conclusion, turbulence is maintained if the Reynolds number is large or, more precisely, if it is greater than its critical value Re_{cr} and, similarly, if the Richardson number Ri is smaller than its critical value Ri_{cr}. This is illustrated in Figure 3.2.

The buoyancy-driven instabilities have been often investigated in laboratory experiments. The best known experiment was that of Bénard (1900) on cellular convection. In 1916, Rayleigh established criteria for the onset of convection in terms of the dimensionless number, called the Rayleigh number Ra ($\mathrm{Ra} = -N^2H^4/\nu\nu_T$), where ν and ν_T are the kinematic and thermal diffusivities and H is the height of the convective layer). When the Rayleigh number is higher than its critical value, well-organized convective rolls can be observed. When Ra increases, the motion becomes more and more complex.

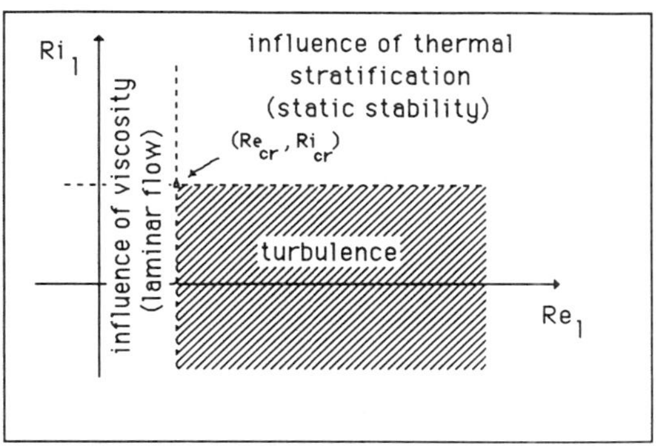

Figure 3.2 Regions of turbulent flow.

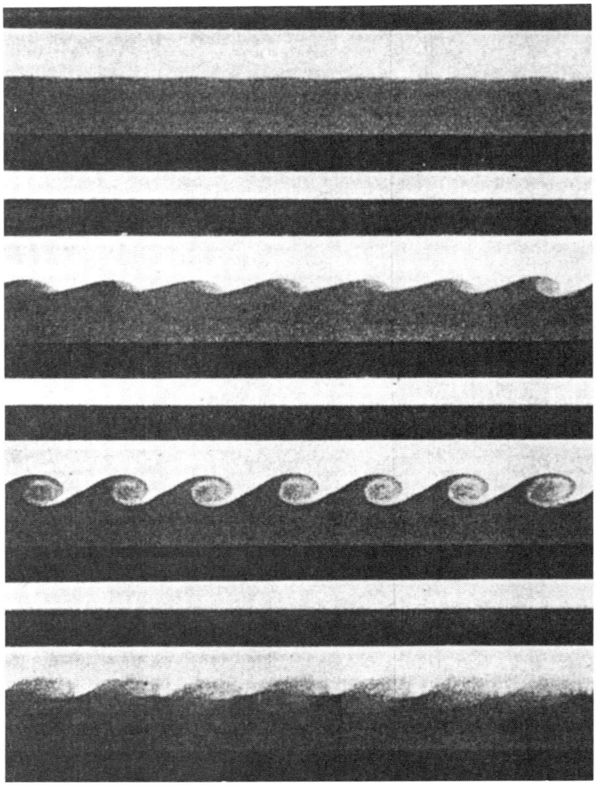

Figure 3.3 Kelvin–Helmholtz waves in a laboratory tube. (After Thorpe, 1979.)

Analogous coherent structures are observed in the atmospheric boundary layer. In more unstable conditions a convection pattern changes to hexagonal cells, called Bénard cells (Figure 3.1). Convective cells were observed with radar by Hardy and Ottersen (1979). The cells have diameters about five times the height of the boundary layer, with the most prominent axis lying along the direction of the mixed-layer mean wind vector.

In stably stratified flow, when the destabilizing influence of the wind shear overcomes the stabilizing effect of the buoyancy force, so-called *Kelvin–Helmholtz instability* takes place. Kelvin–Helmholtz instability, which was discovered in 1868, is initiated when small perturbations on a density discontinuity interface are excited by shear. If the static stability of the flow is not sufficient to dampen them, the perturbations may be amplified. Eventually, the waves may break and dissipate into smaller-scale turbulence. This can happen when the Richardson number is locally below its critical value of 0.25.

Kelvin–Helmholtz waves, shown in Figure 3.3, were generated in the laboratory on the surface between layers of dense brine, dyed by potassium

(a)

(b)

Figure 3.4 Gravity waves on the top of the atmospheric boundary layer: (a) National Center for Atmospheric Research/National Science Foundation photo archive; (b) taken by the author in Chicago in April 1987.

permanganate. The right end of the container was raised through 5° for about 4.4 s, causing a shear flow in which the upper layer, water, flows to the right and the lower layer, fluid brine, moves to the left. The photographs begin at the time when the tube was returned to the horizontal position and are about 0.5 s apart.

In the atmosphere, Kelvin–Helmholtz instability may originate within the stable boundary layer, but has also been observed by radar in the transition layer atop the unstable boundary layer (Gossard and Richter, 1970; Gossard et al., 1971, 1973; Richter et al., 1973). The internal waves, originated through Kelvin–Helmholtz instability, are typically several hundred meters in length. Their velocity of propagation was found to be about the same as the mean velocity of the layer in which they are embedded (Hooke et al., 1973). They may develop either in the shear layer of a low Richardson number within the boundary layer or just above it. With the passage of internal gravity waves, very thin shear layers of small-scale turbulence are observed moving up and down. Sometimes this type of instability is revealed by the cloud patterns on top of the boundary layer (Figure 3.4).

Nonlinear instabilities and interactions between waves can lead to random fluctuations similar to those resulting from turbulence. In such cases it is difficult to interpret the character of the flow, whether it results from genuine turbulence or from internal waves. Moreover, turbulence and waves often interact, and turbulence can be generated as a result of internal waves breaking. Generally, the major difference between turbulence and internal waves is that turbulent flow is highly dissipative and diffusive. On the other hand, wave motion is low-dissipative and nondiffusive and able to propagate with velocities higher than the mean flow.

3.2 STATISTICAL DESCRIPTION OF TURBULENCE

Random properties of turbulent flow can be analyzed in terms of the theory of random processes. The theory is not able to predict an individual realization of the turbulent field. It does, however, allow the evaluation of certain statistical properties.

It is convenient to begin a discussion of atmospheric turbulence by considering its idealized states, such as *stationarity*, *homogeneity*, and *isotropy*. Regardless of the great idealization, a study of simple cases permits an interpretation of some fundamental regularities, which are present in real turbulent flows.

Assumption of *stationarity* implies that all statistical properties of the flow are not dependent on time. In simple terms, "stationarity" means that any part of a measured time record has the same statistics as any other part of the same record.

Characteristics of the turbulent flow in the ABL are generally *nonstationary* mainly because of the diurnal forcing by the sun. The only exception is the vertical velocity, which has a mean value equal to zero. The assumption of stationarity for nonstationary random functions is sometimes acceptable in selected short periods of time, during which changes seem to occur through quasi-stationary states.

Homogeneity means that flow characteristics do not vary in space. Thus stationarity implies statistical invariance with respect to translation on the time axis, while homogeneity implies invariance in space.

Isotropy is usually considered only in conjunction with the assumption of *homogeneity* in all directions. Isotropy implies that the flow is invariant with respect to changes in the coordinate system in space (rotation and reflection).

The statistical structure of a stationary function $u(t)$ is described by the *probability density* $B(u)$. A value of $B(u)$ indicates a probability of finding a value $u(t)$ in an interval between values u and $u + \Delta u$. This probability can be calculated by counting a portion of time $\Delta t/T$ the point $u(t)$ spends in the interval between u and $u + \Delta u$, where T is the length of measurements (Figure 3.5):

$$B(u)\Delta u = \lim_{T \to \infty} \frac{1}{T} \sum_i \Delta t_i \tag{3.15}$$

From this definition it follows that $B(u)$ is always positive and its integral from $-\infty$ to $+\infty$ is equal to 1. The mean value of function $u(t)$ can be found as

$$\bar{u} = \int_{-\infty}^{\infty} uB(u)\, du \tag{3.16}$$

If the value above is equal to the time average of function $u(t)$ (defined in Chapter 2), then u is called *ergodic*.

The mean values of the various powers of u are called *moments*. The mean value is the first moment of u. Moments calculated for powers of the expression $u' = (u - \bar{u})$ are called *central moments*. The second central moment is called *variance* and is often denoted by σ^2. The third central moment, nondimensionalized by σ^3, is called *skewness*. The fourth central moment, nondimensionalized by σ^4, is called *kurtosis*. Meanings of skewness and kurtosis are illustrated in Figure 3.6.

The probability density of a normal or Gaussian distribution has the form

$$B(u) = \frac{1}{\sqrt{2\pi\sigma^2}} \exp\left[-\frac{(u - \bar{u})^2}{2\sigma^2}\right] \tag{3.17}$$

For the normal distribution, skewness is zero and kurtosis is equal to 3.

The concept of one-dimensional probability can be extended for two (or more) variables. Resulting joint probability density $B(u, v)$ is defined analo-

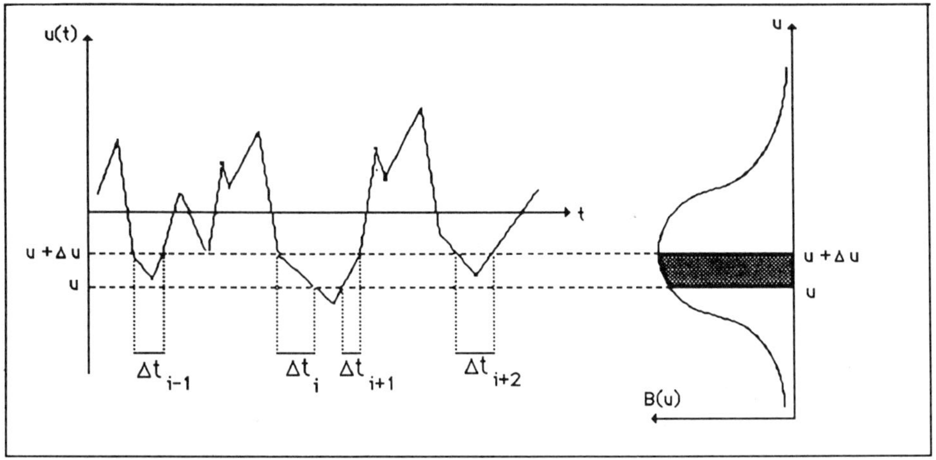

Figure 3.5 Probability density of a stationary function.

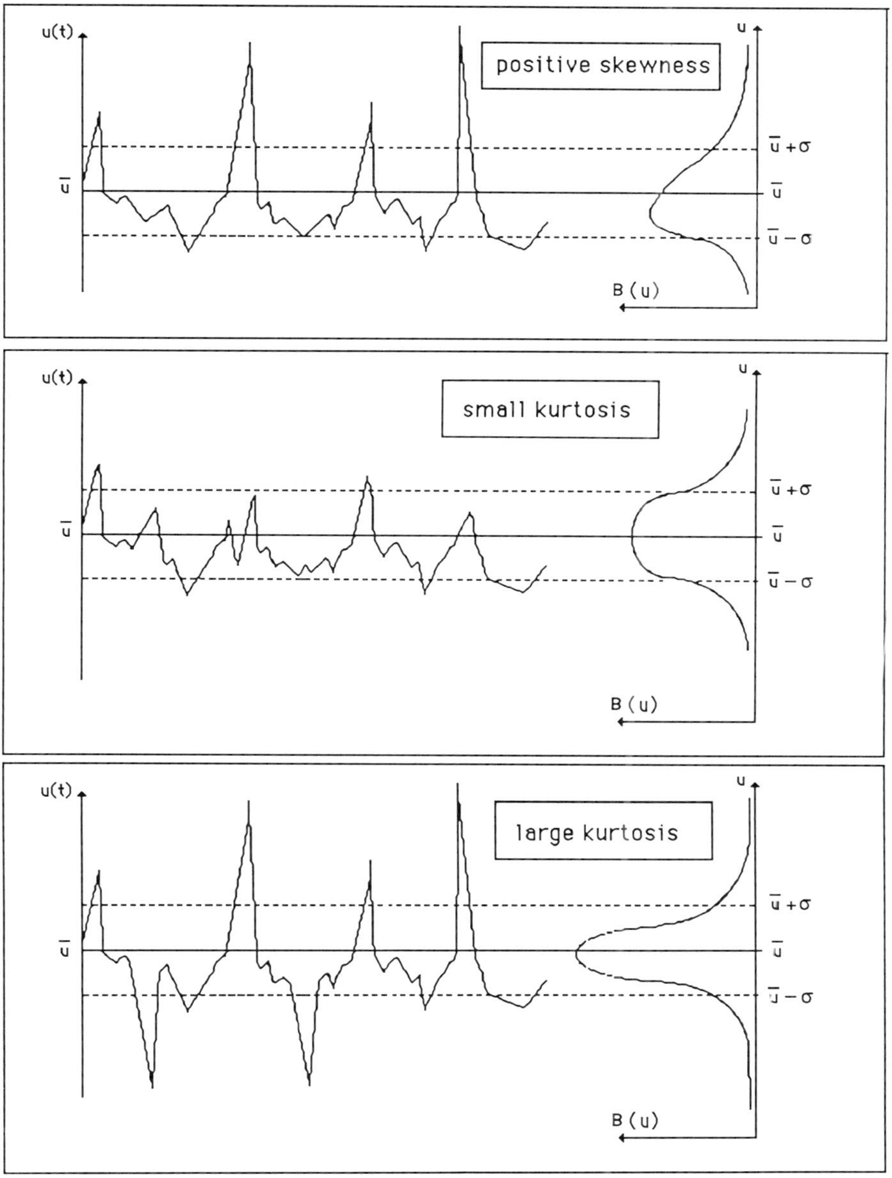

Figure 3.6 Function with positive skewness, small and large kurtosis.

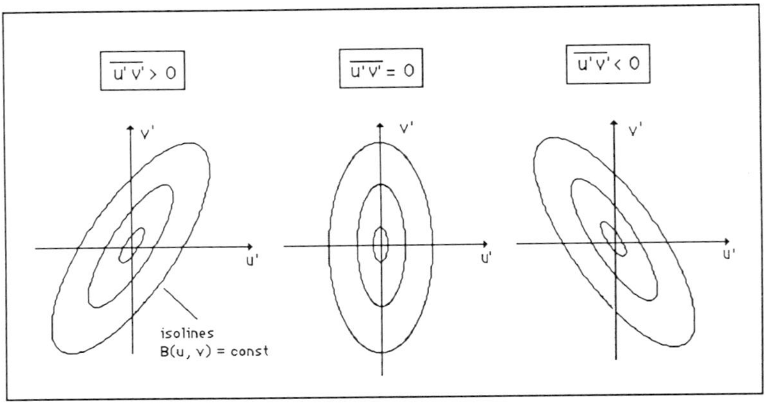

Figure 3.7 Examples of joint probability densities with various correlations.

gously to (3.15) by a fraction of time $\Delta t/T$ the point (u, v) spends in the interval between values $(u, u + \Delta u)$ and $(v, v + \Delta v)$.

The correlation between u and v is defined as

$$\overline{u'v'} = \iint_{-\infty}^{\infty} (v - \bar{v})(v - \bar{v}) B(u, v)\, du\, dv \tag{3.18}$$

and is called *covariance* between u and v. If covariance is equal to zero, $u(t)$ and $v(t)$ are said to be *uncorrelated* (Figure 3.7). Uncorrelated variables are not necessarily independent of each other. Two variables are *statistically independent* only if $B(u, v) = B(u)B(v)$.

The probability density can be considered for variables in different positions in space. The same can be done with statistical moments. For example, the *spatial structure function D* and *spatial correlation function R* are defined as

$$\begin{aligned} D(\mathbf{x}_1, \mathbf{x}_2) &= \overline{[f'(\mathbf{x}_1) - f'(\mathbf{x}_2)]^2} \\ D_i(\mathbf{x}_1, \mathbf{x}_2) &= \overline{[u_i'(\mathbf{x}_1) - u_i'(\mathbf{x}_2)]^2} \\ R(\mathbf{x}_1, \mathbf{x}_2) &= \overline{f'(\mathbf{x}_1) f'(\mathbf{x}_2)} \\ R_{ij}(\mathbf{x}_1, \mathbf{x}_2) &= \overline{u_i'(\mathbf{x}_1) u_j'(\mathbf{x}_2)} \end{aligned} \tag{3.19}$$

where f is a scalar (temperature, humidity), u_i and u_j are components of the wind velocity, a prime denotes a deviation from the mean value and vectors $\mathbf{x}_1$ and $\mathbf{x}_2$ describe the position.

When the field f is homogeneous, the functions D and R depend only on the distance $\mathbf{r} = (r_1, r_2, r_3) = |\mathbf{x}_1 - \mathbf{x}_2|$, and spatial correlation function R is symmetric, $R(\mathbf{r}) = R(-\mathbf{r})$. In the homogeneous and isotropic case, all functions $R_{ij}(\mathbf{r})$ for the i and j components of the wind velocity can be characterized by only two correlation functions, longitudinal R_{uu} and lateral R_{vv}, depending only on distance $r = |\mathbf{r}|$. It is easy to see that $R_{uu}(r) = R_{11}(r_1, 0, 0) = R_{22}(0, r_2, 0) = R_{33}(0, 0, r_3)$ and $R_{vv}(r) = R_{11}(0, r_2, 0) = R_{11}(0, 0, r_3) = R_{22}(r_1, 0, 0) = \cdots$. Also, $R_{vv} = R_{uu} + (r/2)(\partial/\partial r)R_{uu}$ (Hinze, 1959).

Turbulence as a stationary stochastic process is described by the joint probability density for all the variables on which it depends, and for all instants of time and all points in the fluid space. It is, of course, not possible to obtain this function experimentally. However, valuable information can be derived from a few simple statistics of the turbulent flow.

A new direction in research on turbulence has recently been originated. This research includes the theory of strange attractors (e.g., Cvitanović, 1984; Schuster, 1984). The new approach attempts to fill a gap left by the classical theory of turbulence, which deals only with statistical properties of chaotic flows. The classical theory assumes that individual realizations of a turbulent flow are sensitive to initial conditions but that averages do not exhibit such sensitivity because the fluctuations become uncorrelated with time. It does not take into consideration that systems sensitive to initial conditions are capable of creating coherent structures. Coherent structures can be looked at as order within turbulent chaos, implying that many random phenomena can be more predictable than had been thought. The new theory was initiated by a striking discovery that simple deterministic systems can generate chaotic behavior. This was accomplished in 1963 by E. N. Lorenz, who, motivated by the desire to understand the predictability of weather, considered a simplified system of three ordinary equations to describe Rayleigh's convection between two hot plates. The solution of this system behaved in a random fashion and was the first known example of a chaotic or strange attractor.

As an example of the strange attractor, the following equation can be considered:

$$\frac{d^2z}{dt^2} = (1 - z^2)z - 0.25\frac{dz}{dt} + 0.4\cos t \tag{3.20}$$

Two solutions of this equation for two initial conditions, $z = 1.0$ and $z = 1.1$, are shown in Figure 3.8. For the first 15 units of time, the difference between both solutions is very small. After this time the solutions look completely different and can be treated as two realizations of a random process. The small change of coefficients can change a character of the solution. For example, changing the coefficient 0.4 in (3.20) to 0.5 causes the solution to lose its chaotic properties and become periodic.

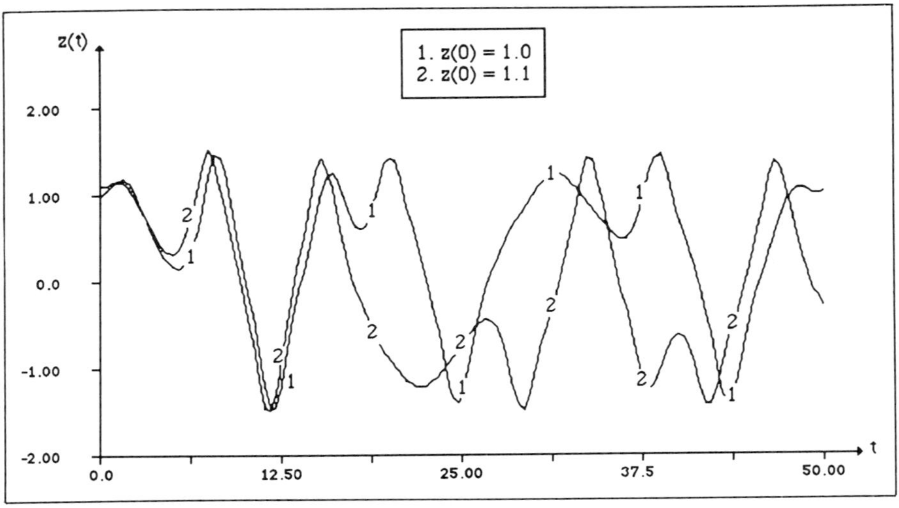

Figure 3.8 Deterministic chaos: two solutions to the ordinary differential equation (3.20) obtained for two initial conditions.

3.3 SCALES OF TURBULENT FLOW

The size of the different eddies is characterized by space scales ℓ. Atmospheric motion occurs over a large range of scales—from thousands of kilometers to millimeters. In 1926, Richardson noticed that smaller-scale motions originated as a result of the instability of larger ones. A *cascade process*, in which eddies of the largest size are broken into smaller and smaller ones, continues down to molecular scales in which the dissipation mechanism turns the kinetic energy of motion into heat. A poetic form of Richardson's notion is well known:

Big whorls have little whorls,
Which feed on their velocity;
Little whorls have smaller whorls,
and so on unto viscosity.

An air velocity record from a fast-moving aircraft flight in the ABL could be interpreted as a spatial record. Based on this record, the correlation function as a function of separation along the flight path can be formed. A measure of the distance over which the signal is correlated defines the *integral scale* $\mathscr{L}$ (or outer scale). This scale is obtained by integration of the normalized correlation function,

$$\mathscr{L} = \int_0^\infty \frac{R(r)}{R(0)}\, dr \tag{3.21}$$

and is interpreted as characteristic of dominant eddies. The integral scale characterizes the eddies carrying the bulk of the turbulent kinetic energy and performing most of the turbulent transport. In principle, there is an integral scale for every flow property, and in general, it depends on the distance from the surface, stability, and the direction of the measurement path (lateral or streamwise).

The turbulent kinetic energy per unit mass E (defined as a half of the averaged sum of the squared velocity fluctuations), the dissipation rate ϵ (defined as a change rate of E in time due to viscosity effects), and the molecular kinematic viscosity ν define another important scale, η, called the *Taylor microscale*:

$$\eta = \left(\frac{\nu E}{\epsilon}\right)^{1/2} \tag{3.22}$$

This scale marks the size where viscous effects become significant.

Turbulent flow in the atmosphere is neither homogeneous nor isotropic, because of the presence of the rigid boundary of the earth's surface. However, the flow can be considered *locally homogeneous and isotropic* in small subregions of the atmosphere far from the earth's surface. Measurements in the atmospheric boundary layer have shown that this is satisfied by the high-frequency fluctuations, which, because of their small size, are not influenced by the surface and have the same properties in all directions. Local homogeneity and isotropy of the atmospheric motion was predicted in 1941 by Kolmogorov. Kolmogorov introduced two hypotheses stating that during the cascade process, in which any large-scale flow in the atmosphere breaks down into motions of smaller scales, the direct influence of larger eddies is lost and smaller eddies tend to have independent properties, universal for all types of turbulent flows. Consequently, they should be isotropic. The form of the first hypothesis is:

> The average properties of the small-scale eddies of any turbulent flow at large Reynolds numbers are uniquely determined by the kinematic viscosity of the fluid ν and the average dissipation rate ϵ.

Kolmogorov's first hypothesis applies in the range determined by the inequality $\mathscr{L} \gg \ell \gg \mu$ (called the *equilibrium range*), where μ is called the *inner scale* or *Kolmogorov's microscale.* From dimensional analysis, μ has the form

$$\mu = \left(\frac{\nu^3}{\epsilon}\right)^{1/4} \tag{3.23}$$

Kolmogorov's microscale defines the size of the viscous eddies dissipating the energy.

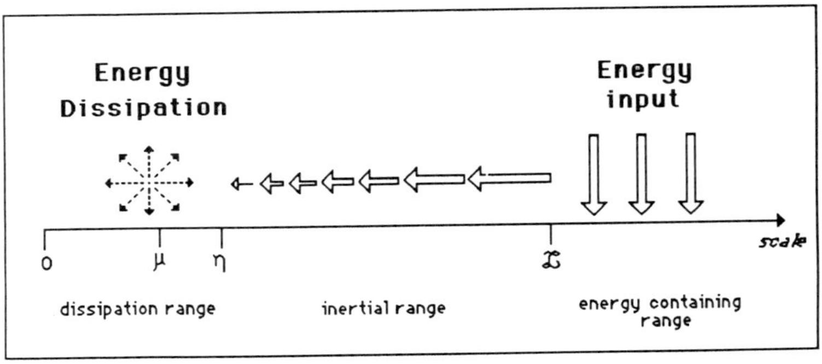

Figure 3.9 Schematic distribution of energy in the atmosphere.

The second hypothesis is as follows:

For sufficiently large Reynolds numbers, there is a subrange of the equilibrium range in which average properties of the flow are determined only by the average rate of dissipation ϵ.

This subrange, defined by the inequality $\mathscr{L} \gg \ell > \eta$, is called the *inertial* subrange, since it is dominated by inertia forces, whose action leads to redistribution of the energy across the turbulent spectrum. In the inertial subrange, energy neither enters the system nor is dissipated (Figure 3.9). It is merely transmitted at rate ϵ from large-scale toward small-scale motion, and then it is dissipated. Therefore, the regime in this subrange must be stationary and have universal character.

In a convective ABL the turbulence dissipation rate per unit mass is about $\epsilon \sim 3 \times 10^{-3}$ m^2 s^{-3}. Using $E \sim 2$ m^2 s^{-2} and $\nu \sim 1.5 \times 10^{-5}$ m^2 s^{-1}, we find that $\eta = (\nu E/\epsilon)^{1/2} \sim 0.1$ m. Hence if $\mathscr{L} = 100$ m, the region between η and $\mathscr{L}$ is three decades long. The dissipative eddy size is $\mu = (\nu^3/\epsilon)^{1/4} = 0.001$ m, and the eddy size range in the convective boundary layer covers roughly five decades. All the scales defined above are illustrated in Figure 3.9. The figure shows energy being generated in the energy-containing range (long wavelengths), transported without energy loss through the inertial subrange, and dissipated because of viscosity in the dissipation range.

3.4 EQUATIONS OF HIGHER MOMENT

Random properties of turbulent motion can only be characterized in terms of statistical moments of the governing parameters. Equations (2.56) or (2.66) describe dynamics of the first moments. First-moment equations include second moments, which are undefined. A new set of equations for second moments can also be derived. A way to derive these equations is quite straightforward. First, equations for fluctuations can be obtained by subtracting averaged equations, term by term, from primitive ones. Multiplying equations obtained this way by the proper quantities and then averaging gives the desired results.

Subtracting the averaged equations (2.56) term by term from the equations (2.51) yields the equations for fluctuations:

$$\frac{\partial u_i'}{\partial t} + \frac{\partial U_i u_\alpha'}{\partial x_\alpha} + \frac{\partial u_i' U_\alpha}{\partial x_\alpha} + \frac{\partial u_i' u_\alpha'}{\partial x_\alpha} = f u_\alpha' \epsilon_{i\alpha 3} + g\frac{\theta'}{\Theta_0}\delta_{i3} - \frac{1}{\rho_0}\frac{\partial p'}{\partial x_i} + \nu\frac{\partial^2 u_i'}{\partial x_\alpha^2} + \frac{\partial \overline{u_i' u_\alpha'}}{\partial x_\alpha}$$

$$\frac{\partial u_\alpha'}{\partial x_\alpha} = 0 \tag{3.24}$$

$$\frac{\partial \theta'}{\partial t} + \frac{\partial u_\alpha' \Theta}{\partial x_\alpha} + \frac{\partial U_\alpha \theta'}{\partial x_\alpha} + \frac{\partial u_\alpha' \theta'}{\partial x_\alpha} = \nu_\theta \frac{\partial^2 \theta'}{\partial x_\alpha^2} - \frac{\partial \overline{u_\alpha' \theta'}}{\partial x_\alpha} + S_\theta'$$

Multiplying the equation for u_i' by u_j' and the equation for u_j' by u_i' and adding the results gives after averaging

$$\underbrace{\frac{\partial \overline{u_i' u_j'}}{\partial t} + U_\alpha \frac{\partial \overline{u_i' u_j'}}{\partial x_\alpha}}_{1} = \underbrace{-\left(\overline{u_j' u_\alpha'}\frac{\partial U_i}{\partial x_\alpha} + \overline{u_i' u_\alpha'}\frac{\partial U_j}{\partial x_\alpha}\right)}_{2} \underbrace{- \frac{\partial \overline{u_i' u_j' u_\alpha'}}{\partial x_\alpha}}_{3}$$

$$\underbrace{- \frac{1}{\rho_0}\left(\overline{u_i' \frac{\partial p'}{\partial x_j}} - \overline{u_j' \frac{\partial p'}{\partial x_i}}\right)}_{4} + \underbrace{\frac{g}{\Theta_0}\left(\overline{u_i'\theta'}\delta_{j3} + \overline{u_j'\theta'}\delta_{i3}\right)}_{5} \tag{3.25}$$

$$\underbrace{+ \nu\left(\overline{u_j' \frac{\partial^2 u_i'}{\partial x_\alpha^2}} + \overline{u_i' \frac{\partial^2 u_j'}{\partial x_\alpha^2}}\right)}_{6} + \underbrace{f\left(\overline{u_j' u_\alpha'}\epsilon_{i\alpha 3} + \overline{u_i' u_\alpha'}\epsilon_{j\alpha 3}\right)}_{7}$$

Term 1 is the total derivative of the covariance $\overline{u_i' u_j'}$, describing its temporal and advective changes. Term 2 represents production due to the interaction of the mean flow with turbulence. Term 3 characterizes transport due to divergence of the covariance. Term 4 describes the interaction between the pressure and wind velocity components. Term 5 represents the buoyancy and can be positive or negative. Term 6 is due to viscous effects. The last term 7 represents the effect of the Coriolis force. This term is negligible for flows with the time scale $\tau \ll f^{-1}$.

Applying an analogous procedure, the equations for covariances involving scalar quantities such as temperature and concentration can be obtained. The equations for the components of the turbulent heat flux have the form

$$\underbrace{\frac{\partial \overline{u_i'\theta'}}{\partial t} + U_\alpha \frac{\partial \overline{u_i'\theta'}}{\partial x_\alpha}}_{1} = \underbrace{-\left[\overline{u_i'u_\alpha'}\frac{\partial \Theta}{\partial x_\alpha} + \overline{\theta' u_\alpha'}\frac{\partial U_i}{\partial x_\alpha}\right]}_{2} \underbrace{- \frac{\partial \overline{u_i'u_\alpha'\theta'}}{\partial x_\alpha}}_{3}$$

$$\underbrace{- \frac{1}{\rho_0}\overline{\theta'\frac{\partial p'}{\partial x_i}}}_{4} + \underbrace{\frac{g}{\Theta_0}\overline{\theta'^2}\delta_{i3}}_{5} \tag{3.26}$$

$$+ \underbrace{\left(\overline{\nu\theta'\frac{\partial^2 u_i'}{\partial x_\alpha^2}} + \overline{\nu_\theta u_i'\frac{\partial^2 \theta'}{\partial x_\alpha^2}}\right)}_{6} + \underbrace{f\overline{u_\alpha'\theta'}\epsilon_{i\alpha 3}}_{7}$$

In the equation obtained, term 1 is the total derivative of covariance $\overline{u_i'\theta'}$. Term 2 represents production due to the interaction of the mean temperature field with turbulence. Term 3 is transport due to divergence of the covariance. Term 4 describes the interaction between the pressure and temperature fields. Term 5 represents the buoyancy effects. Term 6 is due to viscous effects. The last term 7 represents the effect of the Coriolis force. This term is usually negligible. Replacing the temperature in (3.26) by concentration results in an analogical equation for the covariance of a scalar quantity.

The equation for the temperature variance is obtained by multiplying the equation for the temperature perturbation by θ' and averaging. The result is

$$\underbrace{\frac{1}{2}\left(\frac{\partial \overline{\theta'^2}}{\partial t} + U_\alpha \frac{\partial \overline{\theta'^2}}{\partial x_\alpha}\right)}_{1} = \underbrace{-\overline{u_\alpha'\theta'}\frac{\partial \Theta}{\partial x_\alpha}}_{2} \underbrace{- \frac{1}{2}\frac{\partial \overline{u_\alpha'\theta'^2}}{\partial x_\alpha}}_{3} \underbrace{- \nu_\theta \overline{\left(\frac{\partial \theta'}{\partial x_\alpha}\right)^2}}_{4} \tag{3.27}$$

Term 1 is the total derivative of the variance $\overline{\theta'^2}$. Term 2 represents production due to the interaction of the mean flow with turbulence. Term 3 defines transport due to divergence of the covariance. Term 4 is due to viscous effects. The source term has been omitted.

3.5 BUDGETS OF TURBULENT FLUXES AND VARIANCES

Important information about the structure of the atmospheric boundary layer can be obtained by analyzing the terms in equations of higher moment. Budget studies performed for this purpose employ measurements obtained from field experiments as well as the results of numerical simulations. Their results are discussed below. Vertical distributions of the statistical moments are discussed in Chapter 4.

3.5.1 Balance of the turbulent kinetic energy

As mentioned before, turbulent kinetic energy E is defined as half the sum of the squared velocity fluctuations:

$$E = \frac{1}{2}\overline{u_\alpha'^2} = \frac{\overline{u'^2} + \overline{v'^2} + \overline{w'^2}}{2} \tag{3.28}$$

The equation for the turbulent kinetic energy balance is obtained by adding equations (3.25) written for $i = j = 1$, 2, and 3. The result after dividing by 2 and assuming horizontal homogeneity of the mean flow is

$$\frac{1}{2}\frac{\partial \overline{u_\alpha'^2}}{\partial t} = \underbrace{-\left(\overline{u'w'}\frac{\partial U}{\partial z} + \overline{v'w'}\frac{\partial V}{\partial z}\right)}_{\substack{\text{mechanical}\\ \text{production}}} \underbrace{- \frac{1}{2}\frac{\partial \overline{u_\alpha'^2 w'}}{\partial z}}_{\substack{\text{turbulent}\\ \text{transport}}} \underbrace{- \frac{1}{\rho_0}\overline{u_\alpha'\frac{\partial p'}{\partial x_\alpha}}}_{\text{pressure effects}} \underbrace{+ \frac{g}{\Theta_0}\overline{w'\theta'}}_{\substack{\text{buoyant}\\ \text{production}}} \underbrace{- \epsilon}_{\text{dissipation}} \tag{3.29}$$

The descriptions under the terms in (3.29) identify their physical meaning. The term on the left-hand side, and the turbulent transport, pressure, and dissipation terms on the right-hand side, are obtained simply by a summation of analogous terms in (3.25). The first term on the right-hand side is a sum of two terms. The first one is obtained from (3.25) for $i = 1$ and the second for $i = 2$. The analogous term for $i = 3$ vanishes, since it was assumed that $\partial W/\partial z = 0$. In (3.29), there is no Coriolis term, because no work is being done by this force. The total dissipation rate ϵ has the following form:

$$\epsilon = -\nu\overline{u_\beta'\frac{\partial^2 u_\beta'}{\partial x_\alpha^2}} = \underbrace{-\frac{1}{2}\nu\frac{\partial^2 \overline{u_\beta'^2}}{\partial x_\alpha^2}}_{\text{very small}} + \underbrace{\nu\overline{\left(\frac{\partial u_\beta'}{\partial x_\alpha}\right)^2}}_{\text{dissipation}} \tag{3.30}$$

The first term on the right-hand side can be omitted as being small. The second term is always positive, implying that a turbulent dissipation term is always positive and participates in the loss of turbulent energy.

The dissipation rate in the equation (3.30) is expressed by the sum of 9 terms. In isotropic turbulence the expressions of the type $s_{ii} = \overline{(\partial u'_i/\partial x_i)^2}$ are equal for all $i = 1, 2, 3$ and also expressions $s_{kj} = \overline{(\partial u'_k/\partial x_j)^2}$ are equal for all combinations of $k \neq j$. Besides $2s_{11} = 2s_{22} = 2s_{33} = s_{12} = s_{13} = s_{23} = s_{21} = s_{31} = s_{32}$. Because there are 3 terms s_{ii} and 6 related to $2s_{ii}$, the expression for a dissipation rate in the isotropic case simplifies to (Hinze, 1959)

$$\epsilon = 15\nu\overline{(\partial u_1/\partial x_1)^2} = 7.5\nu\overline{(\partial u_1/\partial x_2)^2} \tag{3.31}$$

The relation (3.31) has two very important practical uses. It provides a simple criterion for determining whether or not the structure of dissipating eddies is isotropic. Conversely, assuming isotropy of the small-scale motion, this relation permits determination of the viscous dissipation.

The total pressure term can be expressed in the form

$$\underbrace{-\frac{1}{\rho_0}\overline{u'_\alpha\frac{\partial p'}{\partial x_\alpha}}}_{\text{pressure effects}} = -\frac{1}{\rho_0}\frac{\overline{\partial p'u'_\alpha}}{\partial x_\alpha} + \underbrace{\frac{1}{\rho_0}\overline{p'\frac{\partial u'_\alpha}{\partial x_\alpha}}}_{\text{return-to isotropy term}} \tag{3.32}$$

The second term on the right-hand side vanishes when the field is incompressible. This indicates that this term does not contribute to generating turbulent energy. However, the term exists in equations for each of the components of the turbulent kinetic energy. Hence its role is to redistribute turbulent energy between its components. For this reason the term is named "return to isotropy."

In the case when the transport and pressure terms are horizontally homogeneous, (3.29) simplifies to the form

$$\frac{\partial E}{\partial t} = -\left(\overline{u'w'}\frac{\partial U}{\partial z} + \overline{v'w'}\frac{\partial V}{\partial z}\right) - \frac{\partial\overline{E'w'}}{\partial z} - \frac{1}{\rho_0}\frac{\partial\overline{p'w'}}{\partial z} + \frac{g}{\Theta_0}\overline{w'\theta'} - \epsilon \tag{3.33}$$

The budgets of the turbulent kinetic energy in the convective and stable boundary layers are shown in Figure 3.10. In the convective case (Figure 3.10a), shear production takes place very close to the surface (and also in the inversion level above the mixed layer, which is not shown in the figure). In the surface layer, the momentum flux is negative and the wind shear is positive, resulting in positive shear production. Within the mixed layer, where the vertical gradient of the wind velocity is close to zero, the shear production term vanishes.

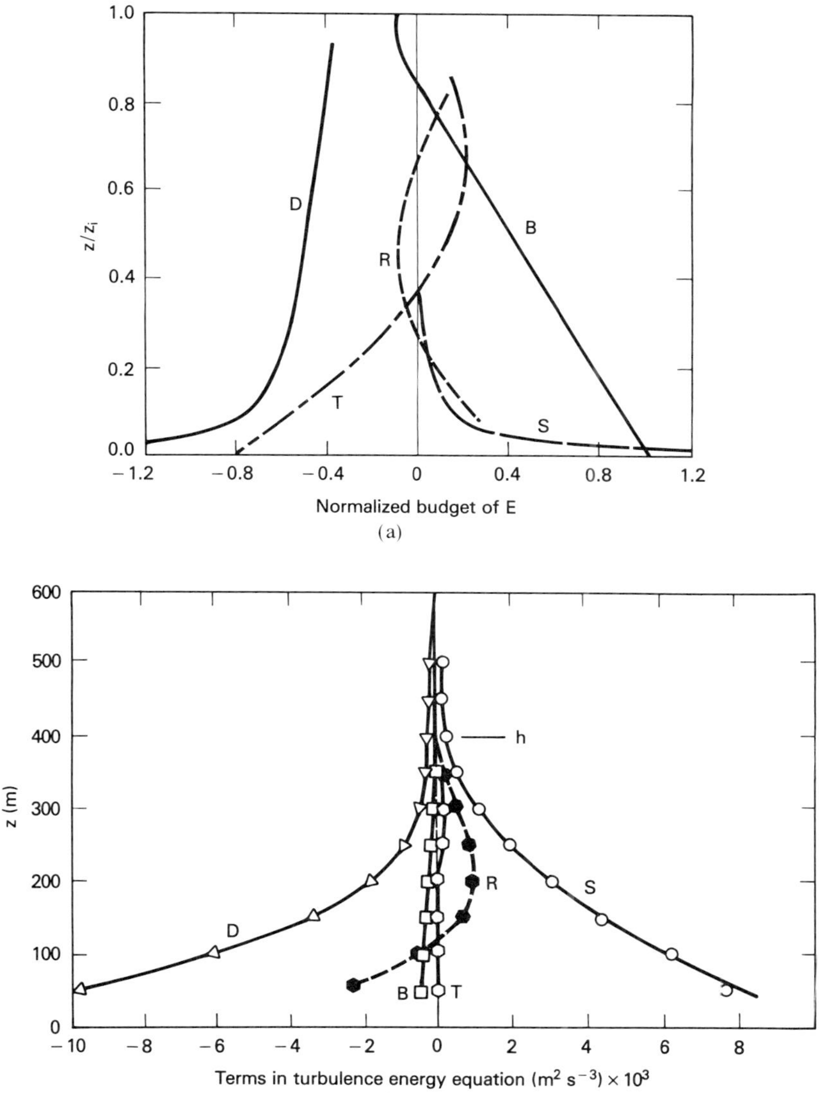

Figure 3.10 Dimensionless turbulent kinetic energy budget: (a) convective case [after Lenschow, Wyngaard, and Pennell, 1980. *Journal of Atmospheric Sciences* published by the American Meteorological Society. (AMTEX data)]; (b) stable case [after Lenschow et al. 1988 (SESAME experiment)]. B, buoyant production; S, mechanical (shear) production; D, dissipation; T, turbulent transport; R, residual.

The buoyant production is almost balanced by the dissipation term. Since the buoyancy force acts vertically, the buoyant production term contributes only to the third component of the turbulent kinetic energy. Turbulent transport is responsible for a return-to-isotropy tendency in the turbulent kinetic energy equation. The transport by pressure fluctuations is difficult to measure and is usually neglected in the turbulent budget.

In the surface layer the turbulent transport term is negative but changes sign in the upper part of the ABL. This implies a loss of energy in the lower half of the mixed layer and a gain in the upper half—the turbulent kinetic energy is exported upward. If there are no turbulent energy fluxes at the top and the bottom of the ABL, the integral of the turbulent transport term becomes zero. This means that this term neither creates nor destroys the turbulent kinetic energy, and its role is to redistribute it with height. At night the picture changes. Any mechanical production is destroyed by dissipation and negative heat flux (Figure 3.10b). The turbulent transport is very small.

It is convenient to express equation (3.33) in a slightly different form by introducing the flux Richardson number:

$$\frac{\partial E}{\partial t} = -\left(\overline{u'w'}\frac{\partial U}{\partial z} + \overline{v'w'}\frac{\partial V}{\partial z}\right)(1 - R_f) - \frac{\partial \overline{E'w'}}{\partial z} - \frac{1}{\rho_0}\frac{\partial \overline{p'w'}}{\partial z} - \epsilon \quad (3.34)$$

The flux Richardson number R_f is defined as a ratio of the buoyant production term to the shear production term:

$$R_f = \frac{(g/\Theta_0)\overline{w'\theta'}}{\overline{u'w'}(\partial U/\partial z) + \overline{v'w'}(\partial V/\partial z)} \quad (3.35)$$

The flux Richardson number provides an important criterion for turbulence. In the surface layer the term in the denominator of (3.35) is negative. In the unstable case, the heat flux is positive and then $R_f < 0$. The first term of the right-hand side in (3.34) is then positive and is a source of turbulent energy. Large negative Richardson numbers can be achieved when the heat flux is positive and the shear production term is zero or negligibly small. This case is referred to as *free convection.* When the Richardson number is zero because of negligible heat flux, the turbulent energy is generated only by shear production. The convection taking place in this case is referred to as *forced convection.* In contrast, in the stable case, the heat flux is negative and $R_f > 0$. Moreover, if $R_f > 1$, the first term on the right-hand side of (3.34) is negative and contributes to the complete suppression of turbulence. However, observations show that turbulence cannot be maintained even at smaller values of R_f. A critical value of 0.25 for R_f is generally accepted.

In the surface layer, equation (3.33) can be expressed in the simplified form

$$-\overline{u'w'}\frac{\partial U}{\partial z} - \frac{\partial \overline{E'w'}}{\partial z} - \frac{1}{\rho_0}\frac{\partial \overline{p'w'}}{\partial z} + \frac{g}{\Theta_0}\overline{w'\theta'} - \epsilon = 0 \quad (3.36)$$

After a simple transformation one can obtain

$$\underbrace{\frac{\kappa z}{u_*}\frac{\partial U}{\partial z}}_{\phi_m} + \underbrace{\frac{\kappa z}{u_*^3}\frac{\partial \overline{E'w'}}{\partial z} + \frac{\kappa z}{u_*^3}\frac{1}{\rho_0}\frac{\partial \overline{p'w'}}{\partial z}}_{\phi_t} - \underbrace{\frac{\kappa z}{u_*^3}\frac{g}{\Theta_0}\overline{w'\theta'}}_{z/L} + \underbrace{\frac{\kappa z}{u_*^3}\epsilon}_{\phi_\epsilon} = 0 \quad (3.37)$$

where κ is a constant, called the von Kármán constant, u_* was defined in (2.77), and all terms ϕ in this equation are dimensionless. The fourth term of the left-hand side can be expressed in the form z/L, where L has a unit of length and is called the Monin–Obukhov length:

$$L = -\frac{u_*^3}{\kappa(g/\Theta_0)\overline{w'\theta'}} \quad (3.38)$$

In the surface layer, momentum and heat fluxes are approximately constant with height, and therefore L is also considered to be constant. The form of equation (3.37) suggests that all dimensionless terms in this equation should depend only on a dimensionless parameter z/L. This issue will be brought up again in Chapter 4.

Under near-neutral conditions, z/L is numerically small. Under these conditions the second and third terms are also small and shear production is balanced locally by dissipation. The first term approaches unity. The rate of dissipation in the neutral surface layer is thus given by $\epsilon = u_*^3/\kappa z$. This is in agreement with experimental data (Wyngaard and Coté, 1971).

Under stable conditions, the budget is dominated by shear production and dissipation, which are proportional to z/L. The turbulent transport is small. The buoyancy term is also relatively small and represents a loss.

Under unstable conditions, all terms are significant. The numerical values of the various terms increase with increasing instability, except for the shear production, which decreases. Over the entire unstable regime, the dissipation exceeds the shear plus thermal production. However, the term representing the vertical flux of turbulent energy is negative.

3.5.2 Heat flux balance

The vertical heat flux balance for a horizontally homogeneous boundary layer has the following form:

$$\frac{\partial \overline{w'\theta'}}{\partial t} = -\underbrace{\overline{w'^2}\frac{\partial \Theta}{\partial z}}_{\substack{\text{mechanical}\\ \text{production}}} - \underbrace{\frac{\partial \overline{w'^2\theta'}}{\partial z}}_{\substack{\text{turbulent}\\ \text{transport}}} + \underbrace{\frac{g}{\Theta_0}\overline{\theta'^2}}_{\substack{\text{buoyant}\\ \text{production}}} - \underbrace{\frac{1}{\rho_0}\overline{\theta'\frac{\partial p'}{\partial z}}}_{\substack{\text{pressure}\\ \text{effect}}} \quad (3.39)$$

where the molecular and Coriolis terms were omitted as small.

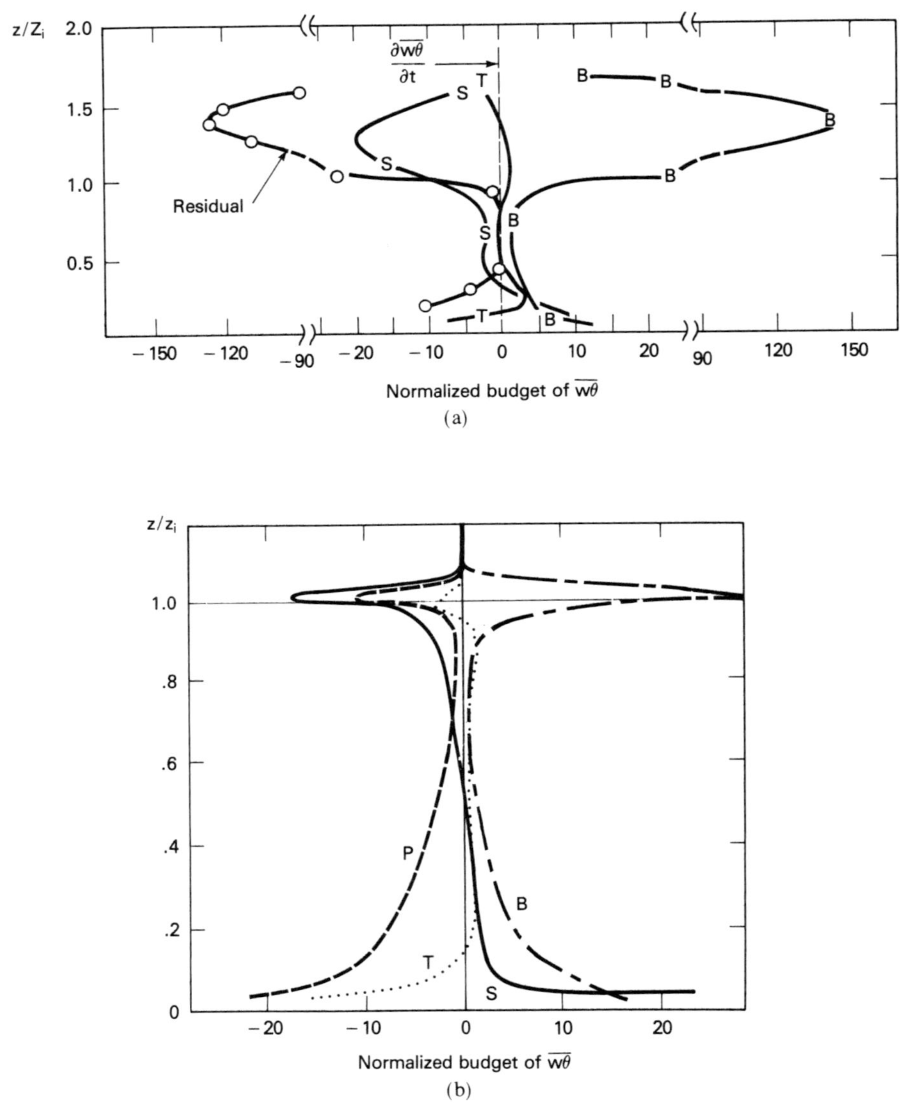

Figure 3.11 Dimensionless heat flux budget in the convective boundary layer: (a) obtained at Boulder BAO tower by Zhou, Lenschow, Stankov, Kaimal, and Gaynor 1985. *Journal of Atmospheric Sciences* published by the American Meteorological Society. (b) as calculated for Wangara Day by the model 33 of Therry and Lacarrére (1983). B, buoyant production; S, shear production; P, pressure effect; T, turbulent transport. Z_i is defined as the height of the inversion base capping the mixed layer. z_i is defined as the height at which the turbulent heat flux is minimum.

The budget for the vertical heat flux in the mixed layer is shown in Figure 3.11. Figure 3.11(a) displays observational data; Figure 3.11(b) is obtained from a numerical simulation of a second-order closure model (numerical models are discussed in Chapter 6). The figure indicates that the vertical heat flux is produced in the surface layer by buoyancy and shear (positive terms). As a result, there is small but regular upward transfer of this flux above the surface layer. At the inversion level, there is a large production of heat flux by buoyancy. Buoyant production at this level is balanced mainly by a negative mechanical production, caused by air parcels penetrating the overlying warmer layer. This negative term forces negative values of heat flux on the top of the mixed layer. The pressure effect decreases the magnitude of the heat flux as a result of the return-to-isotropy tendency.

Convection characterized by the budget in Figure 3.11 is called *penetrative*. The simpler case of *nonpenetrative* convection can also occur. In the laboratory, nonpenetrative convection is obtained by uniform heating from below the fluid, which is insulated at its upper solid boundary. Nonpenetrative

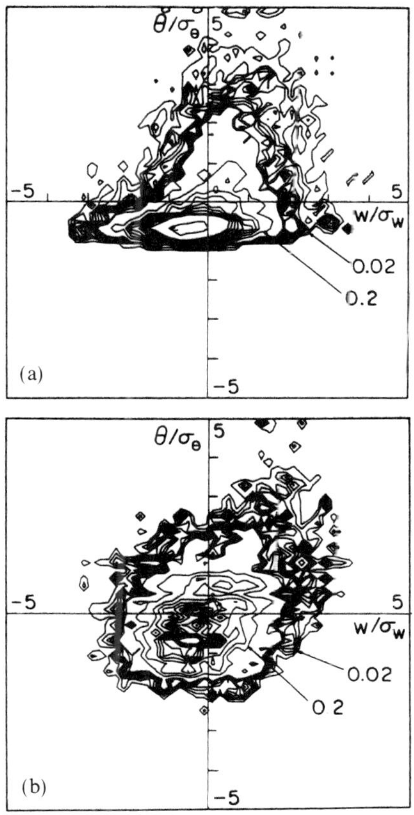

Figure 3.12 Joint probability density for velocity and temperature fluctuations in nonpenetrative convection: (a) $z/z_* = 0.0175$; (b) $z/z_* = 0.50$ (z_*, height of the boundary layer). Laboratory results of Adrian et al. (1986).

Interpretation of the figure:

I quadrant: warm updrafts (thermals)

II quadrant: warm downdraft (circulation forced by mass continuity)

III quadrant: cold downdraft (sinking of locally cooler air)

IV quadrant: cold updrafts (inertially rising cooled thermals)

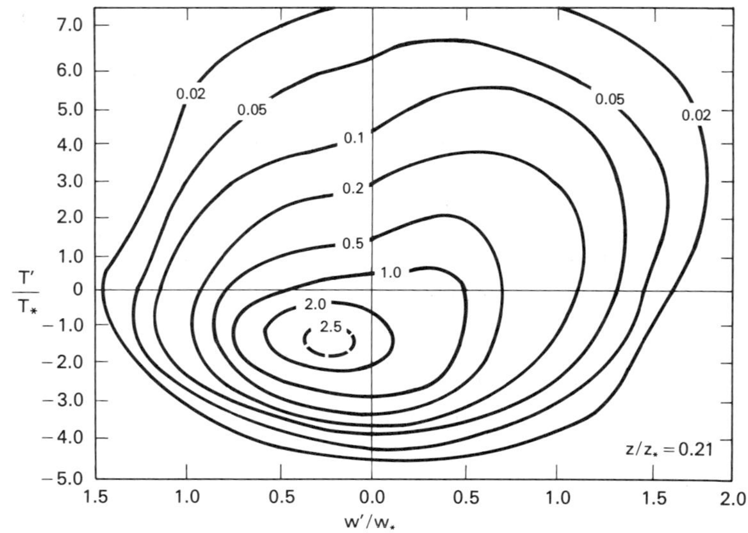

(a)

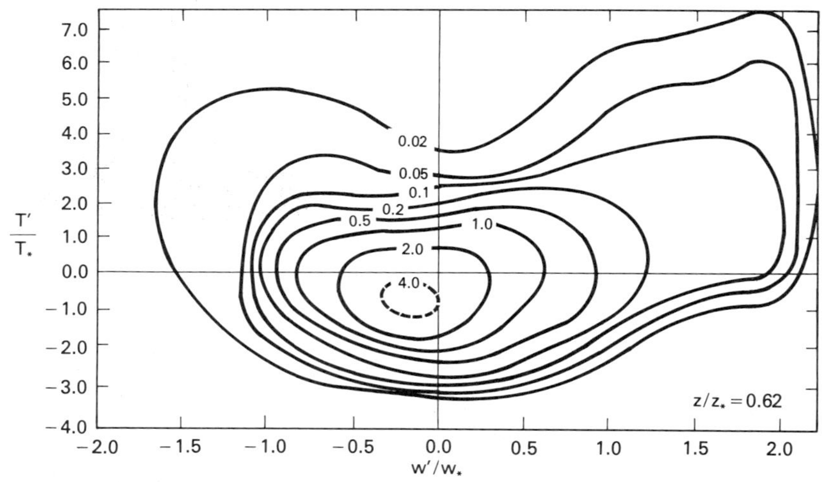

(b)

Figure 3.13 Joint probability distribution for velocity and temperature fluctuations in penetrative convection: (a) $z/z_i = 0.21$; (b) $z/z_i = 0.62$; (c) $z/z_i = 0.99$; (d) $z/z_i = 1.06$. Laboratory results of Deardorff and Willis (1985). Reprinted by permission. Interpretation of the figure as in Figure 3.12. In addition, in the second quadrant there is a contribution from entrained air.

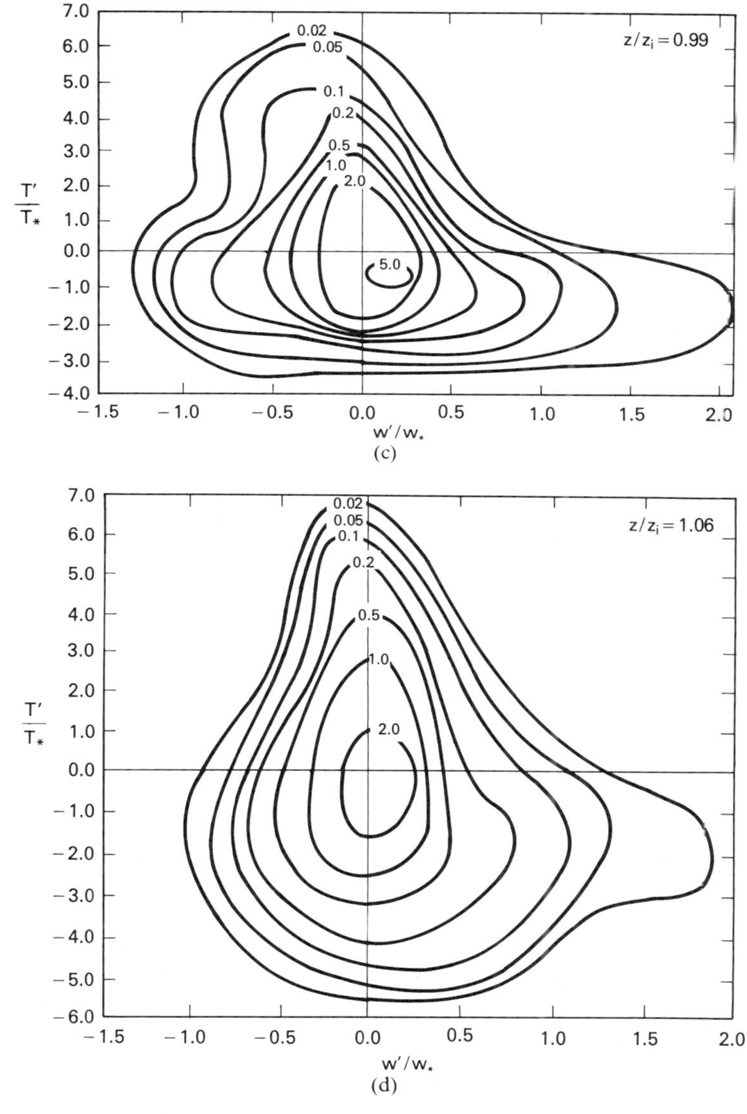

Figure 3.13 Continued

convection rarely, if ever, happens in the atmosphere. However, conditions close to the nonpenetrative regime can be observed early in the morning when convection is weak. This case is described by the so-called *encroachment* slab models (see Section 6.2).

Heating at the surface generates plumes of warmer air called *thermals*. Thermals in the nonpenetrative regime decelerate moving upward, and by the time they reach the upper part of the layer their positive buoyancy changes to slightly negative. In this way ascending thermals are converted into a cool descending flow. Ascending thermals are faster, but they occupy a smaller area than a descending motion, which is slower but more spread out. In general, positive temperature and vertical velocity fluctuations in thermals and negative temperature and vertical velocity fluctuations in descending air (Figure 3.12) cause turbulent heat flux to change linearly with height from a positive value at the bottom to zero at the top (Adrian et al., 1986). From the temperature equation (2.72), it can be concluded that in this case, the boundary layer is unsteady and its temperature increases at a steady rate, given by $Q_0 t/z_i$, where t is time and Q_0 is the surface kinematic heat flux. All other statistical quantities, however, are independent of time and the only effect of unsteadiness is a constant warming of the layer.

In the penetrative convection, thermals tend to overshoot the inversion base. As a result, "wisps" of air from the inversion layer are entrained into the mixed layer. The majority of thermals penetrating the topping inversion layer are characterized by negative temperature and positive vertical velocity fluctuations, while for most of the entrained air, these fluctuations are positive for temperature and negative for vertical velocity (Figure 3.13). The resulting heat flux in the penetrative convection is negative on the top of the mixed layer and also approximately linear.

In the atmosphere thermals are usually visible only on displays of remote sensors (Figure 1.6). Sometimes their presence can be marked by cumulus clouds in the sky. In very special conditions (dry) they can also be seen by eye. It happens when a rising thermal picks up surface dust and debris in a cone of rotating motion. This phenomenon is called a *dust devil*. Vorticity is caused by slight terrain irregularities. Dust devils often extend into the atmosphere for a few hundreds of meters.

In the steady-state mixed layer, not very close to its boundaries, the turbulent transport can be omitted in (3.39):

$$\overline{w'^2}\frac{\partial \Theta}{\partial z} = \frac{g}{\Theta_0}\overline{\theta'^2} - \frac{1}{\rho_0}\overline{\theta'\frac{\partial p'}{\partial z}} \tag{3.40}$$

The pressure term is usually parametrized in the following way (Wyngaard, 1982):

$$\frac{1}{\rho_0}\overline{\theta'\frac{\partial p'}{\partial z}} = \frac{\overline{w'\theta'}}{\tau} \tag{3.41}$$

where τ is a turbulent time scale. Substituting this expression into (3.40) gives

$$\overline{w'\theta'} = -\tau\left(\overline{w'^2}\frac{\partial\Theta}{\partial z} - \frac{g}{\Theta_0}\overline{\theta'^2}\right) \tag{3.42}$$

This indicates that in the convective boundary layer the vertical heat flux is not simply proportional to the temperature gradient. The assumption of proportionality between the turbulent flux and the mean quantity gradient is a basis for the K-models of turbulence considered in Chapter 6. From the arguments above, it follows that this parametrization fails in the convective boundary layer.

3.5.3 Momentum flux balance

The vertical momentum flux balance for a horizontally homogeneous boundary layer has the following form:

$$\frac{\partial\overline{u_i'w'}}{\partial t} = -\underbrace{\overline{w'^2}\frac{\partial U_i}{\partial z}}_{\substack{\text{mechanical}\\ \text{production}}} - \underbrace{\frac{\partial\overline{u_i'w'^2}}{\partial z}}_{\substack{\text{turbulent}\\ \text{transport}}} - \underbrace{\frac{1}{\rho_0}\left(\overline{u_i'\frac{\partial p'}{\partial z}} + \overline{w'\frac{\partial p'}{\partial x_i}}\right)}_{\text{pressure effects}} + \underbrace{\frac{g}{\Theta_0}\overline{u_i'\theta'}}_{\substack{\text{buoyant}\\ \text{production}}}$$

$$+ f\underbrace{\overline{u_\alpha'w'}\epsilon_{i\alpha 3}}_{\text{Coriolis term}} + \underbrace{\nu\left(\overline{w'\frac{\partial^2 u_i'}{\partial x_\alpha^2}} + \overline{u_i'\frac{\partial^2 w'}{\partial x_\alpha^2}}\right)}_{\text{viscous term}} \tag{3.43}$$

where $i = 1, 2$. The viscous term can be omitted as being small.

Momentum flux budgets for a convective afternoon with a strong wind shear are depicted in Figure 3.14. The figure, obtained from a numerical simulation, shows that in the lower part of the mixed layer the mechanical production term is approximately balanced by the pressure term. This result, confirmed also by observations in the atmospheric boundary layer (e.g., Brost et al., 1982), indicates that the pressure covariances destroy all stress flux produced by the mean shear. For $i = 1$, this approximate balance has the form

$$\overline{w'^2}\frac{\partial U}{\partial z} = -\frac{1}{\rho}\left(\overline{w'\frac{\partial p'}{\partial x}} + \overline{u'\frac{\partial p'}{\partial z}}\right) \tag{3.44}$$

Applying a simple parametrization that in analogy to (3.41) the pressure term is equal to $\overline{u'w'}/\tau$, where τ is a time scale, gives

$$\overline{u'w'} = -\tau\overline{w'^2}\frac{\partial U}{\partial z} = -k_m\frac{\partial U}{\partial z} \tag{3.45}$$

where k_m is called the eddy viscosity. The result obtained indicates that in the lower part of the convective boundary layer the momentum flux is propor-

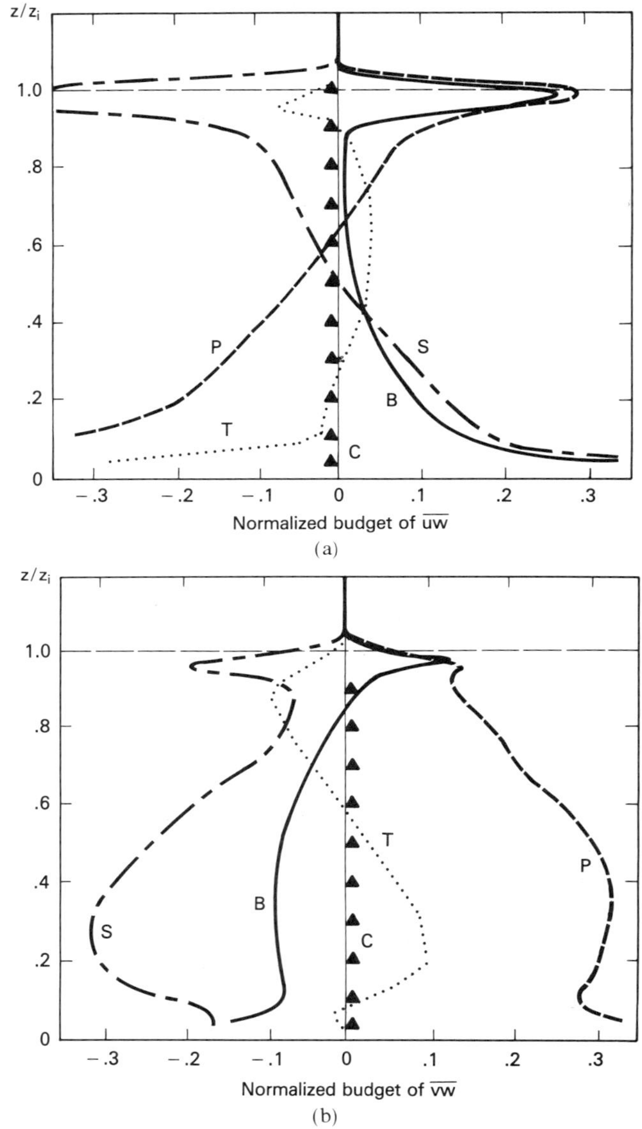

Figure 3.14 Dimensionless (a) eastward and (b) northward momentum flux budget during convective afternoon as calculated by the model of Therry and Lacarrére (1983). S, shear production; B, buoyant production; P, pressure effect; T, turbulent transport; C, Coriolis effect.

tional to the mean shear. This parametrization, however, fails in pure convective cases, when the wind shear is negligible.

3.5.4 Temperature and humidity variance and covariance balances

The temperature variance balance for a horizontally homogeneous boundary layer has the following form:

$$\frac{\partial \overline{\theta'^2}}{\partial t} = \underbrace{-2\overline{w'\theta'}\frac{\partial \Theta}{\partial z}}_{\substack{\text{shear}\\ \text{production}}} \underbrace{-\frac{\partial \overline{w'\theta'^2}}{\partial z}}_{\substack{\text{turbulent}\\ \text{transport}}} \underbrace{-2\nu_\theta \overline{\left(\frac{\partial \theta'}{\partial x_\alpha}\right)^2}}_{\substack{\text{molecular}\\ \text{destruction}}} \tag{3.46}$$

The physical meanings are indicated underneath all terms. The last term in this equation,

$$N = 2\nu_\theta \overline{\left(\frac{\partial \theta'}{\partial x_\alpha}\right)^2} \tag{3.47}$$

is called the destruction rate of the temperature variance and has a meaning and a form analogous to the dissipation rate term, defined by (3.30).

The budget of the temperature variance (observed and numerically simulated) in the mixed layer is shown in Figure 3.15. The figure indicates that the mechanical production is generated only in the surface layer and close to the capping inversion layer, in regions where the vertical gradient of potential temperature does not vanish. The turbulent transport exports temperature variance from these two regions toward the inside of the mixed layer (the results of the numerical simulation shown in the figure indicate a peak whose presence, however, is not confirmed by measurements). Figure 3.15b shows a magnitude of the radiative destruction term [omitted in (3.46) as small] which follows from the source term in (3.24).

The equation of the specific humidity (or concentration of any passive substance) variance has analogous form to (3.46), with the destruction rate analogous to (3.47):

$$N_q = 2\nu_q \overline{\left(\frac{\partial q'}{\partial x_\alpha}\right)^2} \tag{3.48}$$

Also, the equation for the correlation between the temperature and the specific humidity (or concentration of any passive substance) has a form

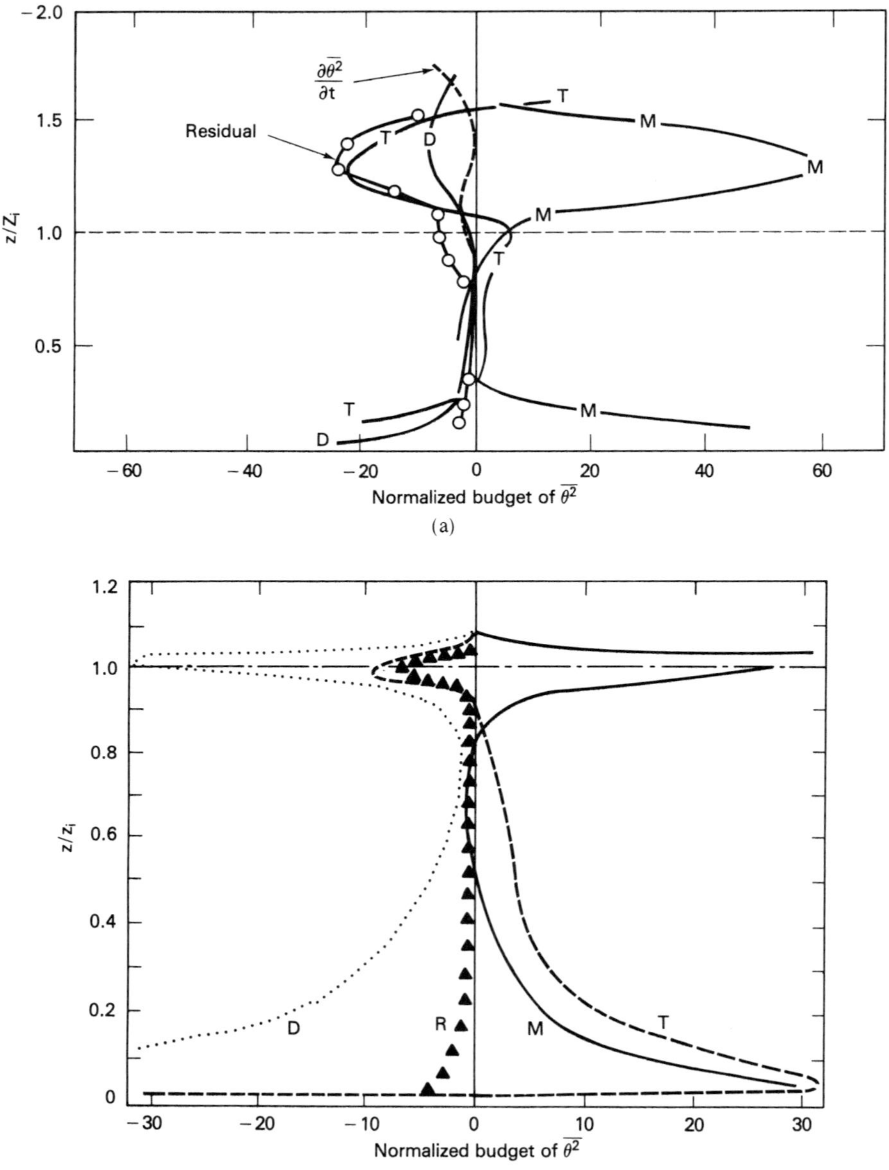

Figure 3.15 Dimensionless virtual potential temperature variance budget in the convective boundary layer: (a) obtained at Boulder BAO tower by Zhov, Lenschow, Stankov, Kaimal, and Gaynor, 1985. *Journal of Atmospheric Sciences* published by the American Meteorological Society. (b) obtained for day 33 of the Wangara experiment from a numerical model of André, De Moor, Lacarrére, Therry, and Du Cachat, 1978. *Journal of Atmospheric Sciences* published by the American Meteorological Society. M, mechanical production; D, molecular destruction; R, radiative destruction; T, turbulent transport. z_i and Z_i defined as in Fig. 3.11.

similar to (3.47):

$$\frac{\partial\overline{\theta'q'}}{\partial t} = \underbrace{-\overline{w'q'}\frac{\partial\Theta}{\partial z} - \overline{w'\theta'}\frac{\partial q}{\partial z}}_{\text{shear production}} - \underbrace{\frac{\partial\overline{w'\theta'q'}}{\partial z}}_{\text{turbulent transport}} - \underbrace{(\nu_\theta + \nu_q)\overline{\frac{\partial\theta'}{\partial x_\alpha}\frac{\partial q'}{\partial x_\alpha}}}_{\text{molecular destruction}} \quad (3.49)$$

The physical meanings are indicated underneath all terms. The last term in (3.49) is the destruction rate analogous to (3.47):

$$N_{\theta q} = (\nu_\theta + \nu_q)\overline{\frac{\partial\theta'}{\partial x_\alpha}\frac{\partial q'}{\partial x_\alpha}} \quad (3.50)$$

This term was obtained by assuming that ν_θ and ν_q are approximately equal and that

$$\frac{1}{2}(\nu_\theta + \nu_q)\frac{\partial^2\overline{\theta'q'}}{\partial x_\alpha \partial x_\alpha} \approx 0 \quad (3.51)$$

The budget of the temperature–humidity covariance in the convective boundary layer is shown in Figure 3.16. The figure indicates that the shear

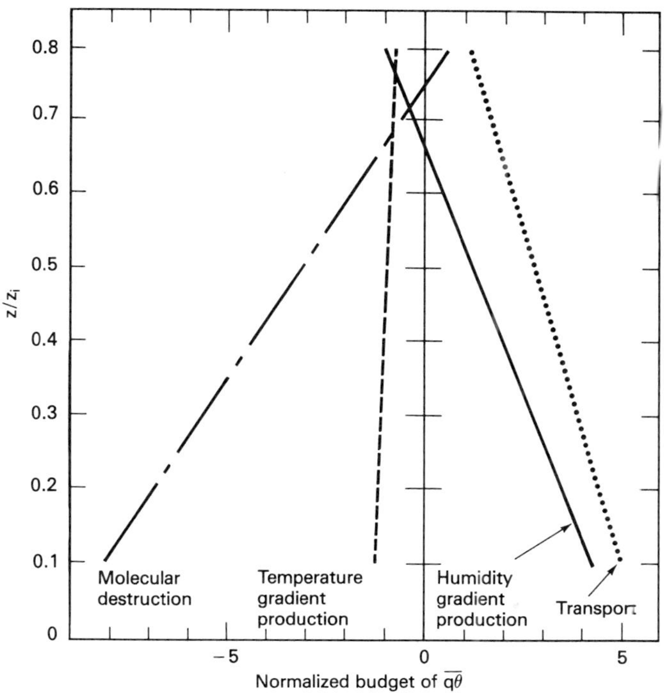

Figure 3.16 Parametrized $\overline{\theta q}$ budget obtained in the AMTEX experiment. (After Wyngaard, Pennell, Lenschow, and LeMone, 1978. *Journal of Atmospheric Sciences* published by the American Meteorological Society.)

production terms are of opposite signs, except for the upper portion of the ABL. The turbulent transport is positive and exports the covariance upward from the surface.

PROBLEMS

3.1. A simple solution of steady-state equations of motion obtained by Ekman in 1905 has the form given by (6.36). Assuming that $a = 0.001$, plot profiles of both components of the wind velocity for the following geostrophic wind vectors $(10, 0)$, $(0, 10)$, $(-10, 0)$, $(0, -10)$, $(3.16, 3.16)$, $(-3.16, -3.16)$, $(3.16, -3.16)$, $(-3.16, 3.16)$. Find the inflection points. Is this solution stable with respect to infinitesimal perturbation?

3.2. Using Figure 3.1, make a $(x - y - z)$ sketch of the horizontal roll vortices in the ABL. Mark the regions where clouds form.

3.3. Show that on the top of the boundary layer, where the momentum stress term can be omitted in (2.69), the wind vector describes an inertial ellipse or circle, clockwise on the northern hemisphere.

3.4. Using (3.40)–(3.42), explain during which condition the turbulent heat flux can be expressed as

$$\overline{w'\theta'} = -k_h\left(\frac{d\Theta}{dz} - \gamma_c\right)$$

3.5. Assuming equations of motion in the form

$$\frac{\partial u}{\partial t} = fv''$$

$$\frac{\partial v}{\partial t} = -fu''$$

where $u'' = u - U_g$, $v'' = v - V_g$, are components of the so-called ageostrophic wind, derive the equation for the changes of the turbulent kinetic energy of the motion. What is the contribution of the Coriolis force to the balance of the turbulent kinetic energy?

3.6. Adopting K-closure (3.45) for U, V, Θ,

$$\overline{u'w'} = -k_m\frac{\partial U}{\partial z}$$

$$\overline{v'w'} = -k_m\frac{\partial V}{\partial z}$$

$$\overline{\theta'w'} = -k_h\frac{\partial \Theta}{\partial z}$$

find the relation between the flux Richardson number R_f (3.35) and the gradient Richardson number Ri defined as

$$\text{Ri} = \frac{g}{\Theta_o}\frac{\partial\Theta/\partial z}{(\partial U/\partial z)^2 + (\partial V/\partial z)^2}$$

3.7. Compare the orders of magnitude of the Coriolis and the production terms in (3.25) and (3.26).

CHAPTER FOUR

SIMILARITY

4.1 DIMENSIONAL ANALYSIS AND SIMILARITY

Statistical characteristics of the turbulent flow in the atmospheric boundary layer are usually expressed in nondimensional form. This form has certain universal properties. It is derived by *dimensional analysis*, which is a technique used in physics to establish a relationship between different quantities. It is based on the assumption that the physical quantities $x_0, x_1, x_2, \ldots, x_n$, necessary and sufficient to describe a certain physical problem are related as

$$x_0 = f(\underbrace{x_1, x_2, \ldots, x_r,}_{\substack{\text{dimensionally} \\ \text{independent} \\ \text{parameters}}} \underbrace{x_{r+1}, \ldots, x_n}_{\substack{\text{dimensionally} \\ \text{dependent} \\ \text{parameters}}}) \tag{4.1}$$

Some of these $n + 1$ physical quantities are dimensionally dependent on others. For example, if the unit of x_1 is a meter, the unit of x_2 is a second, and the unit of x_3 is a meter per second, then x_1 and x_2 are dimensionally independent. However, x_3 is dimensionally dependent on x_1 and x_2. If n physical quantities in (4.1) involve r independent physical units, then (4.1) can be expressed in the form

$$\pi_0 = F(x_1, \ldots, x_r, \pi_1, \pi_2, \ldots, \pi_{n-r}) \tag{4.2}$$

where $\pi_0, \pi_1, \pi_2, \ldots, \pi_{n-r}$ are dimensionless power products of $x_0, x_1, x_2, \ldots, x_n$.

The dimensionless quantities $\pi_0, \pi_1, \pi_2, \ldots, \pi_{n-r}$ are independent of the system of units chosen. This implies that π_0 has to be independent of parameters $x_1, \ldots, x_r$ and proves the *Π theorem*, which states that (4.1) is equivalent to

$$\pi_0 = F(\pi_1, \pi_2, \ldots, \pi_{n-r}) \tag{4.3}$$

The derivation of the logarithmic wind profile in the surface layer can be used as a classical example of the technique described above. In the stationary, horizontally homogeneous and neutral surface layer, stress is practically constant with height and equal to u_*^2. Assuming that the wind shear $x_0 = dU/dz$ is a function of only two parameters, height $x_1 = z$ and friction velocity $x_2 = u_*$, we obtain

$$\frac{dU}{dz} = f(z, u_*) \tag{4.4}$$

The units involved are length and time, and the number of quantities is two ($n = 2$ and $r = 2$). Hence, by dimensional arguments, there are no independent dimensionless parameters and the function π_0 must be constant:

$$\pi_0 = \frac{z}{u_*}\frac{dU}{dz} = \text{const} \tag{4.5}$$

where the constant is traditionally denoted as $1/\kappa$ and κ is called the von Kármán constant. The value of κ is determined experimentally and is usually taken to be 0.4. The equation obtained can be integrated between the two levels, z and z_0, to yield

$$U(z) - U(z_0) = \frac{u_*}{\kappa}\ln\frac{z}{z_0} \tag{4.6}$$

The parameter z_0 is called a *roughness parameter* and is chosen such that $U(z_0) = 0$. It can be interpreted as a property of the surface. The constant z_0 has to be determined from experiments (Table 4.1), with the order of magnitude varying in the range 1 mm to 1 m.

Equation (4.6) indicates that for different values of z_0 and u_*, a family of curves describing a vertical distribution of wind velocity can be obtained. All these curves are *similar*, which means that they coincide into one curve when height z is scaled by z_0 and wind velocity is scaled by u_*.

As a second example, let us consider the structure function D defined by (3.19). The theory presented by Kolmogorov in 1941 (Section 3.3) predicts that

TABLE 4.1 Roughness Parameters for Different Surfaces

Type of Surface	z_0 (cm)
Smooth mud flats	0.001
Smooth snow on short grass	0.005
Snow surface	0.1
Short grass	0.2–9.0
Forest	~ 100

in the inertial subrange the structure function $D(r)$ of wind velocity should be dependent only on dissipation rate ϵ and distance r. In the case of temperature and humidity fields, additional parameters—destruction rate of the temperature variance N, destruction rate of humidity variance N_q, and destruction rate of temperature–humidity covariance N_{Tq}—should be included (Tatarskii, 1961). For these parameters, the dimensional analysis yields Kolmogorov's "2/3 laws":

$$
\begin{aligned}
D_V(r) &= a_V\epsilon^{2/3}r^{2/3}\\
D_T(r) &= a_T\epsilon^{-1/3}Nr^{2/3}\\
D_q(r) &= a_q\epsilon^{-1/3}N_q r^{2/3}\\
D_{Tq}(r) &= a_{Tq}\epsilon^{-1/3}N_{Tq}r^{2/3}
\end{aligned}
\tag{4.7}
$$

where a_V through a_{Tq} are constants.

The structure function is often expressed in the form $D_V(r) = C_V^2 r^{2/3}$, $D_T(r) = C_T^2 r^{2/3}$, $D_q(r) = C_q^2 r^{2/3}$, and $D_{Tq}(r) = C_{Tq} r^{2/3}$, where C_V^2, C_T^2, C_q^2, and C_{Tq} are called the structure parameters. When constants $a_V, \ldots, a_{Tq}$ in (4.8) are replaced by empirically obtained numerical values, the structure parameters have the form

$$
\begin{aligned}
C_V^2 &= 2\epsilon^{2/3}\\
C_T^2 &= 1.6N\epsilon^{-1/3}\\
C_q^2 &= 1.6N_q\epsilon^{-1/3}\\
C_{Tq} &= 1.6N_{Tq}\epsilon^{-1/3}
\end{aligned}
\tag{4.8}
$$

4.2 THE MONIN–OBUKHOV SIMILARITY THEORY

The presence of a vertical turbulent heat flux introduces new similarity parameters and modifies the logarithmic profile (4.6). The effects of stability were considered by Monin and Obukhov, who in 1954 proposed a similarity theory for a stratified surface layer. It is based on the assumptions that the turbulent regime in the surface layer depends on height z, surface kinematic stress $\tau_0 = [(\overline{u'w'})|_0^2 + (\overline{v'w'})|_0^2]^{1/2}$, surface kinematic heat flux $Q_0 = \overline{w'\theta'}|_0$, and also the buoyancy parameter $\beta = g/T_0$. From these parameters, the following scales are obtained:

$$\begin{aligned} &\text{For velocity:} && u_* = \tau_0^{1/2} \\ &\text{For temperature:} && T_* = -\frac{\overline{w'\theta'}|_0}{u_*} \\ &\text{For length:} && L = \frac{u_*^2}{\kappa \beta T_*} \end{aligned} \tag{4.9}$$

In the case when water vapor is present, the virtual temperature fluctuation θ_v' should be used instead of θ' in the definition of T_*. From the definition of virtual temperature given in Section 2.3, it follows that $T_{v_*} = -\overline{w'\theta_v'}|_0/u_*$ and $\overline{w'\theta'}_v \approx \overline{w'\theta'} + 0.61T_0\overline{w'q'}$. In this case a new scale for specific humidity can be introduced and defined as $q_* = -\overline{w'q'}|_0/u_*$.

In the thermally stratified surface layer, (4.4) has to be replaced with

$$\frac{dU}{dz} = f(z, u_*, L) \tag{4.10}$$

A similar relationship can be written for the temperature gradient:

$$\frac{d\Theta}{dz} = f(z, T_*, L) \tag{4.11}$$

According to the comment above, when water vapor is present in the atmosphere, Θ is replaced by a virtual potential temperature Θ_v, and T_* is replaced by T_{*v}. Notice also that the Monin–Obukhov length L was defined before in (3.38).

Equations (4.10) and (4.11) lead to a conclusion that the quantities

$$\phi_m' = \frac{\kappa L}{u_*}\frac{\partial U}{\partial z} \qquad \phi_h' = \frac{\kappa L}{T_*}\frac{\partial \Theta}{\partial z} \tag{4.12}$$

have to be functions of a parameter z/L only. It is convenient to replace (4.12) with the following similarity functions:

$$\begin{aligned} \phi_m &= \zeta \phi_m' \\ \phi_h &= \zeta \phi_h' \end{aligned} \tag{4.13}$$

where ϕ_m, ϕ_m', ϕ_h, and ϕ_h' are called universal functions of a parameter $\zeta = z/L$. Dimensional analysis also leads to the conclusion that the quantities

$$\begin{aligned} &\phi_{12} = \frac{\overline{u'v'}}{u_*^2} \qquad \phi_{11} = \frac{\overline{u'^2}}{u_*^2} \qquad \phi_{22} = \frac{\overline{v'^2}}{u_*^2} \qquad \phi_{33} = \frac{\overline{w'^2}}{u_*^2} \\ &\phi_3' = \frac{L\epsilon}{u_*^3} \quad \left(\text{or } \phi_\epsilon = \frac{z\epsilon}{u_*^3}\right) \qquad \phi_{1\theta} = \frac{\overline{u'\theta'}}{u_* T_*} \qquad \phi_{\theta\theta} = \frac{\overline{\theta'^2}}{T_*^2} \end{aligned} \tag{4.14}$$

must be universal functions of dimensionless height ζ. It is more convenient to use the function ϕ_ϵ than ϕ_ϵ'. Similarly to (4.13), $\phi_\epsilon = \zeta\phi_\epsilon'$.

The behavior of the universal functions can be predicted by dimensional analysis only in an asymptotic sense. It is not difficult to make specific predictions about the universal function for neutral, very stable, and very unstable conditions.

4.2.1 Free convection regime

In the case of a large upward heat flux and light winds ($\zeta \to -\infty$), called the *free convection regime*, it was suggested by Obukhov (1946), Monin and Obukhov (1954), Monin (1959), and Priestley (1955) that τ_0 can be dropped from the list of independent variables. Dimensional analysis shows that in this case,

$$
\begin{aligned}
\phi_m(\zeta) &\sim \phi_h(\zeta) \sim \zeta^{-1/3} \\
\phi_{33}(\zeta) &\sim \zeta^{2/3} \\
\phi_{\theta\theta} &\sim \zeta^{-2/3}
\end{aligned}
\tag{4.15}
$$

It is easy to show that u_* can be eliminated on both sides of (4.15). The prediction connected with temperature distribution was originally made by Prandtl (1932) and later reformulated by Obukhov (1946). Integration of functions $\phi_m(\zeta)$ and $\phi_h(\zeta)$ gives wind velocity and temperature profiles in the exponential form: $(u_2 - u_1)/u_* \sim (\Theta_2 - \Theta_1)/T_* \sim \zeta_2^{-1/3} - \zeta_1^{-1/3}$.

It must be mentioned here that not all functions ϕ follow this prediction. For example, ϕ_{11} and ϕ_{22} have a very different behavior in the surface layer than predicted by dimensional analysis (i.e., $\phi_{11} \sim \phi_{22} \sim \zeta^{2/3}$). This issue is discussed later.

Excluding τ_0 from the list of governing parameters, Wyngaard and Coté (1974) obtained new (free convection) scales:

$$
\begin{aligned}
&\text{Velocity:} && u_f = \left(z\beta\overline{w'\theta'}\big|_0\right)^{1/3} \\
&\text{Temperature:} && t_f = -\frac{\overline{w'\theta'}\big|_0}{u_f} \\
&\text{Length:} && z
\end{aligned}
\tag{4.16}
$$

Since no dimensionless height can be formed from free convection scales, it was concluded that all nondimensional quantities are required to be constant. For example:

$$
\frac{\overline{w'^2}}{u_f^2} = \text{constant} \tag{4.17}
$$

Free convection scales u_f and T_f and surface layer scales u_* and T_* are related:

$$\frac{u_f}{u_*} = \left(\frac{\zeta}{\kappa}\right)^{1/3}$$
$$\frac{t_f}{T_*} = \left(\frac{\zeta}{\kappa}\right)^{-1/3} \tag{4.18}$$

From (4.18) it follows that (4.17) agrees with ϕ_{33} in (4.15).

4.2.2 Neutral regime

For the *neutral regime*, when L is large and ζ is small, all functions $\phi(\zeta)$ must approach constant values:

$$\phi_i(0) = \text{constant} \qquad \text{for } i = m, h, 11, 22, 33, 12, E, \theta\theta, \epsilon, \text{etc.} \tag{4.19}$$

Integration of the relation $\phi_m(0) = 1$, leads to a logarithmic profile of wind in the surface layer. Analogously, integration of the relation $\phi_h =$ constant $= \alpha_\theta$ yields for the near-neutral regime:

$$\Theta(z_2) = \Theta(z_1) + \frac{\alpha_\theta T_*}{\kappa} \ln \frac{z_2}{z_1} \tag{4.20}$$

Observations show that $\alpha_\theta = 0.74$ (Businger et al., 1971).

4.2.3 Very stable regime

In the case of a *very stable regime*, when $\zeta \to \infty$, z is no longer important since turbulence is suppressed by a buoyancy force and has a local character. The nondimensional wind shear and the vertical temperature gradient ϕ'_m, ϕ'_h should asymptotically tend to constant values, which are equivalent to $\phi_m, \phi_h \sim \zeta$. From the same argument, the remaining function, $\phi_{11}, \phi_{22}, \phi_{33}$ should approach constant values. The dissipation rate ϵ, in stable conditions, presumably would scale as u_*^3/L and the nondimensional dissipation rate ϕ_ϵ would tend to become proportional to ζ. Consequently, in the stable surface layer, similarity functions have the forms

$$\begin{aligned}
\phi_m(\zeta) &\sim \zeta \\
\phi_h(\zeta) &\sim \zeta \\
\phi_\epsilon(\zeta) &\sim \zeta \\
\phi_{11}(\zeta) &= \text{const} \\
\phi_{22}(\zeta) &= \text{const} \\
\phi_{33}(\zeta) &= \text{const} \\
\phi_{\theta\theta}(\zeta) &= \text{const}
\end{aligned} \tag{4.21}$$

Integration of functions $\phi_m(\zeta)$ and $\phi_h(\zeta)$ gives linear wind velocity and temperature profiles of the form $(u_2 - u_1)/u_* \sim (\Theta_2 - \Theta_1)/T_* \sim \zeta_2 - \zeta_1$.

4.2.4 Empirical verification

Since the early 1950s, when modern detecting techniques became available, numerous experimental works have been performed to study profiles, fluxes, variances, spectra, and other statistics of the wind and temperature in the surface layer of the atmosphere (see Chapter 8). Most of the results were systematically analyzed, using the framework of the similarity theory.

The results of most of the investigations, appearing in the 1970s and 1980s, are shown in Table 4.2. The visible differences and the scatter of the resulting data are due partly to difficulties of the measurements and partly to variable conditions in the atmosphere.

With regard to the mean wind and temperature gradients, Businger et al.'s (1971) form of ϕ_m and ϕ_h is the most popular (Figure 4.1), but nevertheless evokes some controversies. The first controversy is associated with the value of the von Kármán constant κ. Businger et al. (1971) and later Telford and Businger (1986) obtained $\kappa = 0.35$. Miyake et al. (1970) found no indication that the von Kármán constant differed from the usually accepted value of 0.4. The same result was obtained by Högström (1985, 1986) and Zhang et al. (1988). It has been suggested that the result obtained by Businger et al. (1971) could be influenced by the procedure of reducing wind shear by 10% because of alleged overspeeding (Dyer, 1977).

Another controversy concerns the form of the universal functions in very unstable conditions. Businger et al. (1971) argued that for large $(-\xi)$ values, $\phi_h \sim (-\xi)^{-1/2}$ and $\phi_m \sim (-\xi)^{-1/4}$ (Figure 4.2). The Kansas observations show $-1/2$ power behavior for $-\xi$ up to 2, the largest value found. In this case the local free convection prediction is not verified. Because of the limited $(-\xi)$ range and the difficulty of measuring vanishing temperature gradients this cannot be taken as a decisive test. Carl et al. (1973), as well as Lewellen and Teske (1973), argued that for large ξ $(-\xi > 2)$, ϕ_m approaches the classical limit for free convection. Therefore, Businger's form should be used for $2 > -\xi > 0$ and Carl's form for $-\xi > 2$ (Figures 4.2 and 4.3). Korrell et al. (1982) found from the analysis of Boulder Tower data, the superiority of the Carl curve for $0.6 > \xi > 0$, with the Businger and KEYPS forms overestimating ϕ_m in this region.

On the other hand, free-convection predictions of variances of vertical velocity σ_w^2 and temperature fluctuations σ_θ^2 are confirmed by the observations remarkably well (Figures 4.4 and 4.5). In general, it is found that average quantities containing only w' and θ' obey the free convection similarity quite well, whereas mean profiles or quantities that also include u', v' do not follow this similarity. In the latter case, free convection scaling does not explain all the observations. The regime evidently depends on more parameters.

TABLE 4.2 Form of the Empirical Universal Function of M–O Similarity Theory

(a) Unstable caser

Function	Form	Constants	Range of z/L	Author
ϕ_m	$(1 - az/L)^b$	$a = 16,\ b = -\frac{1}{4}$	$-1;\ 0$	Dyer and Hicks (1970)
		$a = 15,\ b = -\frac{1}{4}$	$-2;\ 0$	Businger et al. (1971)
		$a = 16,\ b = -\frac{1}{3}$	$-10;\ -2$	Carl et al. (1973)
		$a = 16,\ b = -\frac{1}{4}$	$-2;\ 0$	Carl et al. (1973)
		$b = 28,\ b = -\frac{1}{4}$	$-10;\ -0.01$	Dyer and Bradley (1982)
	$\phi_m^4 - 9z/L\phi_m^3 = 1$	(KEYPS formula)		Businger et al. (1971)
	$(1 - 7z/L)^{-1/4} - 0.2$		$-1;\ 0$	Kai (1982)
	$1.06(1 - 15z/L)^{-1/4}$		$-0.05;\ 0$	Bradley et al. (1981)
ϕ_h	$c(1 - az/L)^b$	$a = 16,\ b = -\frac{1}{2},\ c = 1$	$-1;\ 0$	Dyer and Hicks (1970)
		$a = 9,\ b = -\frac{1}{2},\ c = 0.74$	$-2;\ 0$	Businger et al. (1971)
		$a = 9,\ b = -\frac{1}{2},\ c = 1.04$	$-0.5;\ 0$	Bradley et al. (1981)
		$a = 14,\ b = -\frac{1}{2},\ c = 1$	$-10;\ -0.01$	Dyer and Bradley (1982)
	$0.74[1 + 4(z/L)^{3/4}]^{-2/3}$			Businger et al. (1971)
ϕ_{11}	$(12 - 0.5z_i/L)^{2/3}$			Panofsky et al. (1977)
	$[0.036 + 0.003(-z/L)^{-2/3}]^{-1}$		$0.6;\ -0.2$	Bradley and Antonia (1979)
	$62.5(-z/L)^{2/3}$		$-2;\ -0.1$	
	$[0.0003 + 0.0009(-z/L)^{-2/3}]^{-1}$		$-2;\ -0.1$	
ϕ_{22}	$(12 - 0.5z_i/L)^{2/3}$			Panofsky et al. (1977)
	$[-0.002 + 0.021(-z/L)^{-2/3}]^{-1}$		$-2;\ -0.1$	Bradley and Antonia (1979)

ϕ_{33}	$a(1 - bz/L)^c$	$a = 1.56,\ b = 3,\ c = \frac{2}{3}$	$-4; 0$	Panofsky et al. (1977)
		$a = 1,\ b = 14,\ c = \frac{1}{2}$	$-10; -0.01$	Dyer and Bradley (1982)
		$a = 1.7,\ b = 3,\ c = \frac{2}{3}$	$-1; 0$	Kai (1982)
	$1.9(-z/L)^{2/3}$		$-2; -1$	Wyngaard et al. (1971)
	$1.6 + (2z/L)^{2/3}$		$-1.5; 0$	Haugen et al. (1971)
	$[0.089 + 0.171(-z/L)^{-2/3}]^{-1}$		$-2; -0.1$	Bradley and Antonia (1979)
$\phi_{\theta\theta}$	$[a(b - cz/L)^d]^2$	$a = -0.95,\ b = 0,\ e = 1,\ d = -\frac{1}{3}$	$-0.7; 0$	Wyngaard et al. (1971)
		$a = 0.95,\ b = 0.05,\ c = 1,\ d = -\frac{1}{3}$	$-10; -0.1$	Tillman (1972)
		$a = -0.92,\ b = 0,\ c = 1,\ d = -\frac{1}{3}$	$-10; -0.1$	Monji (1973)
	$[0.004 + 1.11(-z/L)^{1/3}]^{-2}$		$-0.6; -0.2$	Bradley and Antonia (1979)
	$[0.13 + 0.84(-z/L)^{1/3}]^{-2}$		$-2; -0.1$	
	$[-0.22 + 1.28(-z/L)^{1/3}]^{-2}$		$-2; -0.1$	
ϕ_ϵ	$[1 + a(z/L)^b]^c$	$a = 0.5,\ b = \frac{2}{3},\ c = \frac{3}{2}$	$-2; 0$	Wyngaard and Coté (1971)
		$a = -16,\ b = 1,\ c = -\frac{1}{4}$	$-0.5; 0$	Garrat (1972)
		$a = 0.75,\ b = \frac{2}{3},\ c = \frac{3}{2}$	$-10; 0$	Kaimal (1978)
		$a = 1.4,\ b = \frac{2}{3},\ c = \frac{3}{2}$	$-1; 0$	Kai (1982)
	$\phi_m - z/L$			McBean and Eliot (1975) Bradley et al. (1981)

TABLE 4.2 (cont.)

(b) Stable case

Function	Form	Constants	Range of z/L	Author
ϕ_m	$1 + az/L$	$a = 5.2$	$-0.03; 1$	Webb (1970)
		$a = 4.7$	0; 2	Businger et al. (1971)
		$a = 5.0$	0.02; 1	Hicks (1976)
	$0.8 + 2z/L$		0; 1	Kai (1982)
ϕ_h	$a + bz/L$	$a = 1, b = 5.2$	$-0.03; 1$	Webb (1970)
		$a = 0.74, b = 4.7$	0; 2	Businger et al. (1971)
		$a = 1, b = 5.0$	0.02; 1	Hicks (1976)
		$a = 1.2, b = 6.1$	0; 1	Kai (1982)
ϕ_{11}	c^2	$c \sim 2.5$	0.05; 1	Takeuchi (1961)
ϕ_{22}	c^2	$c \sim 1.5$	0.05; 1	Takeuchi (1961)
ϕ_{33}	c^2	$c \sim 1.25$	0.05; 1	Takeuchi (1961)
	$1.3(\phi_m - 2.5z/L)^{1/3}$		$-4; 1$	Merry and Panofsky (1976)
	$1.3(1 - 3z/L)^{1/3}$			Kai (1982)
$\phi_{\theta\theta}$	c^2	$c = 1.8$	0; 1	Wyngaard et al. (1971)
		$c = 1.8$	0; 0.5	Tillman (1972)
ϕ_ϵ	$[1 + a(z/L)^b]^c$	$a = 2.5, b = \frac{3}{5}, c = \frac{3}{2}$	0.2	Wyngaard and Coté (1971)
		$a = 4, b = 1, c = 1$	0; 0.75	Garrat (1972)
		$a = 3.7, b = 1, c = 1$		Wyngaard (1975)
		$a = 2.2, b = \frac{3}{5}, c = \frac{3}{2}$	0; 2	Kai (1982)
	$\phi_m - z/L$			Wyngaard (1975)

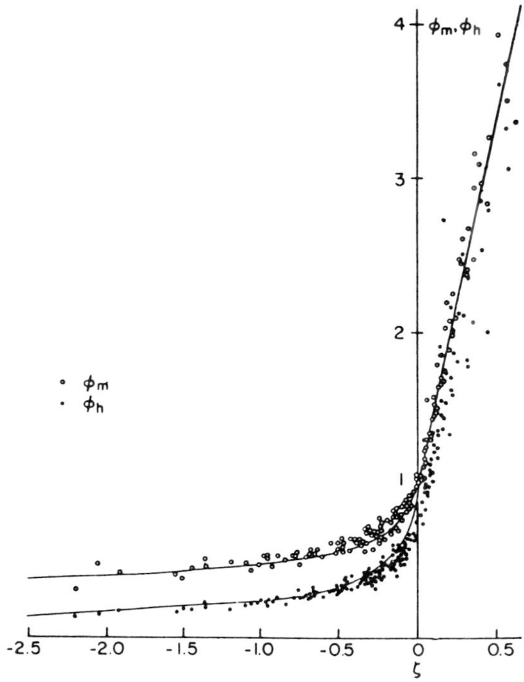

Figure 4.1 Universal functions ϕ_m and ϕ_h. (After Fleagle and Businger, 1980.)

It has been suggested that variances of the horizontal component of wind velocity σ_u^2 and σ_v^2 are influenced considerably by the large convective cells of the scale length z_i, even close to the surface. Therefore, even in the surface layer one would expect σ_u/u_* and σ_v/u_* to be more closely related to z_i/L than to the local stability parameter z/L. This has indeed been confirmed by Panofsky et al. (1977), who have shown that under convective conditions $\sigma_u/u_* \sim \sigma_v/u_* \sim (-z_i/L)^{1/3}$ (Figure 4.6).

The conditions of true free convection occur very seldom in the atmosphere. Since the surface layer height is limited to about one-tenth of the total ABL thickness, the asymptotic condition $-\xi \to \infty$ is probably never realized, although $-\xi$ values in the range 1 to 10 could be expected (see Figure 4.19 in Section 4.4). However, the quasi-free convection regime is known to occur because some predictions are confirmed by observation. The region of validity of free convection formulas varies. Whereas temperature variances behave as predicted for $-\xi > 0.1$, the vertical velocity variance shows the expected behavior only for $-\xi > 0.4$.

In stable air, all measurements suggest the linear form of ϕ_m and ϕ_h. However, since there usually is considerable scatter, the decimal is somewhat questionable. An examination of the stable profiles by Korrell et al. (1982)

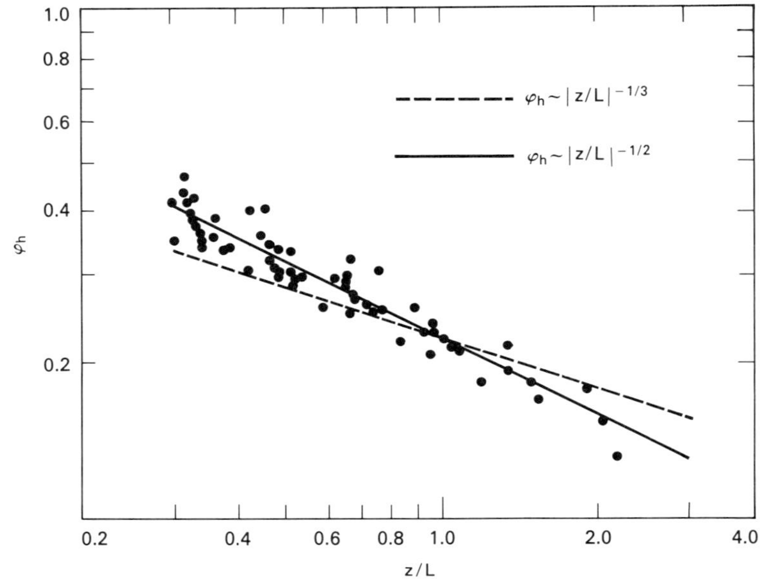

Figure 4.2 Universal function ϕ_h under very unstable conditions according to Businger, Wyngaard, Izumi, and Bradley, 1971. *Journal of Atmospheric Sciences* published by the American Meteorological Society.

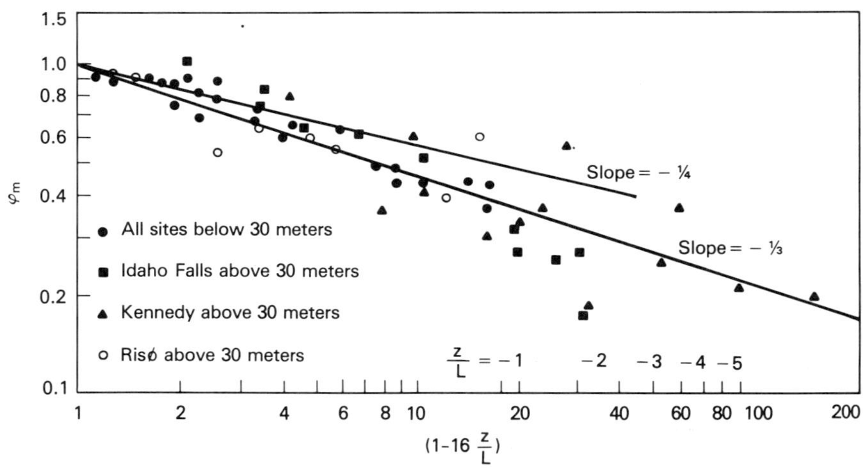

Figure 4.3 Universal function ϕ_m under very unstable conditions according to Carl, Tarbell, and Panofsky, 1973.

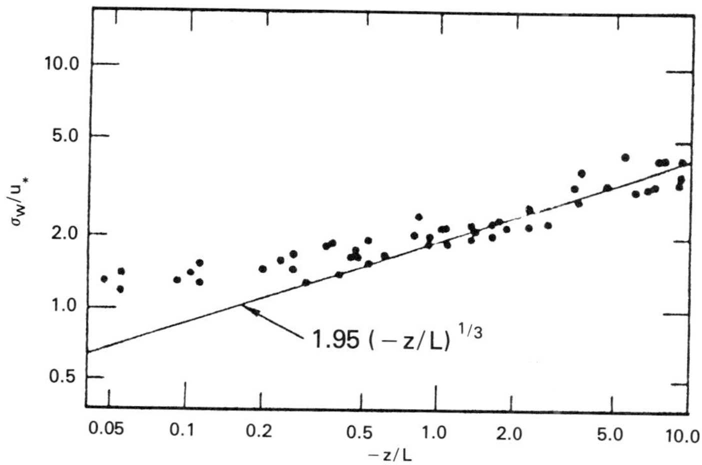

Figure 4.4 Nondimensional standard deviation of the vertical velocity in convective conditions. (Kaimal, Eversole, Lenschow, Stankov, Kahn, and Businger, 1982. *Journal of Atmospheric Sciences* published by the American Meteorological Society.)

shows that the empirical curve of Businger et al. (1971) fits the Boulder Tower data well for $0 < \xi < 0.4$.

Dimensional analysis gives guidance that wind and temperature variances should remain constant in the stable boundary layer (Figure 4.7). Function $\phi_{\theta\theta}$ is generally accepted to be constant (Takeuchi, 1961). However, there is little agreement on the behavior of ϕ_{33} in this case. It has been suggested that parameters like ϕ_{33} might increase for positive ξ. Wesely (1976)

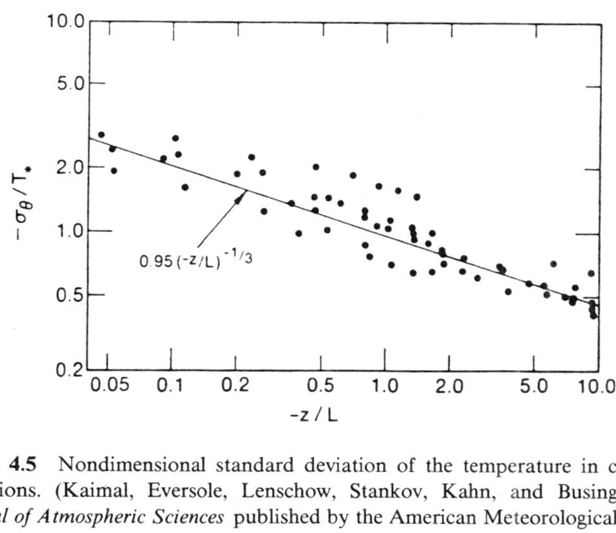

Figure 4.5 Nondimensional standard deviation of the temperature in convective conditions. (Kaimal, Eversole, Lenschow, Stankov, Kahn, and Businger, 1982. *Journal of Atmospheric Sciences* published by the American Meteorological Society.)

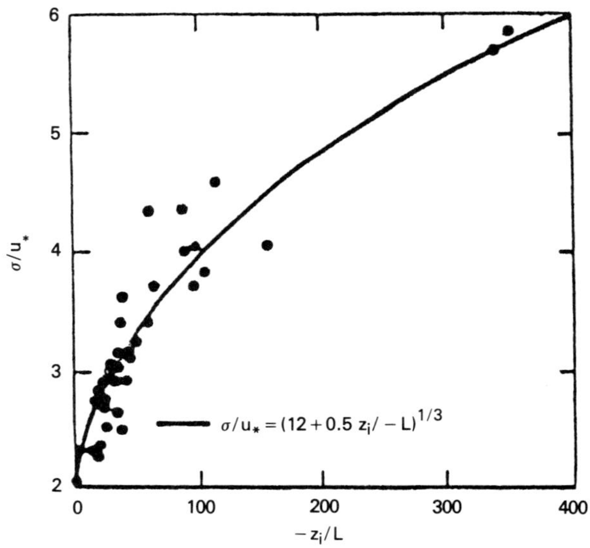

Figure 4.6 Normalized standard deviation of horizontal wind velocity. (From Panofsky et al., 1977.)

indicated no increase of ϕ_{33} under stable conditions. Merry and Panofsky (1976) analyzed many sets of similar ϕ_{33} observations, some of which show strong increases in stable conditions, and concluded that an increase is indeed real. Sethuraman et al. (1978) presented a set of observations of ϕ_{33} obtained over water which showed no evidence of an increase in stable conditions when presented as a function of Ri. Their data also showed relatively constant values of ϕ_{11} and ϕ_{22}. The reported difficulties with ϕ_{33} are partly observational because in the stable surface layer both σ_w and u_* become small and therefore their ratio is poorly determined.

4.3 THE ROSSBY NUMBER SIMILARITY THEORY

Above the surface layer, the equations of motion contain the Coriolis terms. Consequently, a new quantity, the Coriolis parameter f, has to appear on the list of governing parameters in a dimensional analysis. This was taken into consideration by Kazanskii and Monin (1961), who introduced a similarity hypothesis, in which the structure of the thermally stratified atmospheric boundary layer above the surface layer is determined by the following parameters: height z, friction velocity u_*, surface kinematic heat flux Q_0, buoyancy

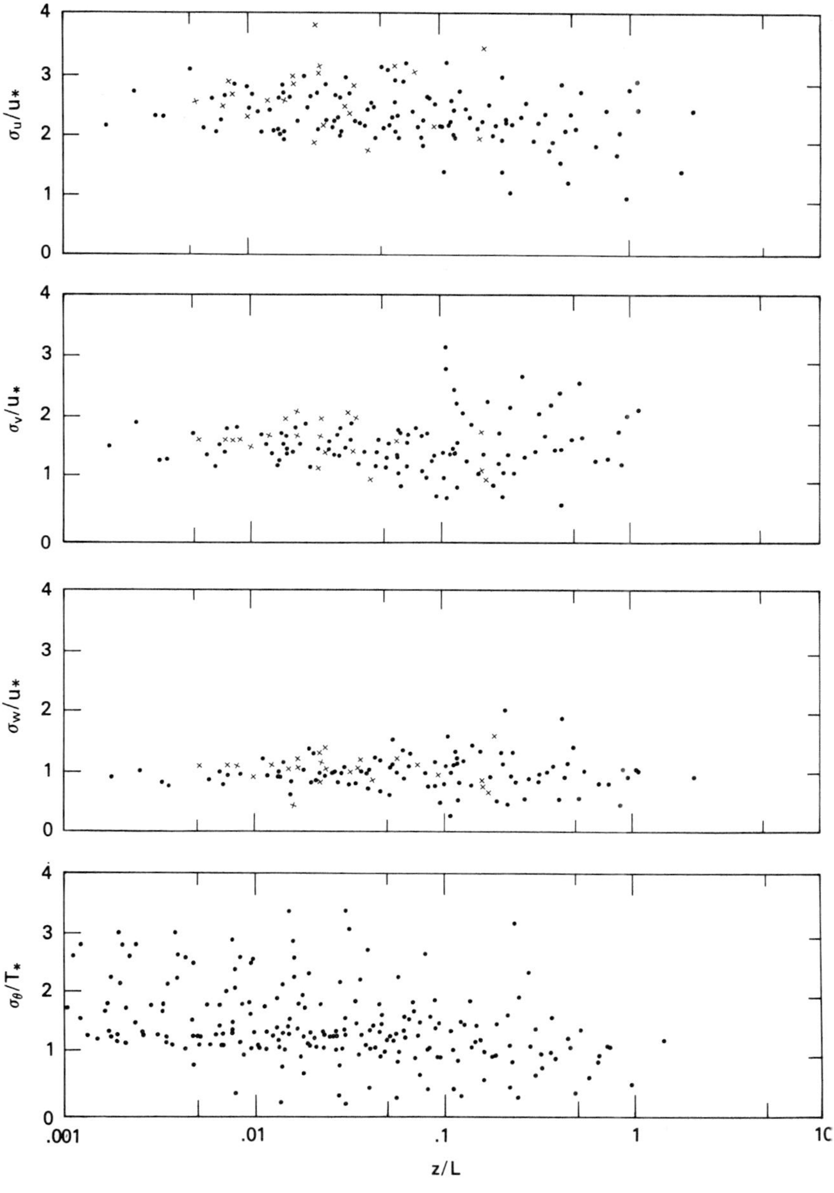

Figure 4.7 Normalized standard deviations of u, v, w, and θ under stable conditions. (After Takeuchi, 1961.)

parameter β, and magnitude of the Coriolis parameter f. From this hypothesis it follows that all universal functions should depend on only two dimensionless parameters, $Z = z/L_E$ and $\mu_0 = L_E/L$, where $L_E = \kappa u_*/f$ (the height of the Ekman layer).

The list of governing parameters does not include a roughness parameter z_0. As shown by Wippermann and Yordanov (1972), with increasing height, solutions of the properly scaled equations of motion lose their dependence on the roughness parameter z_0 (Figure 4.8). In terms of the similarity theory, we can say that these solutions are independent of the *surface Rossby number* $R_0 = G/(fz_0)$, where G is the geostrophic wind. This fact explains the name given to the theory.

By analogy to the surface layer similarity, one can assume that the following relations are valid above the surface layer:

$$
\begin{aligned}
\frac{L_E}{u_*}\frac{dU}{dz} &= \Psi_u(Z, \mu_0) \\
\frac{L_E}{u_*}\frac{dV}{dz} &= \Psi_v(Z, \mu_0) \\
\frac{L_E}{T_*}\frac{d\Theta}{dz} &= \Psi_\theta(Z, \mu_0)
\end{aligned}
\tag{4.22}
$$

Equations (4.22) can be written in the integrated form (Kazanskii and Monin, 1961)

$$
\begin{aligned}
\frac{U - U_g}{u_*} &= F_u(Z, \mu_0) \\
\frac{V - V_g}{u_*} &= F_v(Z, \mu_0) \\
\frac{\Theta - \Theta_t}{T_*} &= F_\theta(Z, \mu_0)
\end{aligned}
\tag{4.23}
$$

Here it is assumed that the wind velocity on the top of the ABL is equal to the geostrophic wind and Θ_t is a value of temperature on the same level.

Recognizing the importance of the actual boundary layer height, Deardorff (1972a) has suggested the use of other dimensionless parameters, $\mu_* = H/L$ and $\mathbf{Z} = z/H$, where H is a generic boundary layer height (for the mixed layer $H = z_i$; for the stable boundary layer $H = h$). It was argued by Sundararajan (1975), Arya (1975b), and Arya and Wyngaard (1975) that L_E is also a relevant scale and should not be dropped from similarity considerations. A more complete and proper dimensional analysis would give $\mu = H/L_E$ as

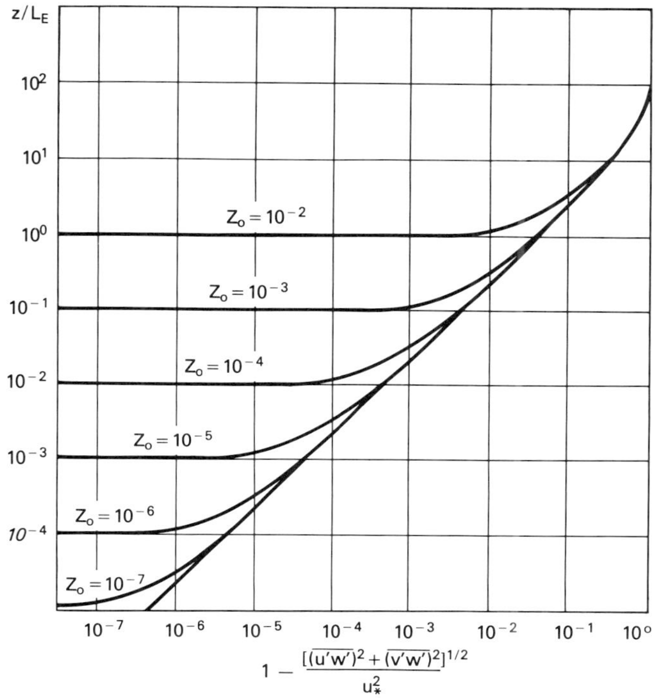

Figure 4.8 Vertical profiles of the nondimensional Reynolds stress (for various values of the nondimensional roughness length $Z_0 = z_0/L_E$ in the case of neutral stratification. (After Wippermann and Yordanov, 1972.)

an additional parameter:

$$\begin{aligned} \frac{U - U_g}{u_*} &= F_u(\mathbf{Z}, \mu, \mu_*) \\ \frac{V - V_g}{u_*} &= F_v(\mathbf{Z}, \mu, \mu_*) \\ \frac{\Theta - \Theta_t}{T_*} &= F_\theta(\mathbf{Z}, \mu, \mu_*) \end{aligned} \tag{4.24}$$

The difference between the Kazanskii–Monin theory and its generalization is the height scale.

In the definition of L_E, f appears in the denominator. For this reason the parameterization based on Kazanskii–Monin theory becomes highly inaccurate and misleading in low latitudes and completely meaningless at the equator. Instead, it is suggested that the height of the upper inversion layer z_i is a more appropriate height scale for the ABL.

The surface layer similarity profiles are obviously useless at extremely large values of z/z_0. The outer layer similarity profiles, on the other hand, cannot be extended to low values of z/z_0 because of the very steep gradient close to the ground. It can be assumed, however, that there exists a level in the atmosphere where both lower and upper profiles can be matched.

The surface layer profiles can be obtained by integration of functions ϕ_m and ϕ_h. For Businger et al.'s (1971) form of ϕ_m and ϕ_h (Table 4.2), the result can be expressed as

$$\begin{aligned} U(z) &= \frac{u_*}{\kappa}\left[\ln\frac{z}{z_0} - \Psi_1\left(\frac{z}{L}\right)\right] \\ V(z) &= 0 \\ \Theta(z) - \Theta_0 &= \frac{\alpha_\theta T_*}{\kappa}\left[\ln\frac{z}{z_0} - \Psi_2\left(\frac{z}{L}\right)\right] \end{aligned} \tag{4.25}$$

where $\kappa = 0.35$, $\alpha_\theta = 0.74$. For the *free convection regime* (Paulson, 1970),

$$\Psi_1\left(\frac{z}{L}\right) = 2\ln\frac{1+x}{2} + \ln\frac{1+x^2}{2} - 2\arctan x + \frac{\pi}{2}$$

$$\Psi_2\left(\frac{z}{L}\right) = 2\ln\frac{1+x^2}{2}$$

where $x = (1 - \gamma z/L)^{1/4}$, $\gamma = 15$ for the wind velocity and 9 for the temperature; for the *stable regime*, $\Psi_1(z/L) = \alpha_\theta \Psi_2(z/L) = -4.7z/L$; and for the *neutral regime*, $\Psi_1(z/L) = \Psi_2(z/L) = 0$.

Matching (4.23) with surface layer profiles (4.25) yields

$$\begin{aligned} \ln\frac{L_E}{z_0} - \frac{\kappa G\cos\alpha_0}{u_*} &= F_u(Z, \mu_0) - \left[\ln Z - \Psi_1\left(\frac{z}{L}\right)\right] \\ -\frac{\kappa G\sin\alpha_0}{u_*} &= F_v(Z, \mu_0) \\ \ln\frac{L_E}{z_0} - \frac{\alpha_\theta(\Theta_t - \Theta_0)}{T_*} &= \frac{\kappa}{\alpha_\theta}F_\theta(Z, \mu_0) - \left[\ln Z - \Psi_2\left(\frac{z}{L}\right)\right] \end{aligned} \tag{4.26}$$

where G is the magnitude of the geostrophic wind and α_0 is the angle between the geostrophic and surface wind vectors.

In the system of equations (4.26), the left-hand sides do not depend on z and the right-hand sides should also be independent of z and be a function of μ_0 only:

$$\begin{aligned} F_u(Z, \mu_0) - \left[\ln Z - \Psi_1\left(\frac{z}{L}\right)\right] &\longrightarrow A(\mu_0) + \ln\kappa \\ F_v(Z, \mu_0) &\longrightarrow B(\mu_0) \\ \frac{\kappa}{\alpha_\theta}F_\theta(Z, \mu_0) - \left[\ln Z - \Psi_2\left(\frac{z}{L}\right)\right] &\longrightarrow C(\mu_0) + \ln\kappa \end{aligned} \tag{4.27}$$

where A, B, and C are dimensionless universal functions of μ_0. Equations (4.26) and (4.27) lead to a system of algebraic equations called *resistance laws:*

$$\ln\left(C_g R_0\right) - \kappa \frac{\cos\alpha_0}{C_g} = A(\mu_0)$$

$$\kappa \frac{\sin\alpha_0}{C_g} = -B(\mu_0) \tag{4.28}$$

$$\frac{\kappa}{\alpha_\theta}\frac{\Delta\Theta}{T_*} - \ln\left(C_g \,\mathrm{Ro}\right) = C(\mu_0)$$

where $C_g = u_*/G$ is the geostrophic drag coefficient; Ro is the Rossby number; A, B, and C are the similarity functions; α_0 is the cross isobar angle; and $\Delta\Theta = \Theta_0 - \Theta_t$, the temperature difference between the bottom and top of the atmospheric boundary layer.

For the generalized relationships (4.24), the resistance laws have the form

$$-\ln\frac{z_0}{H} - \kappa\frac{\cos\alpha_0}{C_g} = a(\mu, \mu_*)$$

$$\kappa\frac{\sin\alpha_0}{C_g} = -b(\mu, \mu_*) \tag{4.29}$$

$$\frac{\kappa}{\alpha_\theta}\frac{\Delta\Theta}{T_*} - \ln\left(C_g \,\mathrm{Ro}\right) = c(\mu, \mu_*)$$

where a, b, and c are the similarity functions. Notice that by substituting $H = u_*/f$ in (4.29), (4.28) results.

The resistance laws were first derived by Kazanskii and Monin (1961) for the barotropic and adiabatic cases. They have been extended to the diabatic barotropic case by Monin and Zilitinkevich (1967) and simultaneously by Blackadar (1967).

It is useful to introduce bulk stability parameters:

$$S = \beta\frac{\Delta\Theta}{fG} \tag{4.30}$$

for Kazanskii–Monin scaling, and

$$S_* = \beta\frac{\Delta\Theta}{G^2}H \tag{4.31}$$

for Deardorff scaling. Expressing them in terms of internal stability parameters results in

$$S = -\frac{\Delta\Theta}{T_*}\frac{C_g}{\kappa^3}\mu_0 \tag{4.32}$$

$$S_* = -\frac{\Delta\Theta}{T_*}\frac{C_g^2}{\kappa}\mu_0 \tag{4.33}$$

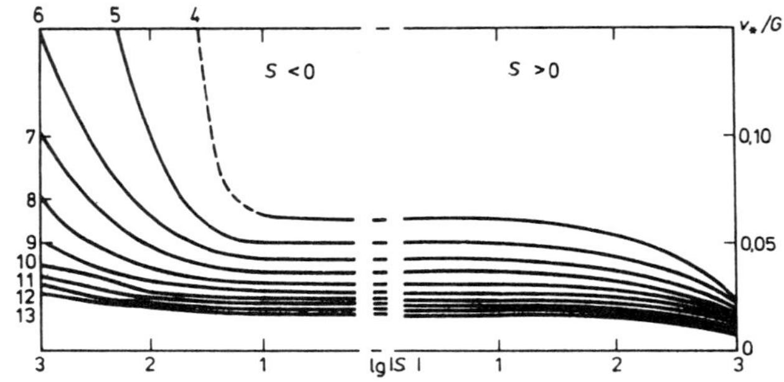

Figure 4.9 Geostrophic drag coefficient $C_g = u_*/G$ as a function of the stability parameter S (numbers by the curves are values of ln Ro). (From Chalikov, 1968.)

Equations (4.28) and (4.32) [or (4.29) and (4.33)] give the relation between the *internal parameters* C_g, α_0, and $\Delta\Theta/T_*$ and the *external parameters* Ro and S (or S_*). If similarity functions A, B, and C or a, b, and c are determined empirically, the system (4.28) or (4.30) can be solved numerically (Figures 4.9 to 4.12). The solution gives the dependence of surface turbulent fluxes on external parameters (synoptic information) and finds application in the parametrization of the ABL in numerical weather prediction models.

There were many theoretical attempts to predict the form of similarity functions A, B, and C or a, b, and c. Many were based on matching surface similarity profiles with analytical solutions of outer-layer equations (e.g., Zilitinkevich and Chalikov, 1968a,b; Chalikov, 1968; Brown, 1978). There also were attempts to determine the similarity functions from observations taken during the Great Plains (Lettau and Davidson, 1957), the Wangara (Clarke et al., 1971), and other experiments. They all reflected a great scatter of data, partly because there are more similarity parameters involved than those

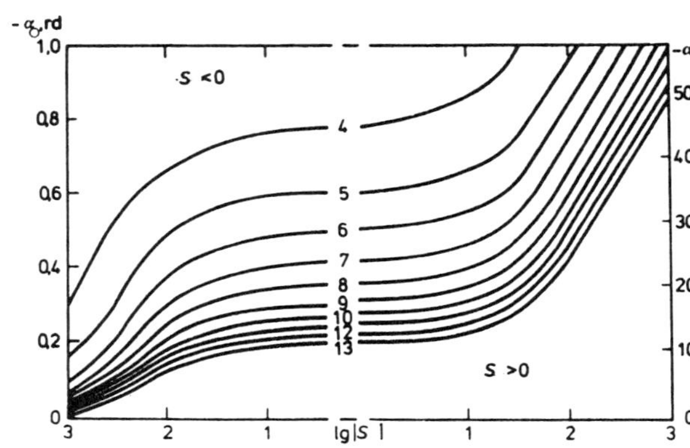

Figure 4.10 Cross-isobar angle α_0 as a function of the stability parameter S (numbers by the curves are values of ln Ro). (From Chalikov, 1968.)

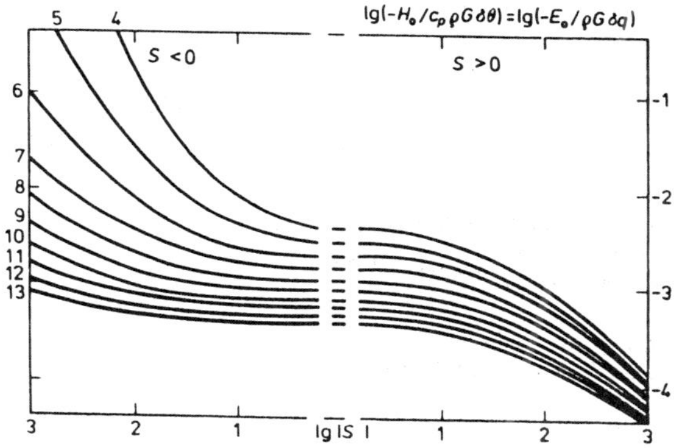

Figure 4.11 Dimensionless fluxes of heat and humidity as a function of the stability parameter S (numbers by curves are values of ln Ro). (After Chalikov, 1968.)

usually considered in the simplest versions of the ABL similarity theory. For example, similarity functions were often assumed to be dependent on only the stability parameters μ, which would be true only if the observed ABL was truly stationary, homogeneous, nonentraining, barotropic, and overlying a flat, horizontal terrain. In reality, the ABL is affected by all these factors and the observations used for determining the various similarity functions are often incomplete. Figures 4.13 and 4.14 illustrate the typical scatter in such a determination.

From the results of the Wangara experiment, Arya (1975) obtained, for $\mu_0 > -50$,

$$
\begin{aligned}
A(\mu_0) &= 1.01 - 0.105\mu_0 - 0.00099\mu_0^2 + 0.00000081\mu_0^3 \\
B(\mu_0) &= 5.14 + 0.142\mu_0 + 0.00117\mu_0^2 - 0.00000033\mu_0^3 \\
C(\mu_0) &= 1.86 - 0.377\mu_0 - 0.00539\mu_0^2 + 0.00000572\mu_0^3
\end{aligned}
\tag{4.34}
$$

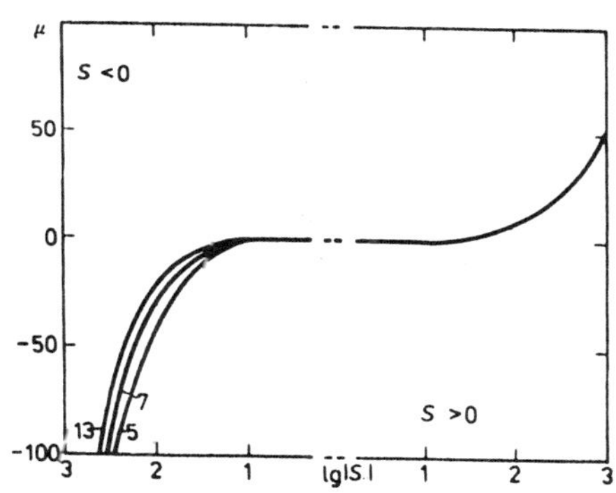

Figure 4.12 Stability parameter μ_0 as a function of stability parameter S (numbers by the curves are values of the ln Ro). (After Chalikov, 1968.)

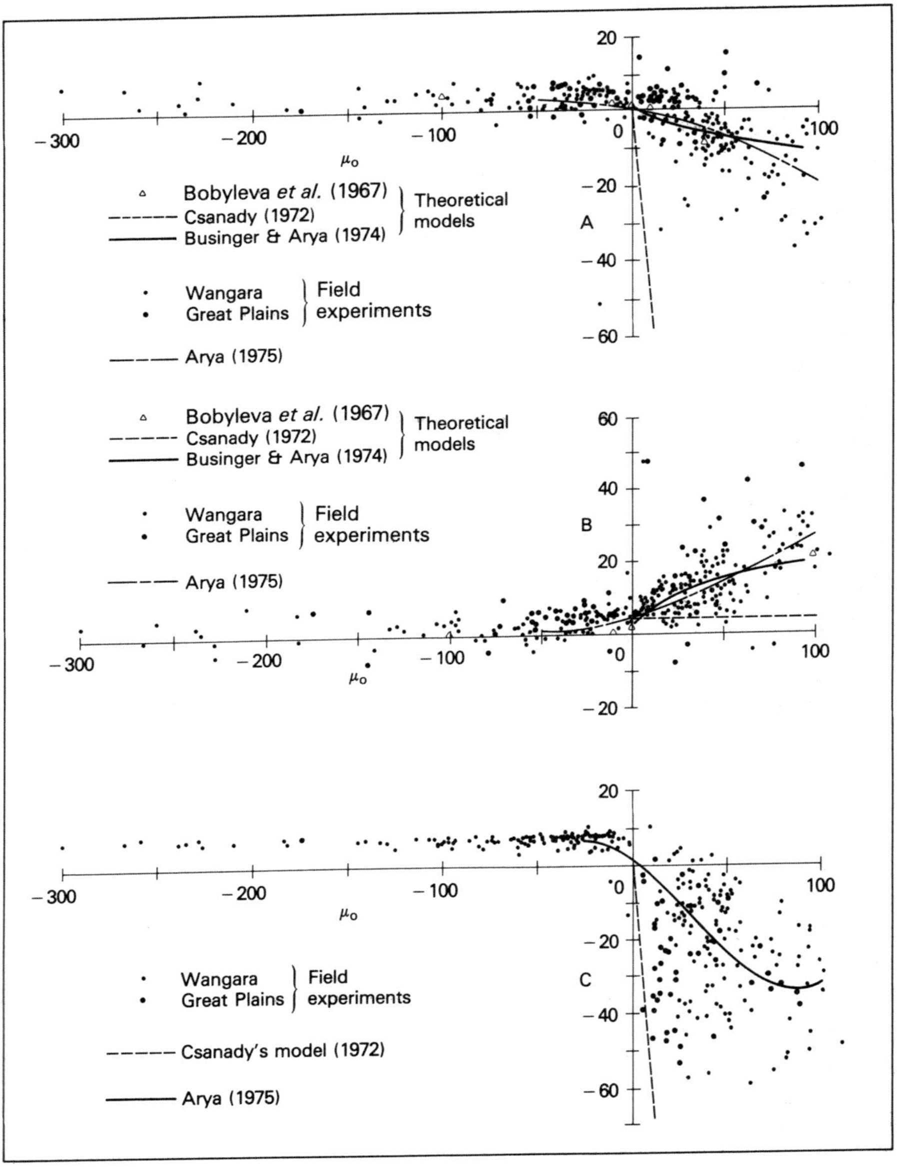

Figure 4.13 Similarity functions A, B, and C. (Arya and Wyngaard, 1975. *Journal of Atmospheric Sciences* published by the American Meteorological Society.)

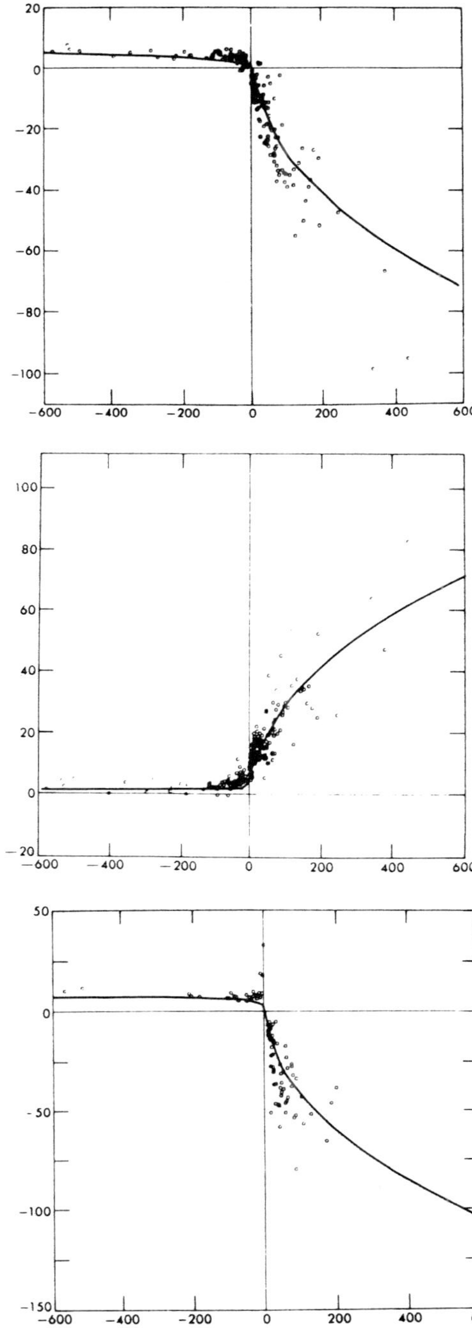

Figure 4.14 Similarity functions *a*, *b*, and *c*. (Yamada, 1976. *Journal of Atmospheric Sciences* published by the American Meteorological Society.)

Curves described by (4.34) are plotted in Figure 4.13. In best-fitting the Wangara data, Yamada (1976) obtained

$$a = \begin{cases} 1.855 - 0.38\dfrac{z_i}{L} & \text{for } 0 \le \dfrac{z_i}{L} \le 35 \\ -2.94\left(\dfrac{z_i}{L} - 19.94\right)^{1/2} & \text{for } 35 \le \dfrac{z_i}{L} \end{cases} \tag{4.35}$$

$$b = \begin{cases} 3.020 + 0.30\dfrac{z_i}{L} & \text{for } 0 \le \dfrac{z_i}{L} \le 35 \\ 2.84\left(\dfrac{z_i}{L} - 12.47\right)^{1/2} & \text{for } 35 \le \dfrac{z_i}{L} \end{cases} \tag{4.36}$$

$$c = \begin{cases} 3.665 - 0.829\dfrac{z_i}{L} & \text{for } 0 \le \dfrac{z_i}{L} \le 18 \\ -4.32\left(\dfrac{z_i}{L} - 11.21\right)^{1/2} & \text{for } 18 \le \dfrac{z_i}{L} \end{cases} \tag{4.37}$$

and additionally, for $h/L < 0$:

$$\begin{aligned} a &= 10.0 - 8.145\left(1.0 - 0.008376\frac{h}{L}\right)^{-1/3} \\ b &= 3.020\left(1.0 - 3.290\frac{h}{L}\right)^{-1/3} \\ c &= 12.0 - 8.335\left(1.0 - 0.03106\frac{h}{L}\right)^{-1/3} \end{aligned} \tag{4.38}$$

Curves described by equations (4.35)–(4.38) are plotted in Figure 4.14.

The boundary layer height z_i, appearing in (4.35)–(4.38), can easily be evaluated for the unstable case, as the height of the upper temperature inversion. For the stable case, the height of the boundary layer is not well defined (e.g., Mahrt et al., 1982; Wetzel, 1982). There are arguments that in the stable case, h is uniquely determined by surface stress, the Earth's rotation, and stability, at least for the idealized steady, homogeneous, and barotropic conditions. From dimensional considerations, h/L_E must then be some universal function of the stability μ_0. Zilitinkevich (1972, 1975a) predicted that the dimensional depth of the idealized, steady-state, stably stratified atmospheric boundary layer obeys

$$\frac{h}{L_E} \sim \mu_0^{-1/2} \tag{4.39}$$

or

$$\frac{h}{L} \sim \mu_0^{1/2} \tag{4.40}$$

This prediction is fairly consistent in the model of Brost and Wyngaard (1978) and the observations of Caughey et al. (1979).

During the last two decades, in the field of boundary layer meteorology there has been great interest in resistance and heat transfer laws. This is caused by the demand for a simple method to combine small-scale boundary layer characteristics (turbulent vertical fluxes) with large-scale parameters in numerical weather prediction models. This interest can be illustrated by reviewing resistance laws modifications, introduced by various authors in order to decrease the substantial scatter of data.

Recognizing the uncertainties involved in measurements of geostrophic winds, some investigators (e.g., Melgarejo and Deardorff, 1974) have defined similarity functions a_h, b_h, and c_h with respect to the actual wind components U_h and V_h at $z = H$ instead of the geostrophic wind components.

Another approach was presented by Deardorff (1972a), who used layer-averaged (over the ABL depth H) values of wind, temperature, and humidity to introduce similarity functions a_m, b_m, and c_m. It appeared that relations involving the layer-averaged parameters are less sensitive to baroclinicity effects.

To suppress the large scatter in empirical plots of similarity functions, the effects of additional parameters had to be taken into consideration. A generalization of the similarity formulations, including the effects of geostrophic wind changing with height (*baroclinicity*), was proposed by Yordanov and Wippermann (1972), Clarke and Hess (1974), and Arya and Wyngaard (1975). As a result of these studies, two new parameters were introduced:

$$M_x = \frac{H}{u_*}\frac{\partial U_g}{\partial x} \qquad M_y = \frac{H}{u_*}\frac{\partial U_g}{\partial y} \tag{4.41}$$

It was shown that the similarity functions can be expressed in the form

$$\begin{aligned} a(\mu, \mu_0, M_x, M_y) &= a(\mu, \mu_0, 0, 0) + \partial a_b \\ b(\mu, \mu_0, M_x, M_y) &= b(\mu, \mu_0, 0, 0) + \partial b_b \\ c(\mu, \mu_0, M_x, M_y) &= c(\mu, \mu_0, 0, 0) + \partial c_b \end{aligned} \tag{4.42}$$

where the corrections on baroclinicity are

$$\begin{aligned} \partial a_b &= M_0 \cos(\beta_0 - \delta) \\ \partial b_b &= M_0 \sin(\beta_0 - \delta) \end{aligned} \tag{4.43}$$

where $M_0 = (M_x^2 + M_y^2)^{1/2}$, $\beta_0 = \arctan(M_y/M_x)$, and δ is an angle dependent on stability. Theoretically determined functions $(\partial a)_b$, $(\partial b)_b$ are shown in Figure 4.15, revealing the substantial effect of baroclinicity.

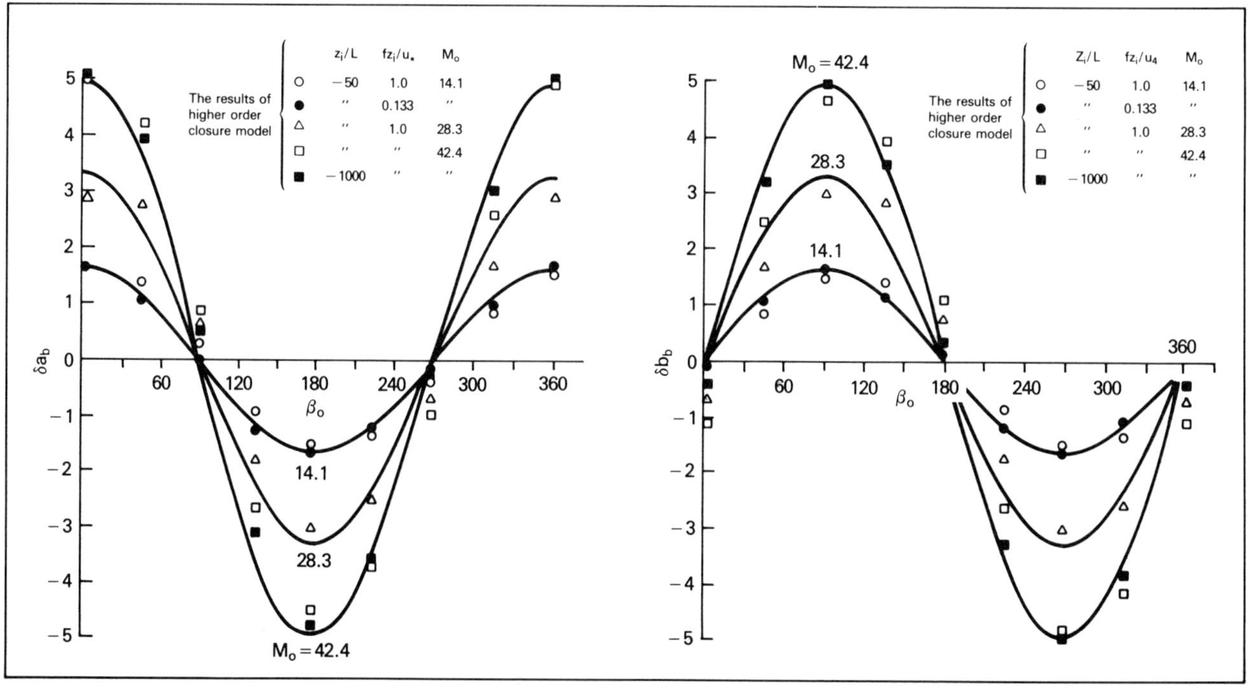

Figure 4.15 Corrections of the similarity functions *a* and *b* due to baroclinicity. β_0, angle between the surface and thermal winds; M_0, magnitude of the thermal wind. (Arya and Wyngaard, 1975. *Journal of Atmospheric Sciences* published by the American Meteorological Society.)

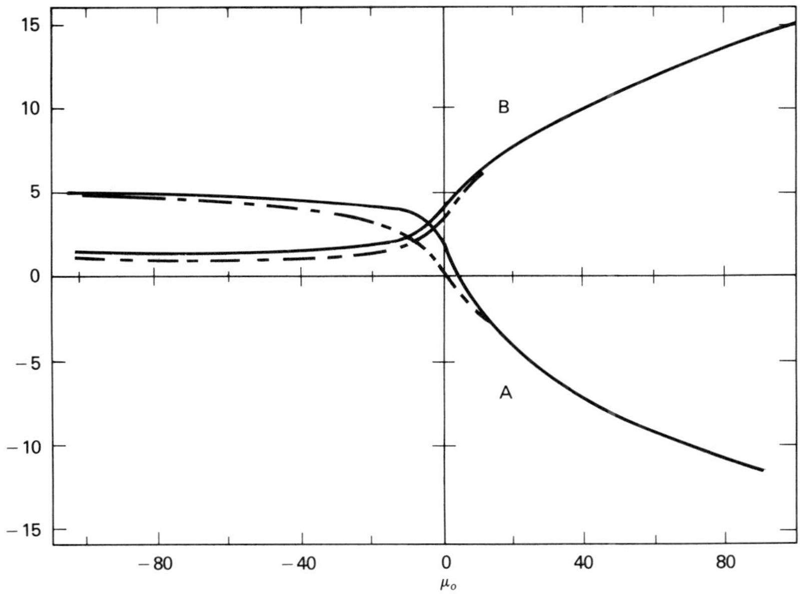

Figure 4.16 Similarity functions A and B from a two-layer model (Brown, 1982) with a secondary flow (solid curve) and without a secondary flow (dashed curve).

Extension of the theory with the inclusion of horizontal roll vortices effects was discussed by Brown (1978). Brown matched analytical profiles in the ABL, including secondary motion, with wind profiles in the surface layer and obtained the theoretical estimation of correction due to secondary motion. Theoretically derived effects resulting from secondary motion are shown in Figure 4.16. As seen from the figure, the effect considered is comparatively minor and achieves maximum values for neutral stratification.

The effects of nonstationarity were studied by Laikhtman and Yordanov (1979) and Yordanov (1980), who generalized resistance and heat transfer laws by introducing two new parameters π_1 and π_2 dependent on a ratio of surface stress components in two adjacent moments of time.

Using the results of the nonstationary ABL model of Wippermann et al. (1973), Yordanov (1980) obtained a theoretical estimation of the diurnal variation of a similarity function (Figure 4.17). The plot of the steady-state dependence of A and B in the form $A + B = 5.5$, obtained by Wippermann (1972a), is also shown. This plot shows that nonstationarity is one of the major sources of scatter of empirical similarity functions.

Sorbjan (1983 and 1984) proposed a generalized form of resistance and heat transfer laws incorporating the effects of a slightly inclined terrain. New similarity functions over a sloping terrain a_s and b_s were shown to be combinations of similarity functions a and b over a horizontal terrain. The

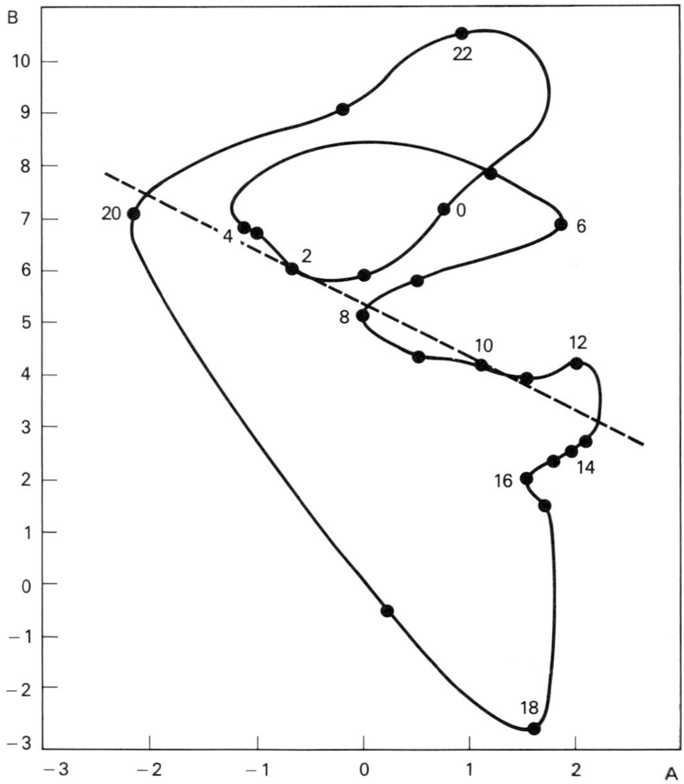

Figure 4.17 Similarity functions B versus A of nonsteady resistance laws obtained from the nonsteady-state ABL model (Yordanov, 1980). The dashed curve $A + B = 5.5$ obtained by Wippermann (1972) for the steady-state resistance laws.

effects of the slope on the similarity functions for stable stratification are shown in Figure 4.18. It was estimated by Sorbjan (1984) that the effect of slope is much larger in this case than that connected with large-scale baroclinicity.

4.4 LOCAL SIMILARITY THEORY

Before discussing further development of the similarity theory it is convenient to review general features of the atmospheric boundary layer expressed in terms of similarity scales. This can be done using the schematic sketch in Figure 4.19. The vertical axis on the sketch emphasizes the vertical structure of the ABL, and the horizontal one shows the impact of the stability. Vertically, the ABL consists of the surface layer with approximately constant turbulent fluxes (the first 10% of the boundary layer) and with the outer layer

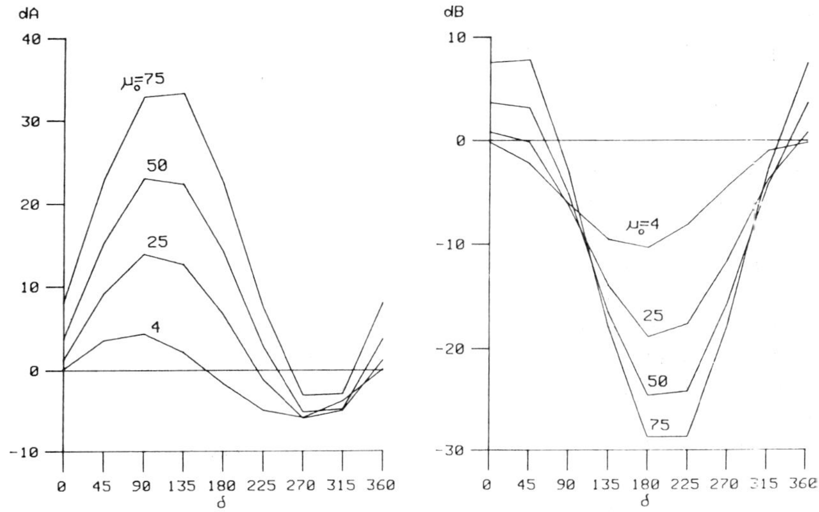

Figure 4.18 Corrections of similarity functions A and B due to the slope effect. δ, angle between the surface stress and the geostrophic wind. (After Sorbjan, 1984. *Journal of Atmospheric Sciences* published by the American Meteorological Society.)

above. With respect to stability, the ABL can be classified as the *convective*, *unstable*, *neutral*, *stable-continuous*, and *stable-sporadic* regimes (Deardorff, 1978b). Lines inclined by 45° to the left and right are parametric curves with z/L = constant.

The regime is called *convective* for $z/L < -0.5$, *unstable* for $-0.5 \leq z/L < -0.02$, *neutral* for $-0.02 \leq z/L \leq 0.02$, *stable-continuous* for $0.02 <$

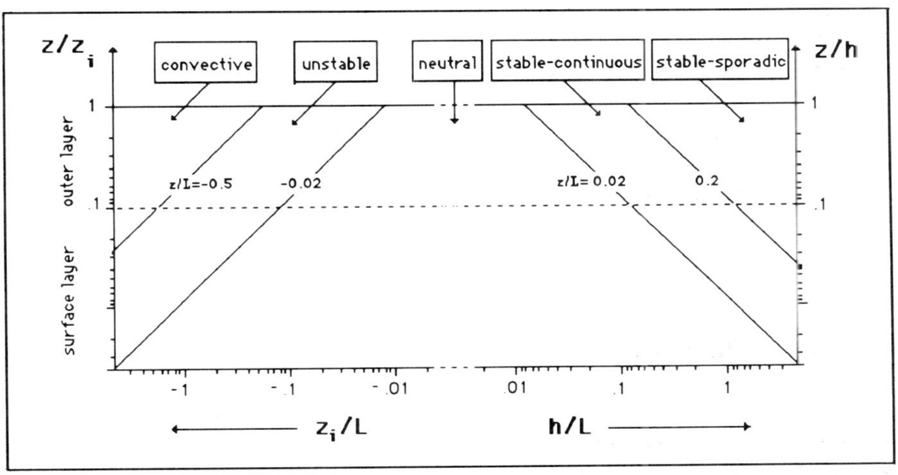

Figure 4.19 Structure of the atmospheric boundary layer.

$z/L \leq 0.2$, and *stable-sporadic* for $z/L > 0.2$. The boundaries between different regimes are somewhat arbitrary.

The structure of the *convective regime* is dominated by buoyancy. The mean wind velocity and temperature profiles are nearly uniform. For this reason, the convective outer layer is called the "mixed layer." The mixed layer is topped by an inversion layer in which temperature increases with height.

In an *unstable regime,* vertical mixing is less effective and causes significant wind shear. A quasi-stationary *neutral* boundary layer is rare in the atmosphere and its structure is not well documented.

In the *stable-continuous* regime turbulence is driven mechanically, and because of the supressing effect of the buoyancy force can exist only close to the ground. Wind velocity has a well-pronounced maximum (low-level jet stream). The height of the maximum wind is usually assumed to be the height of the boundary layer.

In the *stable-sporadic* boundary layer, only a very shallow layer next to the ground surface has a turbulent character. Most of the layer flow is laminar, with patches of turbulent motion resulting from breaking gravity waves. In general, knowledge of processes in the stable boundary layer is more scarce than that of the convective boundary layer, due to the greater measurement problems (smaller turbulence levels) and complications arising from gravity waves and local drainage effects. Furthermore, the stable boundary layer is not stationary, but evolves slowly with time in response to changing surface conditions (Brost and Wyngaard, 1978).

Very close to the ground, where $z/L \sim 0$, wind and temperature profiles are logarithmic. The free convective regime and the stable-sporadic regime never occur next to the ground.

Application of the Monin–Obukhov theory is limited to the lowest 10% of the ABL. The upper part of the boundary layer is considered by the Rossby-number similarity theory which, however, is unable to predict vertical profiles of quantities characterizing the flow. It only allows establishing relationships between surface fluxes and parameters characterizing the flow in the free atmosphere. Vertical distribution of the first- and higher-order moments above the surface layer can be evaluated from the *local similarity theory.* Its name is due to the fact that all scales are height dependent and defined by local values of the governing parameter. Local similarity will be introduced next.

4.4.1 Stable boundary layer

In the stable atmospheric boundary layer, buoyancy forces suppress turbulent eddies, which remain small and independent of the distance from the surface. Therefore, the appropriate scales of turbulent flow in the stable-continuous regime should be based on local values (z dependent) of the Reynolds stress τ and the turbulent heat flux (Nieuwstadt, 1984; Sorbjan, 1986a,b,c):

$$\begin{aligned} &\text{For velocity:} && \mathbf{U}_* = \tau^{1/2} \\ &\text{For temperature:} && \mathbf{T}_* = -\frac{\overline{w'\theta'}}{\mathbf{U}_*} \\ &\text{For concentration:} && \mathbf{C}_* = -\frac{\overline{w'c'}}{\mathbf{U}_*} \\ &\text{For height:} && \Lambda = \frac{\mathbf{U}_*^2}{\kappa\beta\mathbf{T}_*} \end{aligned} \tag{4.44}$$

where

$$\tau = \sqrt{\overline{u'w'}^2 + \overline{v'w'}^2}$$

All the scales above are z dependent. In the surface layer, they take the form of the Monin–Obukhov scales u_*, T_*, and L.

With the help of these scales, the following similarity functions can be formed (Sorbjan, 1986a,b,c):

$$\begin{gathered} \Phi'_m = \frac{\kappa\Lambda}{\mathbf{U}_*^2}\frac{\partial}{\partial z}(U^2 + V^2)^{1/2} \qquad \Phi'_h = \frac{\kappa\Lambda}{\mathbf{T}_*}\frac{\partial}{\partial z}\Theta \qquad \Phi'_c = \frac{\kappa\Lambda}{\mathbf{C}_*}\frac{\partial}{\partial z}C \\ \Phi_{11} = \frac{\overline{u'^2 + v'^2}}{\mathbf{U}_*^2} \qquad \Phi_{33} = \frac{\overline{w'^2}}{\mathbf{U}_*^2} \qquad \Phi_{1\theta} = \frac{\overline{u'\theta'}}{\mathbf{U}_*\mathbf{T}_*} \\ \Phi_{\theta\theta} = \frac{\overline{\theta'^2}}{\mathbf{T}_*^2} \qquad \Phi'_\epsilon = \frac{\Lambda\epsilon}{\mathbf{U}_*^3} \qquad \Phi'_N = \frac{\Lambda N}{\mathbf{U}_*\mathbf{T}_*^2} \end{gathered} \tag{4.45}$$

Local similarity functions are defined in a manner analogous to the Monin–Obukhov similarity functions.

In the stable ABL, turbulence is in the so-called "z-less" state. As a consequence, all similarity functions should approach constant values:

$$\Phi'_m, \Phi'_h, \Phi'_C, \Phi_{11}, \Phi_{33}, \Phi_{1\theta}, \Phi_{\theta\theta}, \Phi'_\epsilon, \Phi'_N = \text{const} \tag{4.46}$$

Values of the universal functions can be obtained by matching stable outer and stable surface layers. The empirical basis of the Monin–Obukhov similarity theory (Table 4.2) gives the values of the constants in (4.46):

$$\Phi_{11} = 9.3 \qquad \Phi_{33} = 2.5 \qquad \Phi_{1\theta} = 3.0 \qquad \Phi_{\theta\theta} = 6.0 \tag{4.47}$$

It is more convenient to introduce five new functions defined in the following way:

$$\Phi_m = Z\Phi'_m \qquad \Phi_h = Z\Phi'_h \qquad \Phi_c = Z\Phi'_c \qquad \Phi_\epsilon = \kappa Z\Phi'_\epsilon \qquad \Phi_N = \kappa Z\Phi'_N \tag{4.48}$$

Functions Φ_m and Φ_h are linear and can be expressed for the small Z (stratification close to neutral) as power series of Z. By taking only first terms, we obtain (Sorbjan, 1986a)

$$\Phi_m = 1 + 4.7Z \qquad \Phi_h = 0.74 + 4.7Z \qquad \Phi_c = 1.0 + 5.2Z \tag{4.49}$$

where the values of constants are obtained by matching them with the similarity functions obtained by Businger et al. (1971) in the surface layer.

From the definition of eddy viscosities and Richardson numbers, the following can also be obtained:

$$K_m = \frac{Z}{\Phi_m} \qquad K_h = \frac{Z}{\Phi_h} \qquad \mathrm{Ri} = \frac{Z\Phi_h}{\Phi_m^2} \qquad R_f = \frac{Z}{\Phi_m} \tag{4.50}$$

In Chapter 3 the transport terms in the budgets of the turbulent kinetic energy and temperature variance were shown to be small in the stable surface layer. This implies that the dissipation and production rates for $\overline{\theta^2}$ are approximately in balance: $N \sim -\overline{w'\theta'}\,\partial\Theta/\partial z$. From similar arguments it can be concluded that the dissipation rate of turbulent kinetic energy is balanced by the sum of buoyant and mechanical energy productions. This yields

$$\begin{aligned} \Phi_N &= c_N \Phi_h \\ \Phi_\epsilon &= c_\epsilon(\Phi_m - Z) \end{aligned} \tag{4.51}$$

where c_N and c_ϵ are parameters. The parameters c in (4.51) are equal to unity when the ABL is in steady state. It can be obtained from (4.51) and (4.49):

$$\begin{aligned} \Phi_\epsilon &= c_\epsilon(1 + 3.7Z) \\ \Phi_N &= c_N(0.74 + 3.7Z) \end{aligned} \tag{4.52}$$

The structure parameters in (4.8) can be transformed into a dimensionless form by using the scales defined in (4.44) and (4.52):

$$\begin{aligned} \frac{(\kappa z)^{2/3} C_v^2}{\mathrm{U}_*^2} &\equiv \Phi_{cv} = 2\Phi_\epsilon^{2/3} \\ \frac{(\kappa z)^{2/3} C_T^2}{\mathrm{T}_*^2} &\equiv \Phi_{cT} = 1.6\Phi_\epsilon^{-1/3}\Phi_N \end{aligned} \tag{4.53}$$

To express the formulas above as functions of height, the local scales first have to be rescaled. It can be done assuming that in the stable, barotropic ABL, over a horizontal terrain the kinematic momentum and heat fluxes have the following distributions:

$$\tau = u_*^2\left(1 - \frac{z}{h}\right)^\alpha \tag{4.54}$$

$$\overline{w'\theta'} = \overline{w'\theta'}\Big|_0\left(1 - \frac{z}{h}\right)^\beta \tag{4.55}$$

and α and β are parameters dependent on the temporal development of the stable ABL, terrain slope, baroclinicity, and so on. Constants α and β must be determined empirically. From (4.44) and (4.54)–(4.55), we also obtain

$$\begin{aligned} \mathbf{U}_* &= u_*\left(1 - \frac{z}{h}\right)^{\alpha/2} \\ \mathbf{T}_* &= T_*\left(1 - \frac{z}{h}\right)^{\beta-\alpha/2} \\ \Lambda &= L\left(1 - \frac{z}{h}\right)^{3/2\alpha-\beta} \end{aligned} \tag{4.56}$$

Data presented by Yokoyama et al. (1979) show that β can vary between 1 and 3. The Minnesota observations (Caughey et al., 1979), taken near sunset, show that $\alpha \sim 2$ and $\beta \sim 3$. The Cabauw data (Nieuwstadt, 1984), collected later in the development of the stable ABL, indicate that both α and β are close to 1.

It can be proven that

$$\beta \geq \alpha \tag{4.57}$$

To show this, consider Φ_h. From (4.46), (4.47), and (4.56) we can derive

$$\frac{L}{T_*}\frac{\partial\Theta}{\partial z} = \frac{c}{\kappa}\left(1 - \frac{z}{h}\right)^{2(\beta-\alpha)} \tag{4.58}$$

where c is a constant.

From (4.58) it follows that a singularity on the top of the ABL is avoided only when $\beta \geq \alpha$.

It can be assumed that the vertical distribution of the concentration flux (e.g., water vapor) can be expressed in the form

$$\overline{w'c'} = \overline{w'c'}\Big|_0\left(1 - \frac{z}{h}\right)^{\gamma} \tag{4.59}$$

where γ is a parameter. Substituting (4.59) into (4.44) and (4.45), (4.49) yields

$$\frac{\kappa z}{c_{*0}}\frac{\partial C}{\partial z} = \left(1 + 5.2\mu_h\frac{z}{h}\right)\left(1 - \frac{z}{h}\right)^{\gamma-2\alpha+\beta} \tag{4.60}$$

where c_{*0} is the surface layer value of C_*.

The diffusivity K_c, nondimensionalized by ku_*h, can be obtained as

$$K_c = \frac{1}{\kappa u_* h}\frac{\overline{w'c'}}{\partial C/\partial z} = \frac{z/h(1 - z/h)^{2\alpha-\beta}}{1 + 5.2\mu_h(z/h)} \tag{4.61}$$

For the particular case of the Minnesota experiment (Caughey et al., 1979), where $\alpha = 2$ and $\beta = 3$, $c_N = 1.5$ and $c_\epsilon = 1.25$, the similarity functions derived take the following form (valid only for the Minnesota data!):

$$
\begin{aligned}
\frac{\kappa z}{u_*}\frac{\partial}{\partial z}(U^2+V^2)^{1/2} &= \left(1+4.7\mu_h\frac{z}{h}\right)\left(1-\frac{z}{h}\right)\\
\frac{\kappa z}{T_*}\frac{\partial}{\partial z}\Theta &= \left(0.74+4.7\mu_h\frac{z}{h}\right)\left(1-\frac{z}{h}\right)^2\\
\frac{\overline{u'^2}+\overline{v'^2}}{u_*^2} &= 9.3\left(1-\frac{z}{h}\right)^2\\
\frac{\overline{w'^2}}{u_*^2} &= 2.5\left(1-\frac{z}{h}\right)^2\\
\frac{\overline{\theta'^2}}{T_*^2} &= 6.0\left(1-\frac{z}{h}\right)^4\\
\frac{\overline{u'\theta'}}{u_*T_*} &= 3.0\left(1-\frac{z}{h}\right)^3\\
\frac{Nh}{u_*T_*^2} &= 4.3\left(1+3.7\mu_h\frac{z}{h}\right)\frac{(1-z/h)^5}{z/h}\\
\frac{\epsilon h}{u_*^3} &= 3.6\left(1+3.7\mu_h\frac{z}{h}\right)\frac{(1-z/h)^3}{z/h}\\
\mathrm{Ri} &= \mu_h\frac{z/h}{[1+4.7\mu_h(z/h)]^2}\left(0.74+4.7\mu_h\frac{z}{h}\right)\\
R_f &= \mu_h\frac{z/h}{1+4.7\mu_h(z/h)}\\
\frac{k_m}{\kappa u_* h} &= \frac{(z/h)(1-z/h)}{1+4.7\mu_h(z/h)}\\
\frac{k_h}{\kappa u_* h} &= \frac{(z/h)(1-z/h)}{0.74+4.7\mu_h(z/h)}
\end{aligned}
\tag{4.62}
$$

Plots of empirical similarity functions Φ_{33} and $\Phi_{\theta\theta}$ are shown in Figures 4.20 and 4.21. The scatter seen in these plots may be due to measurement errors, which are substantial in the stable ABL (since turbulent characteristics are small).

Equations (4.54)–(4.55) are plotted in Figure 4.22 for $\alpha = 2$ and $\beta = 3$. These fit the Minnesota data very well. Similarity functions $\overline{u'^2}$, $\overline{u'\theta'}$, $\overline{\theta'^2}$, $\overline{w'^2}$, Ri, ϵ, and N obtained for these values are plotted in Figures 4.23 to 4.27. In

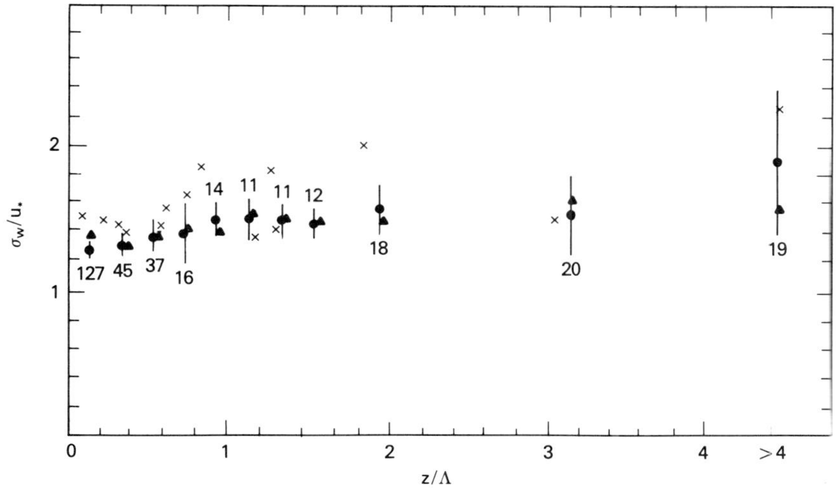

Figure 4.20 Vertical profile of the standard deviation of the vertical velocity nondimensionalized by $\mathbf{U}_*$, in the stable boundary layer. Points, data of Nieuwstadt (1984); crosses, data after Caughey et al. (1979). (Nieuwstadt, 1984; Caughey, Wyngaard, and Kaimal, 1979. *Journal of Atmospheric Sciences* published by the American Meteorological Society.)

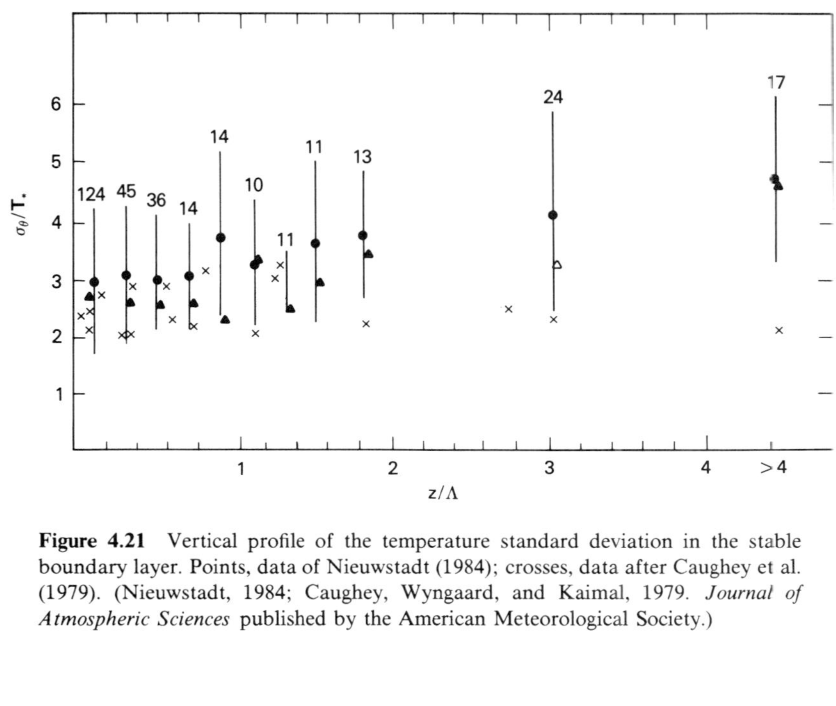

Figure 4.21 Vertical profile of the temperature standard deviation in the stable boundary layer. Points, data of Nieuwstadt (1984); crosses, data after Caughey et al. (1979). (Nieuwstadt, 1984; Caughey, Wyngaard, and Kaimal, 1979. *Journal of Atmospheric Sciences* published by the American Meteorological Society.)

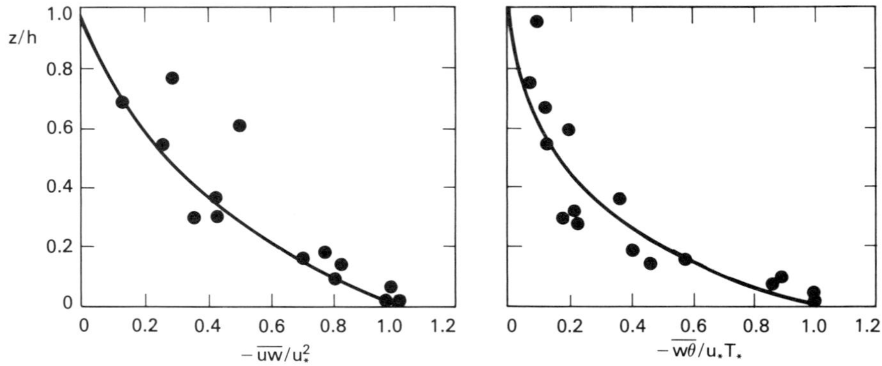

Figure 4.22 Vertical profiles of the dimensionless Reynolds stress and the vertical heat flux in the stable boundary layer. (Caughey, Wyngaard, and Kaimal, 1979. *Journal of Atmospheric Sciences* published by the American Meteorological Society.)

Figure 4.28, structure parameters derived by using (4.8) are also presented. The figures indicate that all characteristics of turbulence decrease with height and vanish on the top of the boundary layer, where the Richardson number reaches its critical value.

4.4.2 Convective boundary layer

The supply of solar energy early in the morning transforms the stable boundary layer into the convective boundary layer. In the convective boundary layer, the potential temperature is almost constant with height (except very close to the surface, Figure 1.4). For this reason, the convective boundary layer is often called the mixed layer. Above the mixed layer, the potential temperature increases with height.

Laboratory experiments (Adrian et al., 1986) indicate that in nonpenetrative convection [described in Sections 3.5.2 and 6.2 (where we discuss encroachment)] the statistical moments have values very close to those in the lower part of the atmospheric boundary layer during penetrative convection (Figure 4.29). Taking this into consideration, it seems convenient to decompose the statistical characteristics of the penetrative regime into nonpenetrative (here and after denoted by b) and residual (here and after denoted by t) components. For example, the decomposition, when applied to the turbulent heat flux, yields

$$\begin{aligned} \overline{w'\theta'}(z) &= \overline{w'\theta'}\Big|_0\left(1 - \frac{z}{z_i}\right) + \overline{w'\theta'}\Big|_i\frac{z}{z_i} \\ &= \overline{w'\theta'_b}(z) + \overline{w'\theta'_t}(z) \end{aligned} \tag{4.63}$$

where the subscript *0* denotes the surface value and the subscript i denotes the

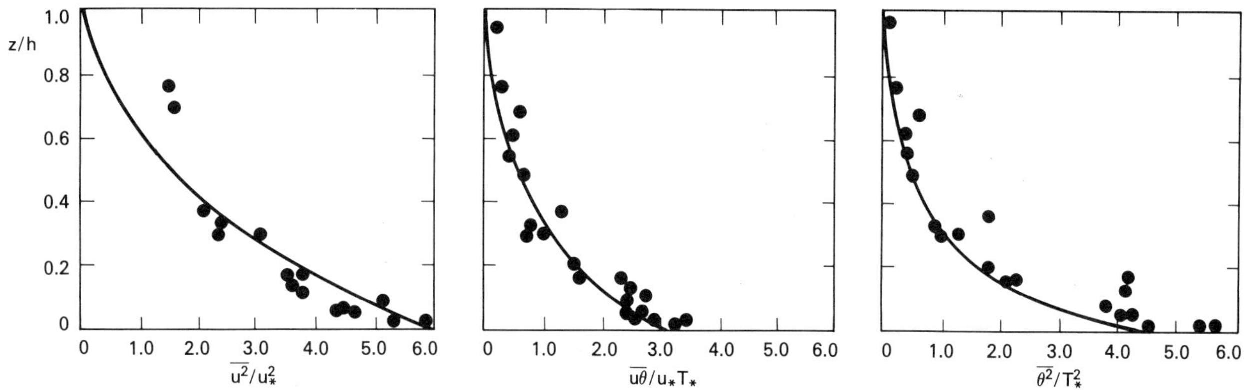

Figure 4.23 Vertical profiles of the dimensionless variances and covariances for wind velocity and temperature in the stable boundary layer. (Caughey, Wyngaard, and Kaimal, 1979. *Journal of Atmospheric Sciences* published by the American Meteorological Society.)

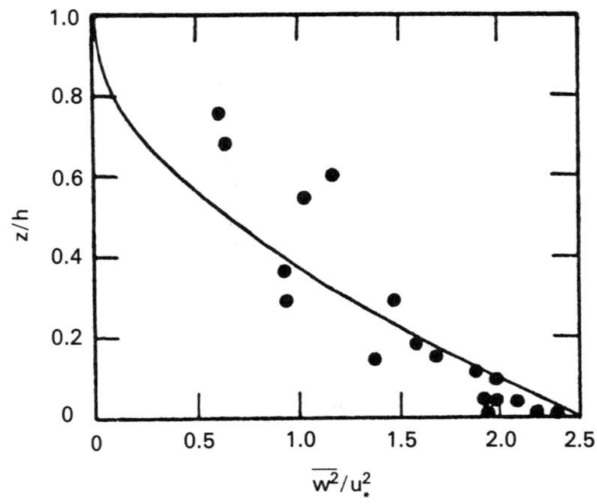

Figure 4.24 Vertical profile of dimensionless vertical velocity variance in the stable boundary layer. (Caughey, Wyngaard, and Kaimal, 1979. *Journal of Atmospheric Sciences* published by the American Meteorological Society.)

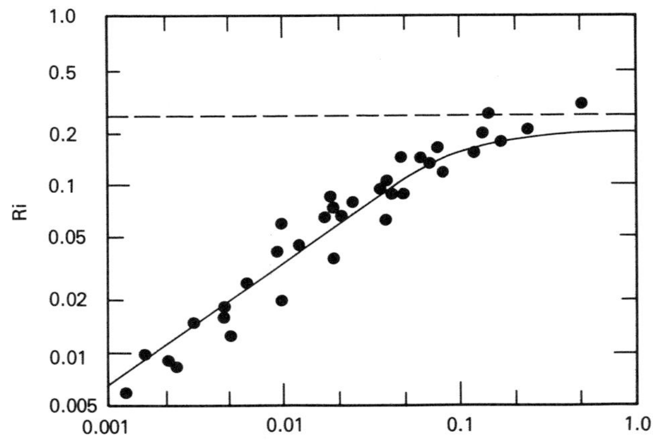

Figure 4.25 Vertical profile of the Richardson number in the stable boundary layer. (Caughey, Wyngaard, and Kaimal, 1979. *Journal of Atmospheric Sciences* published by the American Meteorological Society.)

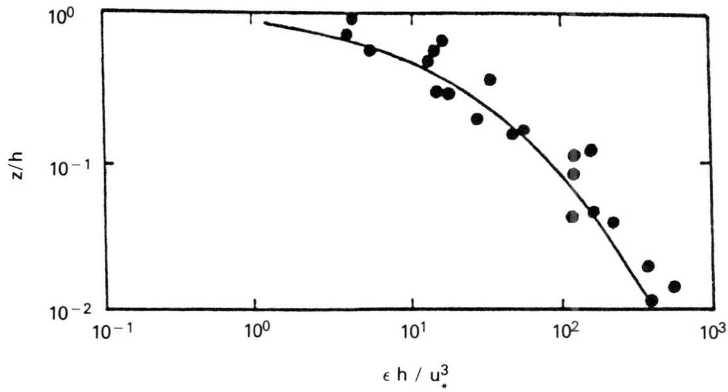

Figure 4.26 Vertical profile of dimensionless dissipation rate of the turbulent kinetic energy in the stable boundary layer. Points obtained from the Minnesota experiment. (Caughey, Wyngaard, and Kaimal, 1979. *Journal of Atmospheric Sciences* published by the American Meteorological Society.)

value at the top of the ABL. The *b* component of the heat flux is positive. Positive fluctuations for both vertical velocity and temperature (ascending positively buoyant air) as well as negative fluctuations for both the vertical velocity and temperature (descending air, which is negatively buoyant) mainly contribute to this component of the heat flux. The *t* component of the heat flux is negative on the top and changes linearly to zero at the surface, due mainly to the contribution from negatively buoyant thermals penetrating the inversion layer and positively buoyant air entraining the mixed layer (see

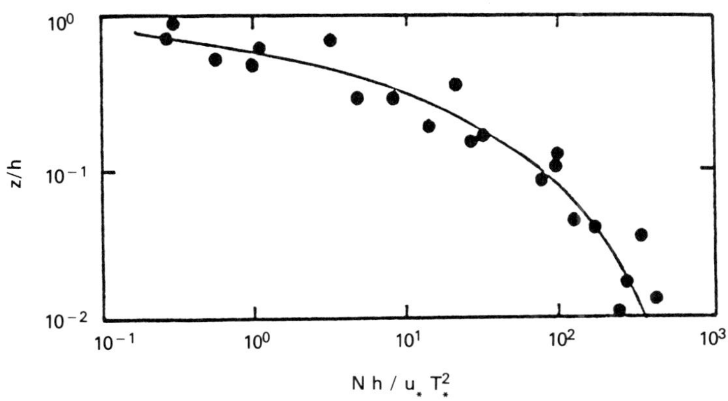

Figure 4.27 Vertical profile of dimensionless destruction rate of the temperature variance in the stable boundary layer obtained from the Minnesota experiment. (Caughey, Wyngaard, and Kaimal, 1979. *Journal of Atmospheric Sciences* published by the American Meteorological Society.)

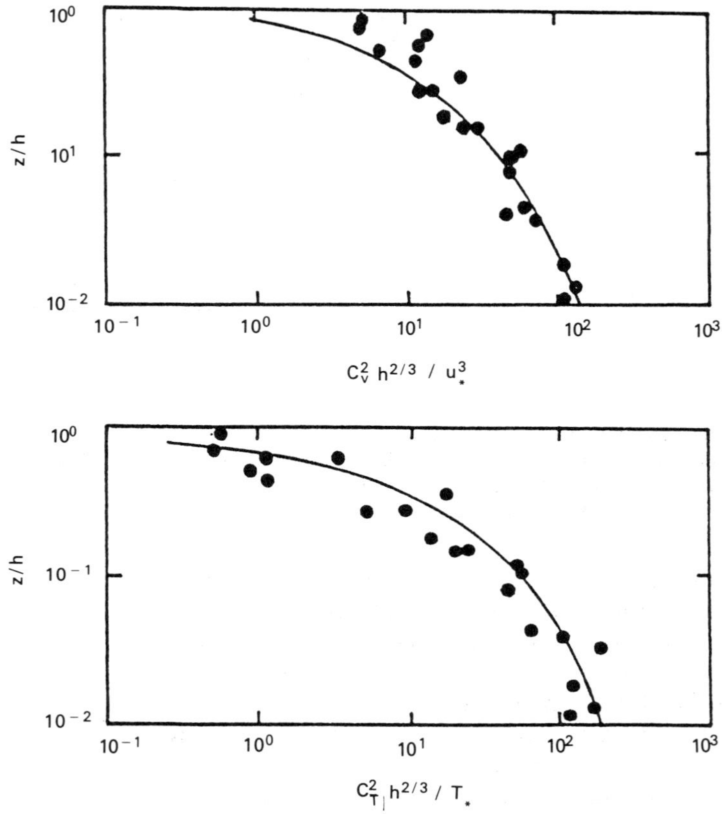

Figure 4.28 Vertical profiles of dimensionless structure parameters in the stable boundary layer. Points obtained from the Minnesota experiment. (Caughey, Wyngaard, and Kaimal, 1979. *Journal of Atmospheric Sciences* published by the American Meteorological Society.)

Figure 3.10). Wilczak and Businger (1983) showed that positive and negative components of the heat flux are linear.

To derive a better understanding of the consequences of the decomposition proposed, consider the equation of the vertical component of turbulent kinetic energy (3.25) in the simplified (steady-state and horizontally homogeneous) form (Sorbjan, 1988)

$$-\frac{1}{2}\frac{d}{dz}\overline{w'^3} - \frac{1}{\rho_0}\frac{d}{dz}\overline{p'w'} + \beta\overline{w'\theta'} - \frac{\epsilon}{3} = 0 \tag{4.64}$$

In the equation above, the first term is the turbulent transport, the second term is the pressure redistribution, the third term is the buoyant production, and the last term is the dissipation rate. Denoting the turbulent

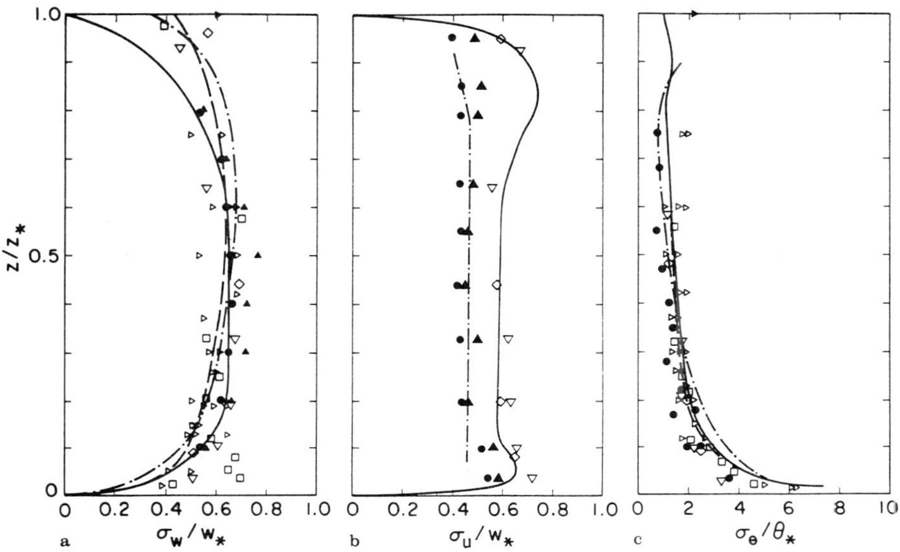

Symbol	Reported by	Observations
●	Willis and Deardorff (1974)	Case S1
▲	Willis and Deardorff (1974)	Case S2
□	Telford and Warner (1964)	Aircraft measurements
▽	Lenschow (1974)	Aircraft measurements as reported by Willis and Deardorff (1974)
◇	Lenschow (1970)	Same as above
▷	Kaimal et al. (1977)	Minnesota experimental data
-·-	Andre et al. (1976)	Clipping approximation-numerical study
---	Zeman and Lumley (1976)	Second order modeling-numerical study

Figure 4.29 Comparison of fluctuation intensities in laboratory experiments (solid line), atmospheric boundary layer data, and computations. (After Adrian et al., 1986. Reprinted by permission.)

transport term by TT, the pressure term by PT, and introducing the decomposition above yields

$$\left(-\mathrm{TT}_b - \mathrm{PT}_b + \beta\overline{w'\theta_b'} - \frac{\epsilon_b}{3}\right) + \left(-\mathrm{TT}_t - \mathrm{PT}_t + \beta\overline{w'\theta_t'} - \frac{\epsilon_t}{3}\right) = 0 \quad (4.65)$$

Observations in the atmosphere (Lenschow et al., 1980), laboratory experiments (Willis and Deardorff, 1974), and the results of second-order closure modeling (e.g., André et al., 1978) indicate that within the mixed layer all terms in (4.64) are approximately linear. Therefore, all terms in (4.65) must be also approximately linear. Terms within the first set of parentheses change linearly with increasing height from nonzero values at the surface to zero at the top of the ABL but become constant when divided by $\beta\overline{w'\theta_b'}$. Inversely, terms within the second set of parentheses increase linearly from zero at the surface to a nonzero value at the top but remain constant when divided by $\beta\overline{w'\theta_t'}$.

The comments above imply that for nonpenetrative convection, characteristics of turbulence involving fluctuations of vertical velocity and temperature can be defined by the following parameters:

$$\beta, \quad \overline{w'\theta_b'}(z), \quad z \quad (4.66)$$

These parameters can be used to construct the following local scales:

For vertical velocity: $$\mathbf{U}_b = \left[\beta z\overline{w'\theta_b'}(z)\right]^{1/3} \quad (4.67)$$

For temperature: $$\Theta_b = -\frac{\overline{w'\theta_b'}(z)}{U_b} \quad (4.68)$$

For height: $$z \quad (4.69)$$

The actual height z is adopted as a height scale. This is consistent with the results of Lenschow and Stephens (1980), who have shown that ascending thermals merge with height and provide the mechanism for increasing the length scale.

Residual components can be assumed to depend on the following governing parameters:

$$\beta, \quad \overline{w'\theta_t'}(z), \quad (z_i + \Delta - z) \quad (4.70)$$

where Δ is the depth of the interfacial layer between the mixed and inversion layers. The parameters above can be used to construct the following local scales:

For vertical velocity: $$\mathbf{U}_t = \left[\beta(z_i + \Delta - z)\overline{w'\theta_t'}(z)\right]^{1/3} \quad (4.71)$$

For temperature: $$\Theta_t = -\frac{\overline{w'\theta_t'}(z)}{U_t} \quad (4.72)$$

For height: $$z_i + \Delta - z \quad (4.73)$$

The distance $z_i + \Delta - z$ is assumed to be a height scale, since the identity of entrained air would depend on the distance from the inversion layer. It can be noticed that the assumed height scale on the top of the mixed layer is equal to Δ. A similar length scale in this region was proposed by Wyatt (1978).

Nondimensionalizing of terms within the first parentheses of (4.65) by the appropriate scales yields

$$\frac{z}{\mathbf{U}_b^3}\frac{d\overline{w'^3}}{dz}\bigg|_b = c_{3b} = \text{const}$$
$$\frac{z}{\mathbf{U}_b^3}\frac{d\overline{w'\pi'}}{dz}\bigg|_b = c_{\pi b} = \text{const} \tag{4.74}$$
$$\frac{z}{\mathbf{U}_b^3}\epsilon|_b = c_{\epsilon b} = \text{const}$$

while the terms within the second set of parentheses become

$$\frac{(z_i + \Delta - z)}{\mathbf{U}_t^3}\frac{d\overline{w'^3}}{dz}\bigg|_t = c_{3t} = \text{const}$$
$$\frac{(z_i + \Delta - z)}{\mathbf{U}_t^3}\frac{d\overline{w'\pi'}}{dz}\bigg|_t = c_{\pi t} = \text{const} \tag{4.75}$$
$$\frac{(z_i + \Delta - z)}{\mathbf{U}_t^3}\epsilon|_t = c_{\epsilon t} = \text{const}$$

where π is the kinematic pressure, $\pi = p/\rho$, and all coefficients c are universal constants, satisfying $c_{3b} + c_{\pi b} + c_{\epsilon b} = 1$ and $c_{3t} + c_{\pi t} + c_{\epsilon t} = 1$.

The local scales (4.67)–(4.68) and (4.71)–(4.72) can be expressed in an explicit form. Using (4.63), it can be obtained that

$$\mathbf{U}_b = w_*\left(\frac{z}{z_i}\right)^{1/3}\left(1 - \frac{z}{z_i}\right)^{1/3} \tag{4.76}$$

$$\Theta_b = \Theta_*\left(\frac{z}{z_i}\right)^{-1/3}\left(1 - \frac{z}{z_i}\right)^{2/3} \tag{4.77}$$

and

$$\mathbf{U}_t = w_* R^{1/3}\left(\frac{z}{z_i}\right)^{1/3}\left(1 - \frac{z}{z_i} + D\right)^{1/3} \tag{4.78}$$

$$\Theta_t = \Theta_* R^{2/3}\left(\frac{z}{z_i}\right)^{2/3}\left(1 - \frac{z}{z_i} + D\right)^{-1/3} \tag{4.79}$$

where Θ_* and w_* are the mixed layer scales introduced by Deardorff (1970) and defined as

$$w_* = \left(\beta z_i \overline{w'\theta'}\Big|_0\right)^{1/3}$$

$$\Theta_* = -\frac{\overline{w'\theta'}\Big|_0}{w_*} \tag{4.80}$$

and the parameters R and D are defined as

$$R = \frac{\overline{w'\theta'}\Big|_i}{\overline{w'\theta'}\Big|_0}$$

$$D = \frac{\Delta}{z_i} \tag{4.81}$$

When $z/z_i \rightarrow 0$, the results obtained are consistent with the free-convection similarity in the surface layer.

Substituting (4.76)–(4.79) into (4.74)–(4.75) and adding the b and t contributions gives

$$\frac{z_i}{w_*^3}\frac{d\overline{w'^3}}{dz} = c_{3b}\left(1 - \frac{z}{z_i}\right) + c_{3t}R\frac{z}{z_i}$$

$$\frac{z_i}{w_*^3}\frac{d\overline{w'\pi'}}{dz} = c_{\pi b}\left(1 - \frac{z}{z_i}\right) + c_{\pi t}R\frac{z}{z_i} \tag{4.82}$$

$$\frac{z_i}{w_*^3}\epsilon = c_{\epsilon b}\left(1 - \frac{z}{z_i}\right) + c_{\epsilon t}R\frac{z}{z_i}$$

The first two formulas in (4.82) can be integrated, yielding the following result:

$$\frac{\overline{w'^3}}{w_*^3} = c_{3b}\frac{z}{z_i} + 0.5(c_{3t}R - c_{3b})\left(\frac{z}{z_i}\right)^2$$

$$\frac{\overline{w'\pi'}}{w_*^3} = c_{\pi b}\frac{z}{z_i} + 0.5(c_{\pi t}R - c_{\pi b})\left(\frac{z}{z_i}\right)^2 \tag{4.83}$$

The similarity function obtained for the third moment of vertical velocity is compared with the observations of the AMTEX experiment in Figure 4.30

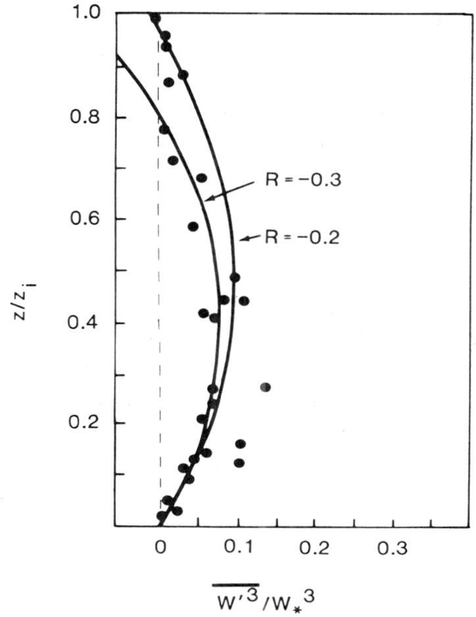

Figure 4.30 Vertical profile of the dimensionless third moment of vertical wind velocity in the convective boundary layer. Data from the AMTEX experiments. (Lenschow, Wyngaard, and Pennell, 1980. *Journal of Atmospheric Sciences* published by the American Meteorological Society.)

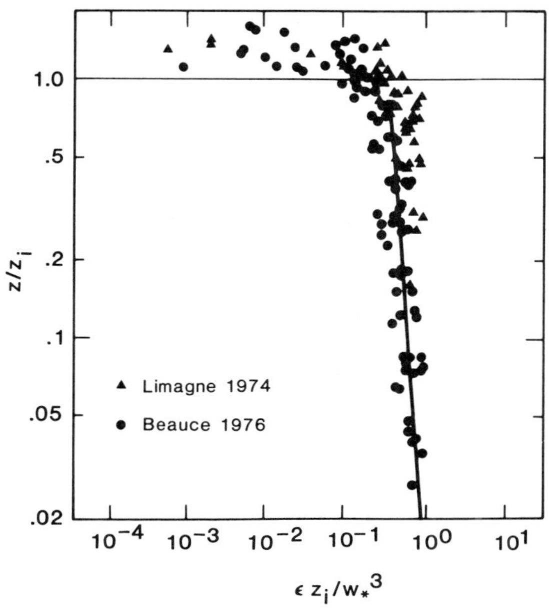

Figure 4.31 Vertical profile of normalized dissipation rate of turbulent kinetic energy in the convective boundary layer. (After Guillement et al., 1983.)

(Lenschow et al., 1980). The local similarity curve was plotted for $c_{3b} = 0.4$ and $c_{3t} = 2$ and $D = 0$. The curve obtained for $R = -0.3$ has slightly negative values at the top. The similarity function for the dissipation rate, shown in Figure 4.31, coincides with the experimental data for $c_{\epsilon b} = 0.8$ and $c_{\epsilon t} = 2.5$.

The approach discussed above can be used to describe vertical profiles of other characteristics of turbulence by decomposing them and then relating to the b and t components of the heat flux. In this way, any statistical moment of a turbulent field can be expressed as a combination of two basis functions found in terms of dimensional analysis. This procedure provides a useful parametrization of the convective boundary layer. As an example let us consider the vertical velocity and temperature variances as well as the destruction rate of the temperature variance. Since z is the only height scale accepted for the nonpenetrative convection, it is therefore impossible to construct the dimensionless height parameter in this case. As a consequence, universal functions for the b component of these characteristics of turbulence must be independent of height (Sorbjan, 1988).

$$\frac{\overline{w_b'^2}}{\mathrm{U}_b^2} = c_{wb}$$
$$\frac{\overline{\Theta_b'^2}}{\Theta_b^2} = c_{\theta b} \tag{4.84}$$
$$\frac{N_b z}{\mathrm{U}_b \Theta_b^2} = c_{Nb}$$

where c_{wb}, $c_{\theta b}$, $c_{\epsilon b}$, c_{Nb} are coefficients independent of height (they can depend on R).

Analogously as a consequence of the accepted assumption that $z_i + \Delta - z$ is the only height scale for the residual process, universal functions for the corresponding t components of the considered characteristics of turbulence should also be independent of height:

$$\frac{\overline{w_t'^2}}{\mathrm{U}_t^2} = c_{wt}$$
$$\frac{\overline{\Theta_t'^2}}{\Theta_t^2} = c_{\theta t} \tag{4.85}$$
$$\frac{N_t(z_i + \Delta - z)}{\mathrm{U}_t \Theta_t^2} = c_{Nt}$$

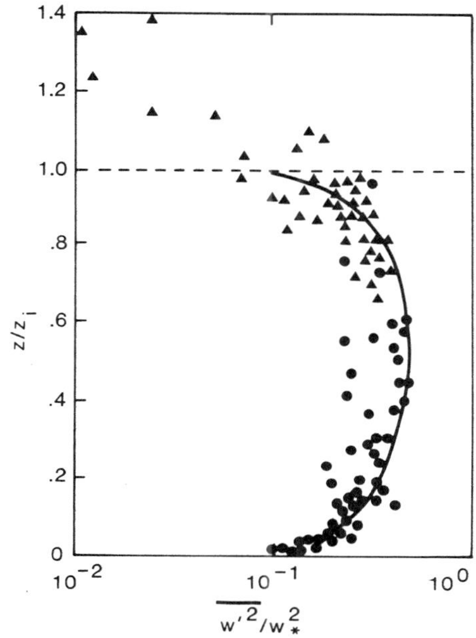

Figure 4.32 Vertical profiles of a dimensionless vertical velocity variance in the convective boundary layer. Data from the 1976 Ashchurch experiments are shown by triangles, the 1973 Minnesota data are shown by circles. The similarity curve is plotted for $R = -0.2$. (After Caughey and Palmer, 1979.)

Combining b and t contributions gives us

$$\frac{\overline{w'^2}}{w_*^2} = c_{wb}\left(\frac{z}{z_i}\right)^{2/3}\left(1 - \frac{z}{z_i}\right)^{2/3} + c_{wt}R^{2/3}\left(\frac{z}{z_i}\right)^{2/3}\left(1 - \frac{z}{z_i} + D\right)^{2/3}$$

$$\frac{\overline{\theta'^2}}{\theta_*^2} = c_{\theta b}\frac{(1 - z/z_i)^{4/3}}{(z/z_i)^{2/3}} + c_{\theta t}R^{4/3}\frac{(z/z_i)^{4/3}}{(1 - z/z_i + D)^{2/3}} \tag{4.86}$$

$$\frac{Nz_i}{\theta_*^2 w_*} = c_{Nb}\frac{(1 - z/z_i)^{5/3}}{(z/z_i)^{4/3}} + c_{Nt}R^{5/3}\frac{(z/z_i)^{5/3}}{(1 - z/z_i + D)^{4/3}}$$

The equations obtained are dependent on parameters R and D. For $z \to 0$, the expressions obtained approach functions predicted by the Monin–Obukhov similarity theory.

Vertical wind velocity and temperature variances are depicted in Figures 4.32 and 4.33. Vertical wind velocity variance reaches maximum for z/z_i near 0.4. Temperature variance continues to decrease with height above the surface layer, reaches a minimum for z/z_i near 0.8, and increases above. Curves in the figures were obtained for one R value, $R = -0.2$, and for the following values of universal constants: $c_{wb} = 1$, $c_{wt} = 0.5$, $c_{\theta b} = 2$, and $c_{\theta t} = 8$. All similarity curves derived in this section were plotted for $D = 0$.

Figure 4.34 illustrates the distribution of the destruction rate of the temperature variance. The curve in this figure was plotted for $c_{Nb} = 2$,

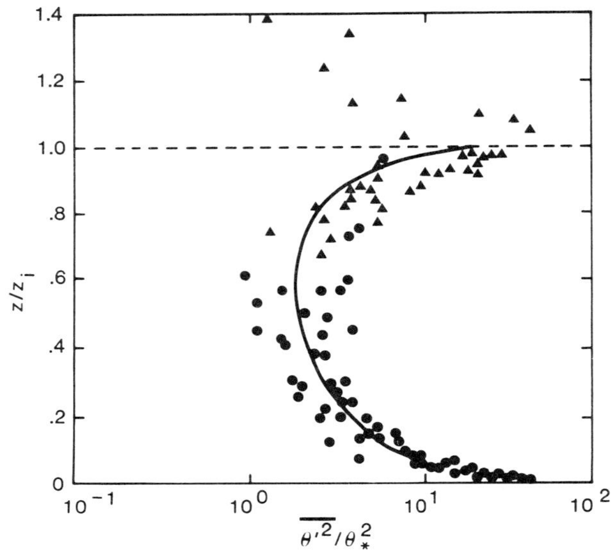

Figure 4.33 Vertical profiles of a dimensionless temperature variance in the convective boundary layer. Data from the 1976 Ashchurch experiments are shown by triangles, the 1973 Minnesota data are shown by circles. The similarity curve is plotted for $R = -0.2$. (After Caughey and Palmer, 1979.)

$c_{Nt} = -5$, and for $R = -0.2$. The same approach can be used to derive equations for other statistical moments: namely, for temperature and humidity,

$$\frac{z_i}{\Theta_*}\frac{d\Theta}{dz} = c_{Tb}\frac{(1 - z/z_i)^{2/3}}{(z/z_i)^{4/3}} + c_{Tt}R^{2/3}\,\frac{(z/z_i)^{2/3}}{(1 - z/z_i + D)^{4/3}}$$
$$\frac{z_i}{q_*}\frac{dq}{dz} = c_{qb}\frac{(1 - z/z_i)^{2/3}}{(z/z_i)^{4/3}} + c_{qt}R_qR^{-1/3}\,\frac{(z/z_i)^{2/3}}{(1 - z/z_i + D)^{4/3}} \tag{4.87}$$

for the third moments of the vertical velocity and temperature,

$$\frac{z_i}{\Theta_*^2 w_*}\frac{\partial}{\partial z}\overline{w'\theta'^2} = c_{Pb}\frac{(1 - z/z_i)^{5/3}}{(z/z_i)^{4/3}} + c_{Pt}R^{5/3}\,\frac{(z/z_i)^{5/3}}{(1 - z/z_i + D)^{4/3}}$$
$$\frac{z_i}{w_*^2\Theta_*}\frac{\partial}{\partial z}\overline{w'^2\theta'} = c_{Rb}\frac{(1 - z/z_i)^{4/3}}{(z/z_i)^{2/3}} + c_{Rt}R^{4/3}\,\frac{(z/z_i)^{4/3}}{(1 - z/z_i + D)^{2/3}} \tag{4.88}$$

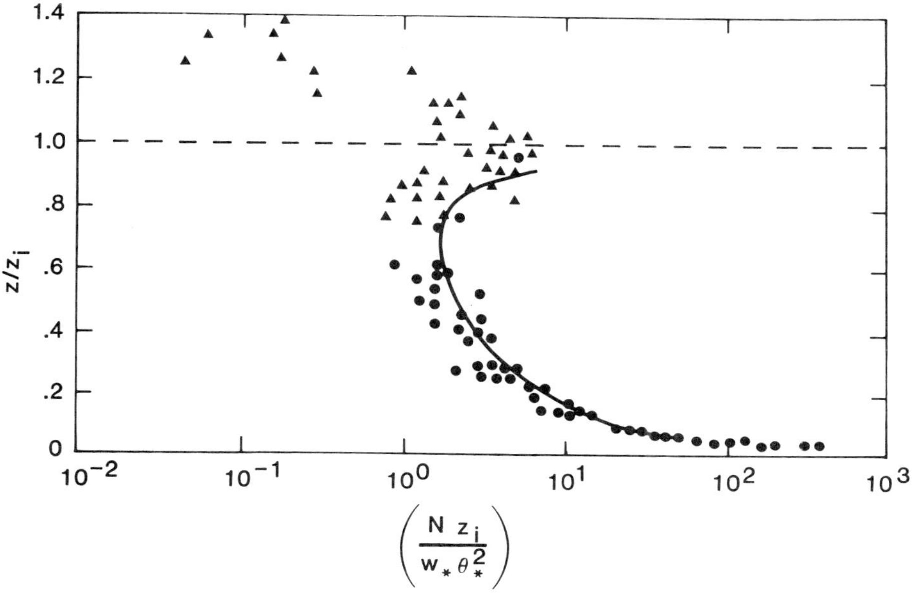

Figure 4.34 Vertical profiles of a dimensionless temperature variance destruction in the convective boundary layer. Data from the 1976 Ashchurch experiments are shown by triangles, the 1973 Minnesota data are shown by circles. The similarity curve is plotted for $R = -0.2$. (After Caughey and Palmer, 1979.)

for second moments of humidity and temperature,

$$\begin{aligned} \frac{\overline{q'^2}}{q_*^2} &= c_{q2b}\frac{(1 - z/z_i)^{4/3}}{(z/z_i)^{2/3}} + c_{q2t}R_q^2R^{-2/3}\frac{(z/z_i)^{4/3}}{(1 - z/z_i + D)^{2/3}} \\ \frac{\overline{q'\theta'}}{q_*\Theta_*} &= c_{q\theta b}\frac{(1 - z/z_i)^{4/3}}{(z/z_i)^{2/3}} + c_{q\theta t}R_qR^{-1/3}\frac{(z/z_i)^{4/3}}{(1 - z/z_i + D)^{2/3}} \end{aligned} \tag{4.89}$$

for the third moment of the vertical velocity, humidity, and temperature.

$$\frac{\overline{w'q'\theta'}}{w_*q_*\Theta_*} = c_{Qb}\frac{(1 - z/z_i)^{5/3}}{(z/z_i)^{1/3}} + c_{Qt}R_qR^{2/3}\frac{(z/z_i)^{5/3}}{(1 - z/z_i + D)^{1/3}} \tag{4.90}$$

and finally, for the destruction rates of humidity variance and humidity–temperature covariance:

$$\begin{aligned} \frac{N_qz_i}{q_*^2w_*} &= c_{Nqb}\frac{(1 - z/z_i)^{5/3}}{(z/z_i)^{4/3}} + c_{Nq}R_q^2R^{-1/3}\frac{(z/z_i)^{5/3}}{(1 - z/z_i + D)^{4/3}} \\ \frac{N_{q\theta}z_i}{w_*\Theta_*q_*} &= c_{Nq\theta b}\frac{(1 - z/z_i)^{5/3}}{(z/z_i)^{4/3}} + c_{Nq\theta t}R_qR^{2/3}\frac{(z/z_i)^{5/3}}{(1 - z/z_i + D)^{4/3}} \end{aligned} \tag{4.91}$$

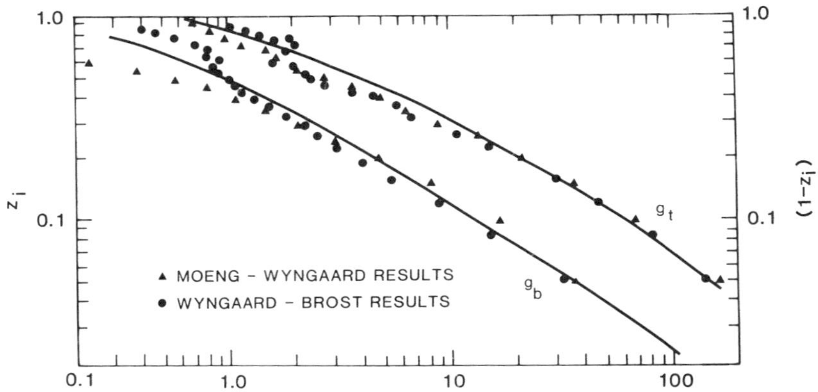

Figure 4.35 Dimensionless function $g_b = (z_i/\theta_*)(d\theta_b/dz)$ and $g_t = (z_i/R\theta_*)(d\theta_t/dz)$. Triangles, model of Moeng and Wyngaard (1984); circles, model of Wyngaard and Brost (1984); solid line, local similarity prediction. (Moeng and Wyngaard, 1984. *Journal of Atmospheric Sciences* published by the American Meteorological Society.)

where q_* is the humidity scale and parameter R_q is as a ratio of humidity fluxes on the top and the bottom of the ABL.

It can be noted that functions (4.87) are quite close to results obtained by Moeng and Wyngaard (1984). Moeng and Wyngaard decomposed the field of a passive, conservative scalar into "bottom-up" and "top-down" components and evaluated both of them using a numerical model. They also argued

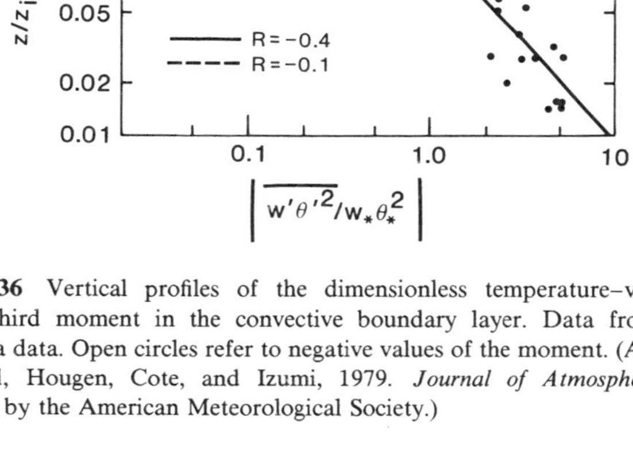

Figure 4.36 Vertical profiles of the dimensionless temperature–vertical wind velocity third moment in the convective boundary layer. Data from the 1973 Minnesota data. Open circles refer to negative values of the moment. (After Kaimal, Wyngaard, Hougen, Cote, and Izumi, 1979. *Journal of Atmospheric Sciences* published by the American Meteorological Society.)

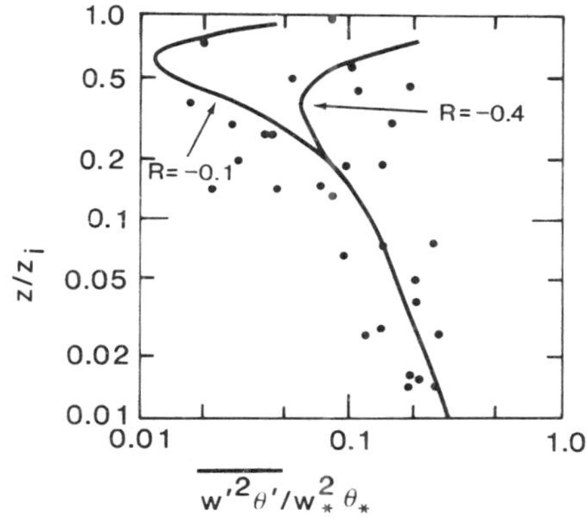

Figure 4.37 Vertical profiles of the dimensionless temperature–vertical wind velocity third moment in the convective boundary layer. Data from the 1973 Minnesota data. (After Kaimal, Wyngaard, Hougen, Cote, and Izumi, 1979. *Journal of Atmospheric Sciences* published by the American Meteorological Society.)

that their results apply to a specific humidity field as well as to the temperature field. Both b and t contributions to the temperature gradient in (4.87) are compared with bottom-up and top-down components in Figure 4.35. The best fit was obtained for $c_{Tb} = 0.6$ and $c_{Tt} = -2$.

The local similarity functions for the third moments $\overline{w'^2\theta'}$, $\overline{w'\theta'^2}$ were obtained by a numerical integration of their gradients, given by equations

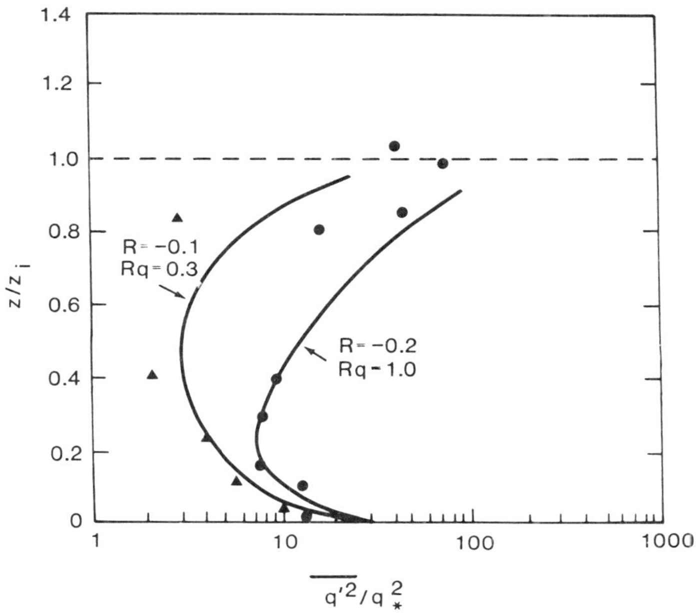

Figure 4.38 Vertical profiles of dimensionless humidity variance in the convective boundary layer. Data from the AMTEX experiment. Similarity curves plotted for different values of parameters R and R_q. (Moeng and Wyngaard, 1984. *Journal of Atmospheric Sciences* published by the American Meteorological Society.)

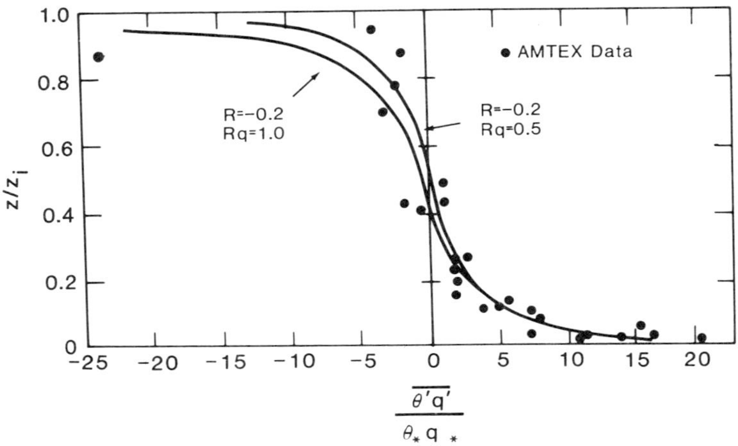

Figure 4.39 Vertical profiles of the dimensionless temperature–humidity covariance in the convective boundary layer. Data from the AMTEX experiment. Similarity curves plotted for different values of parameters R and R_q. (After Wyngaard, Pennell, Lenschow, and LeMone, 1978. *Journal of Atmospheric Sciences* published by the American Meteorological Society.)

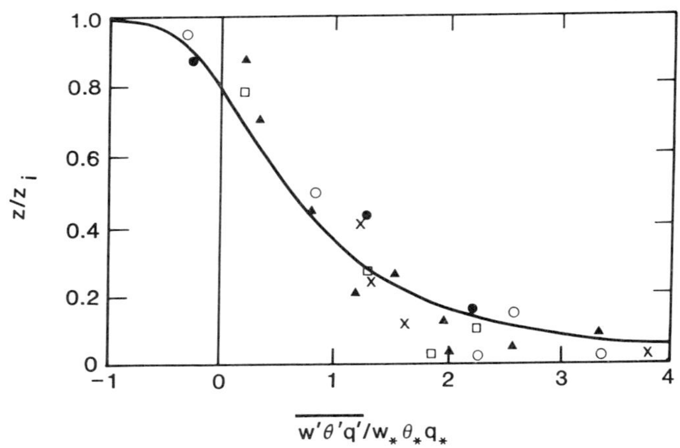

Figure 4.40 Vertical profile of the dimensionless vertical velocity-humidity–temperature covariance in the convective boundary layer. Data from the AMTEX experiment. Different symbols indicate different runs. (After Wyngaard, Pennell, Lenschow, and LeMone, 1978. *Journal of Atmospheric Sciences* published by the American Meteorological Society.)

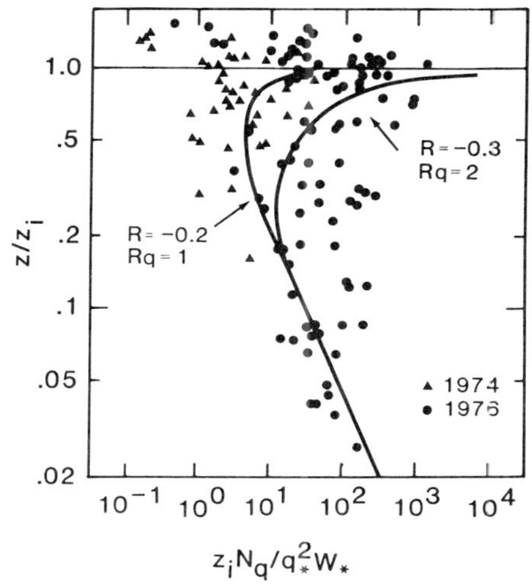

Figure 4.41 Vertical profiles of dimensionless humidity variance destruction rate in the convective boundary layer. Data from the Limagne and Beauce experiment. (After Guillement et al., 1983.)

(4.88). The results are compared with the observations of the AMTEX experiment in Figures 4.36 and 4.37. The curves in the figures are plotted for $c_{Rb} = -0.2$ and $c_{Rt} = 2$, $c_{Pt} = -0.67$ and $c_{Pt} = 2$.

In Figures 4.38 to 4.42, statistics involving the humidity fluctuations are shown. The humidity variance behaves similarly to the temperature variance, reaching a pronounced maximum on the top of the mixed layer. The curves in Figures 4.38 to 4.42 were plotted for the following values of the universal constants: $c_{q2b} = 2$, $c_{q2t} = 8$, $c_{q\theta b} = 1.5$, $c_{q\theta t} = 1.5$, $c_{Qb} = 1.5$, $c_{Qt} = -4$, $c_{Nqb} = 2$, $c_{Nqt} = -5$, $c_{Nq\theta b} = 0.6$, and $c_{Nq\theta t} = -4$.

Expressions (4.8) can be used to derive equations for the structure parameters. For example, using the expression in (4.82) for the dissipation

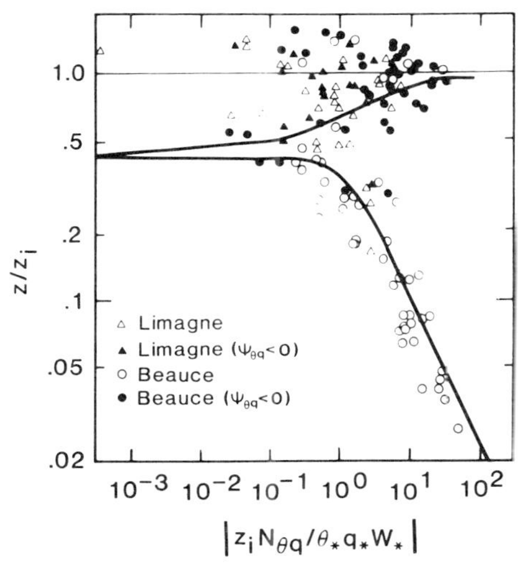

Figure 4.42 Vertical profiles of dimensionless humidity–temperature covariance destruction rate in the convective boundary layer. Data from the Limagne and Beauce experiment. The similarity curve is plotted for $R = -0.2$, $R_q = 1$. (After Guillement et al., 1983.)

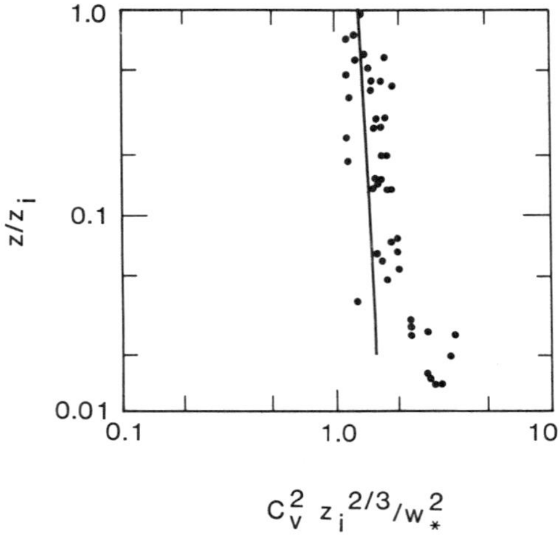

Figure 4.43 Vertical profile of the dimensionless structure parameter for wind velocity. Data from the Minnesota experiment. (After Kaimal, Wyngaard, Hougen, Cote, and Izumi, 1979. *Journal of Atmospheric Sciences* published by the American Meteorological Society.)

rate, it can be obtained that the equation of the structure parameter for wind velocity has the following form:

$$C_V^2 \frac{z_i^{2/3}}{w_*^2} = 2\left[c_{\epsilon b}\left(1 - \frac{z}{z_i}\right) + c_{\epsilon t} R \frac{z}{z_i}\right]^{2/3} \tag{4.92}$$

In a similar way, the other structure parameters can also be derived.

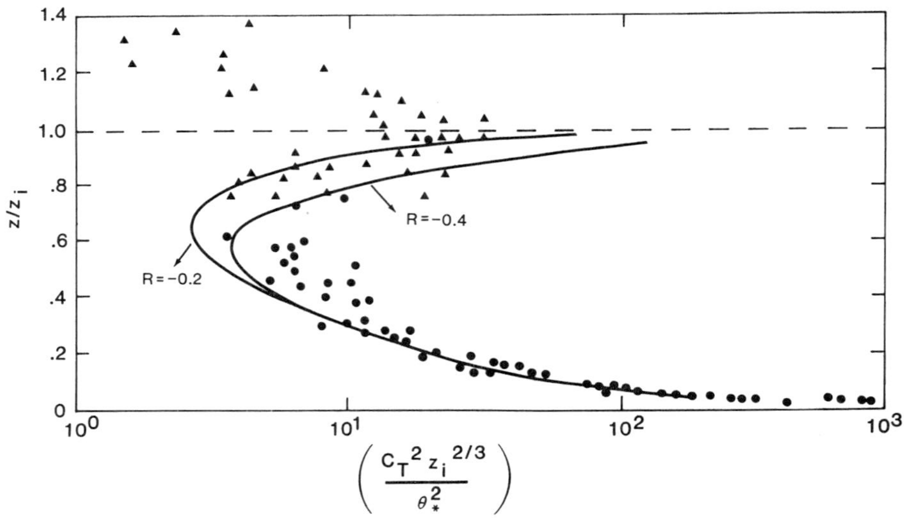

Figure 4.44 Vertical profiles of a dimensionless temperature structure parameter in the convective boundary layer. Data from the 1976 Ashchurch experiments are shown by triangles, the 1973 Minnesota data are shown by circles. The similarity curve is plotted for $R = -0.2$ and $R = -0.4$. (After Caughey and Palmer, 1979.)

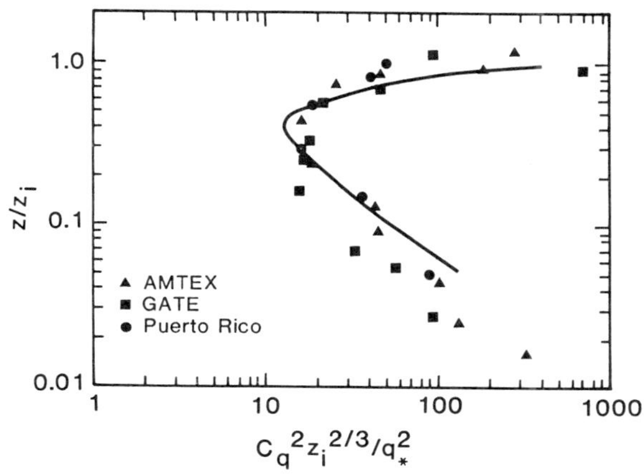

Figure 4.45 Vertical profile of dimensionless humidity structure parameter in the convective boundary layer. Data from the AMTEX, GATE, and Puerto Rico experiments. The similarity curve plotted for $R = -0.2$, $R_q = -1$. (After Wyngaard and LeMone, 1980. *Journal of Atmospheric Sciences* published by the American Meteorological Society.)

Equation (4.92) is plotted for $C_{\epsilon t}R = 0.8$ in Figure 4.43, together with the Minnesota data. It can be seen that the local similarity prediction fits the empirical points for $z/z_i > 0.05$. Structure parameters for temperature and humidity covariances are shown in Figures 4.44 and 4.45.

PROBLEMS

4.1. Use dimensional analysis to derive:

(a) An expression for the period of the pendulum T, assuming that it depends on the length of the pendulum l and the gravity acceleration g. (*Hint:* Assume that $T = l^a g^b$, and find a and b analyzing units on both sides of the equation).

(b) The Monin–Obukhov length L in (3.38), assuming that L depends on $\overline{w'\theta'}|_0$, u_*, and β.

(c) Eddy diffusivity k_m and dissipation rate ϵ, assuming that they depend on mixing length l and turbulent kinetic energy E.

4.2. Given are observations of wind velocity in the surface layer:

Height (m)	Speed (m/s)
1.6	6.34
0.4	4.46

Assuming neutral stratification, find u_*, z_0, and k_m.

4.3. Define:

(a) u_*, T_* (b) u_f, t_f (c) w_*, Θ_* (d) $\mathbf{U_*, T_*}$ (e) $\mathbf{U_b, \Theta_b, U_t, \Theta_t}$

4.4. Using similarity functions derived by (Businger et al., 1971) and listed in Table 4.2, prove that profiles of wind velocity and temperature (or specific humidity) can be expressed by (4.25).

4.5. The bulk transfer coefficients C_D, C_H, and C_E are defined as (over the sea)

$$\tau_0 = u_*^2 = C_D(U_r - U_o)^2 \qquad \text{(for momentum transfer)}$$

$$\overline{w'\theta'}\big|_0 = C_H U_r(\Theta_r - \Theta_0) \qquad \text{(for heat transfer)}$$

$$\overline{w'q'}\big|_0 = C_E U_r(q_r - q_0) \qquad \text{(for water vapor transfer)}$$

where U_r and Θ_r are values of wind velocity and potential temperature at the reference level within the surface layer and U_0 and Θ_0 are their surface values. If the bulk transfer coefficients are known, turbulent fluxes can be determined from observations of the wind, temperature, and humidity.

(a) Find expressions for C_D, C_H, and C_E for the neutral case.

(b) Find expressions for C_D, C_H, and C_E in the diabatic case.

4.6. Analyze (4.25) and indicate on the sketch which wind velocity and temperature profiles are characteristic for neutral, stable, and convective conditions.

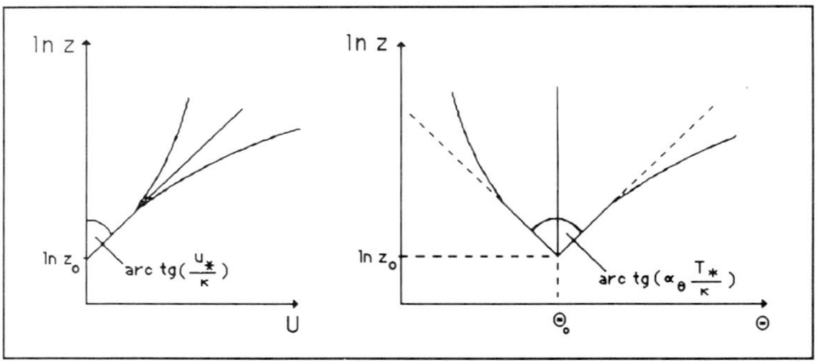

4.7. Zilitinkevich (1970) suggested relating temperature and specific humidity at the ground surface ($z = 0$) with those on the top of the viscous sublayer ($z = z_0$) using expressions of the form

$$\Theta_0 = \Theta_s + 0.0962\frac{T_*}{\kappa}\mathrm{Re}_g^{0.45}$$

$$q_0 = q_s + 0.0962\frac{q_*}{\kappa}\mathrm{Re}_g^{0.45}$$

where Re_g is the surface Reynolds number, defined as $\mathrm{Re}_g = u_* z_0/\nu$; ν is the kinematic viscosity of air ($\sim 1.5 \times 10^{-5}$ m^2/s); and the specific humidity scale q_* is defined in a manner analogous to T_* in (4.9). Find temperature and specific humidity differences in the viscous sublayer assuming that $z_0 = 10$ cm, $u_* = 0.25$ m/s, $T_* = 0.1$ K, and $q_* = 0.01$ g/kg.

4.8. When the ground cover is high (forest, cornfield, etc.), (4.25) are valid but only when vertical coordinate z is replaced by an expression $z - D$, where D is called *zero-plane displacement* ($D \sim 0.6$ to $0.7h$, where h is the height of the vegetation). The modified formulas apply for $z > z_0 + D$. Within a dense, horizontally and vertically uniform ground cover ($z < D$), the following expression can be used to estimate the wind velocity (e.g., Pielke, 1984):

$$U(z) = U(D) \exp\left[a\left(\frac{\ln z}{\ln D} - 1\right)\right]$$

where a is a coefficient dependent on leaf-area index—a ratio of leaf area per area of ground surface. Plot a wind velocity profile in neutral conditions, assuming that $D = 100$ cm, $z_0 = 1$ cm, $a = 2$, $u_* = 0.3$ m/s, and $U(D) =$ 2 m/s. Connect smoothly the two lines for $z \sim D$.

4.9. Given are observations of wind velocity and temperature in the surface layer:

Height (m)	Speed (m/s)	Temperature (°C)
4	5.50	30.29
2	4.74	30.17

Find u_* and T_* assuming that $z_0 = 2.4$ cm.

4.10. Find the proper sign (or value) for the following fluxes in the surface layer (assume that the surface wind velocity is aligned along the x-axis):

	Stable	Neutral	Unstable
$\overline{w'\theta'}$			
$\overline{u'w'}$			
$\overline{v'w'}$			

4.11. Assuming that $\Theta_t = 290$ K, $\Theta_0 = 300$ K, $G = 10$ m/s, $f = 10^{-4}$ s^{-1}, and $z_0 = 0.1$ m, find the surface heat flux, the friction velocity u_*, and the cross-isobar angle α. Use Figures 5.9 to 5.12.

4.12. Given are observations of wind velocity and temperature on $N \geq 2$ levels within the surface layer. Write a computer program evaluating u_*, T_*, and z_0.

Hint: Substitute in equations listed in (4.25) that $y_u = u_*/\kappa \ln z_0$, $x_u = u_*/\kappa$, $y_\theta = \Theta_0 + 0.74T_*/\kappa \ln z_0$, $x_\theta = 0.74T_*/\kappa$. Use the equations obtained and the least-squares method to find y_u, y_θ, and z_0 (for the fixed value of L). Next, substitute the equations obtained for y_u and y_θ in (4.9) and derive a nonlinear equation for L. Solve it and find L. From definitions of y_u and y_θ find u_* and T_*.

4.13. Complete this table using the following symbols: $+$, $-$, $+0$, -0, 0, $+\infty$, $-\infty$.

	L	z/L	T_*	θ_*	t_f	R_f
Extremely stable						
Neutral						
Extremely unstable						

4.14. Assuming that

$$\overline{w'\theta'} = -k_h \frac{\partial \Theta}{\partial z}$$

find k_h employing (4.63) and (4.87). In which region of the convective boundary layer is K-parametrization incorrectly defined?

4.15. The Bowen ratio is defined as a ratio of the sensible and latent heat fluxes, $\text{Bo} = c_p \rho \overline{w'\theta'}|_0 / (L\rho \overline{w'q'}|_0)$ (ρ is the air density, c_p the specific heat at constant pressure, and L the latent heat).

(a) Using (1.1), derive a formula allowing determination of evaporation if the net radiation, the heat flux to the ground, and Bowen ratio are provided.

(b) Using formulas listed in Problem 4.5, evaluate the Bowen ratio from measurements of temperature and humidity.

CHAPTER FIVE

SPECTRAL ANALYSIS

5.1 GENERAL CONSIDERATIONS

In many theoretical and practical problems, it is important to know whether the atmospheric motion extends uniformly over all scales, or whether there are certain preferred scales of motion separated by "gaps." This question is usually studied by considering the *spectra* of atmospheric variables. The spectrum measures the distribution of the variance of a variable over wavelengths or periods. If the variable is a velocity component, the spectrum also describes the distribution of kinetic energy over wavelengths or periods.

Additional important information on the dynamics of atmospheric motions can be obtained by analysis of turbulent fluxes. Fluxes are covariances between velocity components and some other quantity, such as temperature or moisture. Sometimes it is convenient to determine how motions of different scales contribute to a certain covariance. This question is studied by considering the *cospectra* of atmospheric variables. Analysis of cospectral densities can be used to determine over how long a period the covariance has to be averaged in order to include the contributions of all relevant eddies.

Much of the mathematics involved in spectral analysis makes use of Fourier techniques. Some basic spectral formulations are elaborated in the Appendix.

5.2 ENERGY SPECTRA

Two types of spectra can be considered: *time spectra* and *space spectra.* In the first case, observations are made at one point at frequent time intervals and the spectrum is plotted as a function of period or its reciprocal, frequency. In the second case, observations are made at many places simultaneously. The spectrum is then plotted as a function of wavelength or its reciprocal, wavenumber.

To discuss a time spectrum, let us consider a stationary random process (mean value constant with time). A correlation function R for such a process, defined in the following way:

$$R(\tau) = \overline{u'(t)u'(t+\tau)} \tag{5.1}$$

(where u' is the deviation of the function u from its mean value), depends only on a lag time τ. If u is a wind velocity, evidently $R(0)$ is twice the kinetic energy. Using the Fourier transform, we can define

$$R(\tau) = \tfrac{1}{2}\int_{-\infty}^{\infty} \Phi(\omega) e^{i\omega\tau}\, d\omega \tag{5.2}$$

where

$$\Phi(\omega) = \frac{1}{\pi}\int_{-\infty}^{\infty} R(\tau) e^{-i\omega\tau}\, d\tau \tag{5.3}$$

Notice that from (5.2), it follows that for $\tau = 0$,

$$R(0) = \tfrac{1}{2}\int_{-\infty}^{\infty} \Phi(\omega)\, d\omega \tag{5.4}$$

Function $\Phi(\omega)$ shows how turbulent kinetic energy is distributed as a function of frequency $\omega = 2\pi/T = 2\pi n$, where T is a period of a sinusoidal oscillation and n is the frequency in hertz.

Because of stationarity $R(\tau) = R(-\tau)$ (R is an even function). From this and from (5.3) we obtain

$$\begin{aligned}\Phi(\omega) &= \frac{1}{\pi}\int_{-\infty}^{\infty} R(\tau)(\cos\omega\tau - \sin\omega\tau)\, d\tau \\ &= \frac{2}{\pi}\int_{0}^{\infty} R(\tau)\cos\omega\tau\, d\tau\end{aligned} \tag{5.5}$$

It is easy to show that $\Phi(\omega)$ is real and even. This allows us to rewrite (5.2) in the form

$$R(\tau) = \int_{0}^{\infty} \Phi(\omega)\cos\omega\tau\, d\omega \tag{5.6}$$

For $\tau = 0$, equation (5.6) yields

$$R(0) = \int_{0}^{\infty} \Phi(\omega)\, d\omega \tag{5.7}$$

It can be seen that $\Phi(\omega)\, d\omega$ [the area under the curve $\Phi(\omega)$] is the contribution of eddies with the frequency ω to the turbulent kinetic energy. The function $\Phi(\omega)$ is called the energy spectrum density. Changing from frequency ω expressed in radians per second to frequency $n = \omega/2\pi$ in cycles per second, a new spectral density $S(n) = 2\pi\Phi(2\pi n)$ can be introduced:

$$\sigma^2 = R(0) = \int_{0}^{\infty} \Phi(\omega)\, d\omega = 2\pi\int_{0}^{\infty} \Phi(2\pi n)\, dn = \int_{0}^{\infty} S(n)\, dn \tag{5.8}$$

where σ^2 is the variance.

The result above can also be expressed in a slightly changed form:

$$\sigma^2 = \int_0^\infty S(n)\, dn = \int_0^\infty nS(n)\, d \ln n \qquad (5.9)$$

This indicates that the area under the curve $S(n)$ plotted against n is the same as the area under the curve $nS(n)$ plotted against $\ln n$. The function nS is traditionally adopted for the description of atmospheric spectra. This is done for several good reasons. First, this relation is very useful for wide frequency ranges. Moreover, the quantity $nS(n)$ has the dimension of the variance. If variance σ^2 is finite, then from (5.9) it follows that $nS(n) \to 0$ for $n \to \infty$. From this and the obvious fact that $nS(n) \to 0$, when $n \to 0$, it can be concluded that $nS(n)$ vanishes at both the small and large frequencies and must have its maximum somewhere in between.

Reciprocals of the frequency at maxima obtained from the plot of the function $nS(n)$ are interpreted as the principal time scales of the flow. Similarly, the minima in this plot are viewed as *spectral gaps,* defining domains with relatively small kinetic energy.

5.3 CROSS SPECTRA

Consider now the covariance function

$$R_{uv}(t, t+\tau) = \overline{u'(t)v'(t+\tau)} \qquad (5.10)$$

In the stationary case, the covariance function

$$R_{uv}(t, t+\tau) = R_{uv}(\tau) \qquad (5.11)$$

is no longer an even function. Applying the Fourier transform, the following is obtained:

$$\begin{aligned} \Psi(\omega) &= \frac{1}{\pi}\int_{-\infty}^{\infty} R_{uv}(\tau) e^{-i\omega\tau}\, d\tau \\ &= C(\omega) - iQ(\omega) \end{aligned} \qquad (5.12)$$

where the even part is called the *cospectrum*:

$$\begin{aligned} C(\omega) &= \frac{1}{\pi}\int_{-\infty}^{\infty} R_{uv}(\tau) \cos\omega\tau\, d\tau \\ &= \frac{1}{\pi}\int_0^{\infty} [R_{uv}(\tau) + R_{uv}(-\tau)] \cos\omega\tau\, d\tau \end{aligned} \qquad (5.13)$$

and the odd part is called the *quadrature spectrum*:

$$\begin{aligned} Q(\omega) &= \frac{1}{\pi}\int_{-\infty}^{\infty} R_{uv}(\tau) \sin\omega\tau\, d\tau \\ &= \frac{1}{\pi}\int_0^{\infty} [R_{uv}(\tau) - R_{uv}(-\tau)] \sin\omega\tau\, d\tau \end{aligned} \qquad (5.14)$$

Function $\Psi(\omega)$ is complex and can be expressed in the form

$$\Psi(\omega) = |\Psi(\omega)| e^{-i\theta(\omega)} \tag{5.15}$$

where the *amplitude* of the cross spectrum is

$$|\Psi(\omega)|^2 = [C(\omega)]^2 + [Q(\omega)]^2 \tag{5.16}$$

and the *phase angle* between $u(t)$ and $v(t)$ at the frequency ω is

$$\theta(\omega) = \arctan \frac{Q(\omega)}{C(\omega)} \tag{5.17}$$

Normalizing the square of the cross-spectrum amplitude, a new quantity called *coherence* can be defined:

$$\Gamma(\omega) = \frac{[C(\omega)]^2 + [Q(\omega)]^2}{\Phi_u(\omega)\Phi_v(\omega)} \tag{5.18}$$

If $u = v$, then $\Phi_u(\omega) = \Phi_v(\omega)$ and $C(\omega) = \Phi_u(\omega)$, $Q(\omega) = 0$; thus $\Gamma(\omega) = 1$. If $R_{uv}(\tau) = 0$, then $\Psi(\omega) = 0$ and $\Gamma(\omega) = 0$. The coherence is interpreted as a spectral correlation coefficient. Spectral analysis of fluxes of various properties and analysis of two time series in the frequency domain are the two main applications of cross-spectrum analysis in meteorology.

5.4 SPATIAL SPECTRA

A space covariance function for wind velocity can be defined analogously to (5.10), substituting time t by a position vector $\mathbf{x}$. In the case of homogeneity, the space covariance function R_{ij} depends only on the distance vector $\mathbf{r} = (r_1, r_2, r_3)$ between two points in space:

$$R_{ij}(\mathbf{r}) = \overline{u_i'(\mathbf{x})u_j'(\mathbf{x}+\mathbf{r})} \tag{5.19}$$

Space spectral density and space covariance function are related in the following way:

$$\begin{aligned} \mathscr{B}_{ij}(\mathbf{k}) &= \frac{1}{\pi^3} \iiint_{-\infty}^{\infty} R_{ij}(\mathbf{r}) e^{-i\mathbf{k}\cdot\mathbf{r}}\, d\mathbf{r} \\ R_{ij}(\mathbf{r}) &= \frac{1}{2^3} \iiint_{-\infty}^{\infty} \mathscr{B}_{ij}(\mathbf{k}) e^{i\mathbf{k}\cdot\mathbf{r}}\, d\mathbf{k} \end{aligned} \tag{5.20}$$

where $\mathbf{k} = (k_1, k_2, k_3)$ is a wavenumber vector, $d\mathbf{k} = dk_1\, dk_2\, dk_3$ and $d\mathbf{r} = dr_1\, dr_2\, dr_3$. Adding all covariances for which $i = j$ and assuming that $\mathbf{r} = \mathbf{0}$

yields

$$R_{\alpha\alpha}(\mathbf{0}) = 2E = \frac{1}{2^3}\iiint_{-\infty}^{\infty} \mathscr{B}_{\alpha\alpha}(\mathbf{k})\, d\mathbf{k}$$
$$= \frac{1}{2^3}\int_0^{\infty}\left[\iint_K \mathscr{B}_{\alpha\alpha}(\mathbf{k})\, d\Sigma\right] dk = 2\int_0^{\infty} \psi_E(k)\, dk \tag{5.21}$$

where E is the kinetic turbulent energy, K is a sphere defined by $k_1^2 + k_2^2 + k_3^2 = k^2$, $d\Sigma$ is elementary area of the sphere K, and ψ_E is called the *three-dimensional spectrum*, defined by integral over K in (5.21).

In addition to (5.20), one-dimensional spectra can be also defined. One-dimensional spectra depend only on the wavenumber in a particular direction. Usually, longitudinal (streamwise) wavenumber k_1 is chosen because it is the only one that can be related to observations in a simple way.

The one-dimensional spectra which are most often measured are *longitudinal* and *lateral spectra.* These are defined in the following way:

$$\mathscr{H}_{11}(k_1) = \frac{1}{\pi}\int_{-\infty}^{\infty} R_{11}(r,0,0)e^{-ik_1r}\, dr$$
$$\mathscr{H}_{22}(k_1) = \frac{1}{\pi}\int_{-\infty}^{\infty} R_{22}(r,0,0)e^{-ik_1r}\, dr \tag{5.22}$$

where R_{11} and R_{22} are *longitudinal* and *lateral* correlations (see section 3.2). From (5.20) we can easily obtain that

$$R_{11}(r,0,0) = \frac{1}{2}\int_{-\infty}^{\infty} \mathscr{H}_{11}(k_1)e^{ik_1r}\, dk_1$$
$$\mathscr{H}_{11}(k_1) = \frac{1}{2^2}\iint_{-\infty}^{\infty} \mathscr{B}_{11}(\mathbf{k})\, dk_2\, dk_3 \tag{5.23}$$

A similar expression can be derived for $\mathscr{H}_{22}(k_1)$.

One-dimensional spectra are easier to measure; however, they are not wholly appropriate for the description of turbulence, as they give misleading information about three-dimensional fields. One-dimensional spectrum in a three-dimensional field at wavenumber k can contain contributions from components of all wavenumbers larger than k (Figure 5.1). This phenomenon is called *aliasing.* Aliasing is not a serious problem at high wavenumbers because small eddies tend to have about the same size in all directions.

Measurements indicate that values of $R_{11}(r,0,0)$ are positive. Values of $R_{22}(r,0,0)$, however, become negative for some values of r (Figure 5.2). This is because the mean value of u_2 is equal to zero and, therefore, there is no net mass flux across the plane x_1, x_3. As a result,

$$\iint_{-\infty}^{\infty} u_2(x_1,0,x_3)\, dx_1\, dx_3 = 0 \tag{5.24}$$

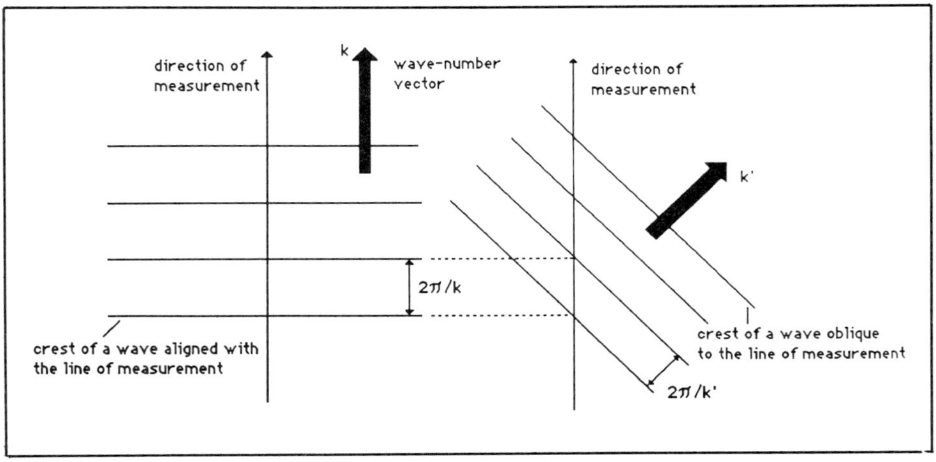

Figure 5.1 Aliasing in a one-dimensional spectrum.

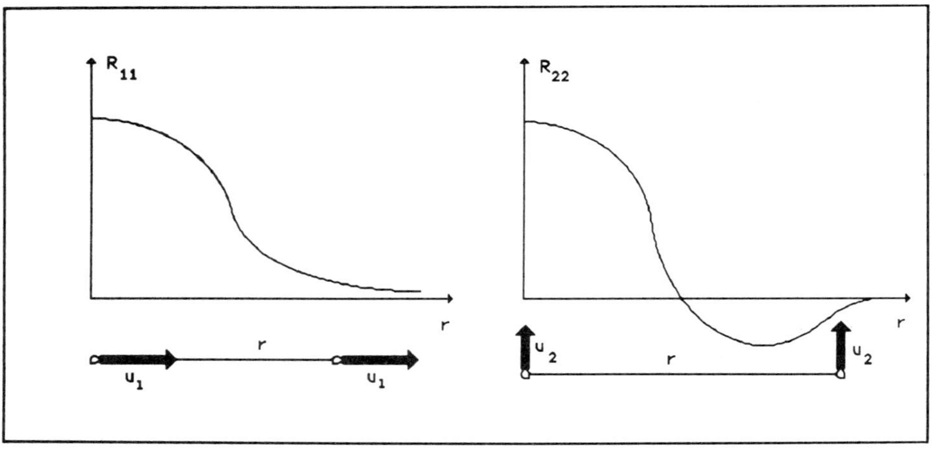

Figure 5.2 Longitudinal and lateral correlations.

If the integral is multiplied by u_2 at some given point and then averaged, we obtain

$$\iint_{-\infty}^{\infty} R_{22}(r_1, 0, r_3)\, dr_1\, dr_3 = 0 \tag{5.25}$$

This means that R_{22} must be negative somewhere in the x_1–x_3 plane (Tennekes and Lumley, 1972).

It can be shown that in isotropic turbulence, longitudinal spectra $\mathscr{H}_{11}$ are related to lateral spectra $\mathscr{H}_{22}$ and to three-dimensional spectra ψ_E in the following way (Monin and Yaglom, 1971):

$$\psi_E(k) = \frac{k^3}{2}\frac{d}{dk}\left[\frac{1}{k}\frac{d\mathscr{H}_{11}(k)}{dk}\right]$$
$$\mathscr{H}_{22}(k) = \frac{1}{2}\left[\mathscr{H}_{11}(k) - k\frac{d\mathscr{H}_{11}(k)}{dk}\right] \tag{5.26}$$

Notice that because of isotropy k_1 could be substituted by k in (5.26). According to Kolmogorov (1941), in the inertial subrange, energy neither enters the system nor is dissipated. It is merely transmitted at rate ϵ from smaller wavenumbers (large-scale motion) to larger wavenumbers (small-scale motion), where it is dissipated. As a result, three-dimensional spectral density function $\psi_E(k)$ would be dependent on the viscous dissipation ϵ and the wavenumber k only. Dimensional analysis yields

$$\psi_E(k) = \alpha\epsilon^{2/3}k^{-5/3} \tag{5.27}$$

where α is the universal constant. From (5.26) it follows that if $\psi_E(k) \sim k^{-5/3}$, then $\mathscr{H}_{11}(k) \sim k^{-5/3}$ and $\mathscr{H}_{22}(k) \sim k^{-5/3}$. In addition, $\mathscr{H}_{22}(k) = 4/3\mathscr{H}_{11}(k)$. Three-dimensional spectral density function $\Psi_E(k)$ in the entire range of wavenumbers is depicted schematically in Figure 5.3.

The simplicity and beauty of Kolmogorov's approach are apparent. Unfortunately, there is a minor flaw, noticed in 1944 by Landau and Lifschitz. It is related to the phenomenon of *intermittency*. A random time function is called intermittent when it consists of random intervals with nonzero values alternating with intervals during which it is approximately equal to zero (Leslie, 1973). Turbulent flow may become intermittent when the production of turbulence occurs only locally, caused either by the breaking waves and other instabilities in stable layers or by the convective plumes in unstable shear layers.

The dissipation of turbulent kinetic energy in the intermittent case is observed to occur in patches. Therefore, the standard Kolmogorov theory is valid only in regions of active turbulence. As a result, local dissipation instead

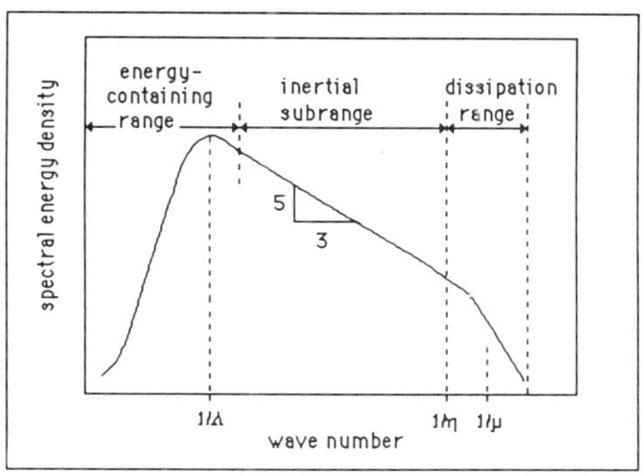

Figure 5.3 Schematic of the spectral energy density ψ_E.

of averaged, should be taken into consideration in the dimensional analysis; that is (Wymgaard, 1973),

$$\psi_\lambda \sim \epsilon_\lambda^{2/3} k^{-5/3} \tag{5.28}$$

where the local average over a region of linear dimensions $\lambda \sim k^{-1}$ is denoted by subscript λ. The spectrum ψ_E should be obtained from the ensemble average of individual estimates, denoted by $\langle\ \rangle$:

$$\psi_E \sim \langle \epsilon_\lambda^{2/3} \rangle k^{-5/3} \tag{5.29}$$

where $\psi_E = \langle \psi_\lambda \rangle$.

This result differs from (5.27), where the average dissipation rate $\epsilon = \langle \epsilon_\lambda \rangle$ to the two-thirds power appears instead of the average of the dissipation rate to the two-thirds power $\langle \epsilon_\lambda^{2/3} \rangle$. As a result, Kolmogorov's law has to be corrected. The corrected law can be written in the form

$$\psi_E \sim \epsilon^{2/3} k^{-5/3-m} \tag{5.30}$$

where m is a coefficient. In fully developed turbulence, this correction is on the order of 3%.

Intermittency can be described in terms of the theory of *fractals* (Mandelbrot, 1982). Fractal theory considers an intermittent cascade process in which only a fraction β of the volume occupied by cascading eddies will be filled by turbulence of a smaller scale. Therefore, the entire domain of turbulence consists of regions of active turbulence and regions with no turbulence. Fraction β is the same for all scales down to molecular scales, where the dissipation mechanism turns kinetic energy of motion into heat.

It has also been suggested that at very strong thermal stability the dissipation rate in (5.27) should be replaced by the Brunt-Våisålå frequency N. Dimensional analysis in this case yields that $\psi_E \sim N^2 k^{-3}$, which should be valid for $k < N^{3/2} E^{-1/2}$ (*buoyancy subrange*, i.e., the lower end of the inertial subrange).

In the inertial subrange, the longitudinal spectrum $\mathcal{H}_u(k)$ and lateral spectra $\mathcal{H}_v(k)$ and $\mathcal{H}_w(k)$ for three components of the wind velocity have the form

$$\begin{aligned} \mathcal{H}_u(k) &= \alpha_u \epsilon^{2/3} k^{-5/3} \\ \mathcal{H}_v(k) &= \tfrac{4}{3} \alpha_u \epsilon^{2/3} k^{-5/3} \\ \mathcal{H}_w(k) &= \tfrac{4}{3} \alpha_u \epsilon^{2/3} k^{-5/3} \end{aligned} \tag{5.31}$$

Arguments leading to (5.27) can break down if there is any substantial transfer of energy directly into or out of the inertial subrange. In this case there will be a jump in spectral density corresponding to the energy input or output. Nevertheless, the spectral curve will have the same slope $(-\frac{5}{3})$ at the new energetic level corresponding to the new rate of energy supply to larger wavenumbers k.

The three-dimensional and one-dimensional temperature spectrum can be defined analogous to (5.20)–(5.21) (Hinze, 1959). For the one-dimensional temperature spectrum, Corrsin (1951) proposed the following inertial subrange form:

$$\mathscr{H}_\theta(k) = \alpha_T \epsilon^{-1/3} N k^{-5/3} \tag{5.32}$$

where α_T is the universal constant analogous to α_u and N is the destruction rate of the temperature variance.

Similarly, for one-dimensional cospectra (real parts of one-dimensional cross spectra), Wyngaard and Coté (1972) have derived the following formulas:

$$\begin{aligned}
\mathscr{H}_{uw}(k) &= -\alpha_{uw}\epsilon^{1/3}\frac{dU}{dz}k^{-7/3} \\
\mathscr{H}_{w\theta}(k) &= -\alpha_{w\theta}\epsilon^{1/3}\frac{d\Theta}{dz}k^{-7/3} \\
\mathscr{H}_{u\theta}(k) &= \alpha_{u\theta}\frac{dU}{dz}\frac{d\Theta}{dz}k^{-3}
\end{aligned} \tag{5.33}$$

The form of the last equation is not unique; another possible combination has the form (Sorbjan, 1986c)

$$\mathscr{H}_{u\theta}(k) = \alpha_{u\theta}\epsilon^{1/4}\left(\frac{dU}{dz}\right)^{1/4}\frac{d\Theta}{dz}k^{-5/2} \tag{5.34}$$

Observations confirm the slope prescribed by (5.34).

The notation will now be changed in order to follow the tradition in meteorological literature. In meteorology, frequency, not wavenumber, is measured, and therefore space spectra must be replaced by frequency spectra. To relate time series to spatial distributions, Taylor in 1938 suggested that if the mean velocity U carrying the turbulent flow is greater than the turbulent fluctuation u', then u' at the fixed point x_0 may be treated as arising from an unchanging pattern of turbulent flow moving over this point: $u'(x_0, t_0 - t) = u'(x_0 - x, t_0)$, where $x = Ut$. Multiplying this expression by $u'(x_0, t_0)$ and averaging gives $R_u(t) = R_u(x)$. As a result of *Taylor's hypothesis*, the correlation between simultaneous events at two different points along the flow direction is the same as the correlation between two moments of time at a single point. From the same argument, the u spectrum can be expressed as [compare with (5.8)]

$$\int_0^\infty \mathscr{H}_u(k)\,dk = R_u(x=0) = R_u(t=0) = \int_0^\infty S_u(n)\,dn \tag{5.35}$$

where the relation between the wavenumber k (m^{-1}) and, the frequency n (s^{-1}) is $k = 2\pi n/U = \omega/U$, and moreover,

$$\frac{2\pi}{U}\mathscr{H}_u\left(\frac{2\pi n}{U}\right) = S_u(n)$$

or

$$k\mathscr{H}_u(k) = nS_u(n) \tag{5.36}$$

It is traditional in atmospheric sciences to plot frequency spectra $S(n)$ not against n but against the nondimensional frequency f, defined as $f = nz/U = kz/2\pi$.

Expressions (5.31)–(5.34) are rewritten in the following way:

$$
\begin{aligned}
nS_u(n) &= \left(\frac{z}{2\pi}\right)^{2/3} \alpha_u \epsilon^{2/3} f^{-2/3} \\
nS_v(n) &= \left(\frac{z}{2\pi}\right)^{2/3} \alpha_v \epsilon^{2/3} f^{-2/3} \\
nS_w(n) &= \left(\frac{z}{2\pi}\right)^{2/3} \alpha_w \epsilon^{2/3} f^{-2/3} \\
nS_\theta(n) &= \left(\frac{z}{2\pi}\right)^{2/3} \alpha_T \epsilon^{-1/3} N f^{-2/3}
\end{aligned}
\tag{5.37}
$$

and

$$
\begin{aligned}
nC_{uw}(n) &= -\left(\frac{z}{2\pi}\right)^{4/3} \alpha_{uw} \epsilon^{1/3} \frac{dU}{dz} f^{-7/3} \\
nC_{w\theta}(n) &= -\left(\frac{z}{2\pi}\right)^{4/3} \alpha_{w\theta} \epsilon^{1/3} \frac{d\Theta}{dz} f^{-7/3} \\
nC_{u\theta}(k) &= \left(\frac{z}{2\pi}\right)^{3/2} \alpha_{u\theta} \epsilon^{1/4} \left(\frac{dU}{dz}\right)^{1/4} \frac{d\Theta}{dz} f^{-3/2}
\end{aligned}
$$

where the functions S and C are frequency spectral and cospectral densities.

Estimates by Kaimal et al. (1972) show that $\alpha_u = 0.5 \pm 0.05$. Constants α_v and α_w are equal to $\frac{4}{3}\alpha_u$, as a consequence of isotropy. Estimates of $2\alpha_T$ are near 0.8 [the factor 2 is included because the quoted authors used the destruction rate $N/2$ for half of the temperature variance in (5.37)]. Kaimal et al. (1972) found $2\alpha_T = 0.82 \pm 0.08$, which agrees with 0.79 ± 0.1 obtained by Wyngaard and Coté (1971) and with 0.83 obtained by Paquin and Pond (1971). For the stable surface layer, Wyngaard and Coté (1972) found that $\alpha_{uw} = 0.12$ and $\alpha_{w\theta} \sim 0.20$.

5.5 SPECTRAL CURVE FITTING

To enable the comparison of different experimental and laboratory data, analytical expressions fitting measured data are usually presented. It has been customary in micrometeorology to use certain mathematical expressions to describe the spectra in the entire frequency range. For example, for wind velocity spectra these expressions have the following form:

$$
\frac{nS(n)}{u_*^2} = \frac{Af}{(1 + Bf)^{5/3}}
$$

or

$$\frac{nS(n)}{u_*^2} = \frac{Af}{1 + Bf^{5/3}} \tag{5.38}$$

A more general expression including both forms is

$$\frac{nS(n)}{u_*^2} = \frac{Af^c}{(1 + Bf^a)^b} \tag{5.39}$$

where A, B, a, b, and c are constants. The constants A and B do not influence the shape of a spectrum, only its position. Spectra in the forms such as those above are characterized by a low-frequency slope of c on a log-log plot and by a high-frequency slope $(c - ab)$. The spectra are more or less pointed, depending on values a, b, and c.

Five criteria are used to find the five coefficients A, B, a, b, and c. The first criterion is that each of the functions (5.38)–(5.39) should be consistent with Kolmogorov's law for the inertial subrange. This leads to the establishment of two requirements: for a slope and for a value of universal constant. The first two formulas have the proper slope for a large f. The slope for the third formula must be

$$c - ab = -\tfrac{2}{3} \tag{5.40}$$

In the third formula, the following condition for a universal constant of the spectral density of the u component of the wind velocity is obtained from (5.37):

$$\alpha_u (2\pi)^{-2/3} \left(\frac{\epsilon z}{u_*^3} \right)^{2/3} = \frac{Ab}{B} \tag{5.41}$$

Another criterion arises if one demands that the observed position f_m of the spectral maximum coincides with the maximum given by the formula

$$f_m = \left(\frac{c}{B(ab - c)} \right)^{1/a} \tag{5.42}$$

Moreover, the observed value of the spectral maximum $(nS)_{\max}$ should be equal to the maximum given by the formula. This leads to the fourth criterion:

$$(nS)_{\max} = A \left(\frac{1.5c}{B} \right)^{c/a} (1 + 1.5c)^{-b} \tag{5.43}$$

Finally, one may require that the integrals of the observed and theoretical spectra are the same. This requirement involves some problems, as one neither can nor should integrate an observed spectrum over all frequencies.

Integration of functions $nS(n)/u_*^2$ (5.39) gives (for $a = c = 1$ or for $b = c = 1$):

$$A\int_0^\infty \frac{f\,d\ln f}{(1+Bf)^p} = \frac{A}{B(p-1)}$$
$$A\int_0^\infty \frac{f\,d\ln f}{1+Bf^p} = \frac{A\pi}{pB^{1/p}\sin(\pi/p)} \tag{5.44}$$

In derivation of (5.44) it was employed that $d\ln n = d\ln f$.

Each integral should be equal to variance σ^2. As a consequence, the fifth requirement for the energy spectrum is in the form for the first formula, in (5.38),

$$\frac{\sigma^2}{u_*^2} = \frac{1.5A}{B} \tag{5.45}$$

and for the second formula,

$$\frac{\sigma^2}{u_*^2} = \frac{1.98A}{B^{3/5}} \tag{5.46}$$

A similar approach can be used to obtain analytical expressions for other spectral and cospectral densities.

Using (5.42) and (5.45)–(5.46) allows the expression of (5.36) in the form

$$\frac{nS_i(n)}{\sigma_i^2} = \frac{f/f_m}{[1+1.5(f/f_m)]^{5/3}} \tag{5.47a}$$

$$\frac{nS_i(n)}{\sigma_i^2} = \frac{0.644f/f_m}{1+1.5(f/f_m)^{5/3}} \tag{5.47b}$$

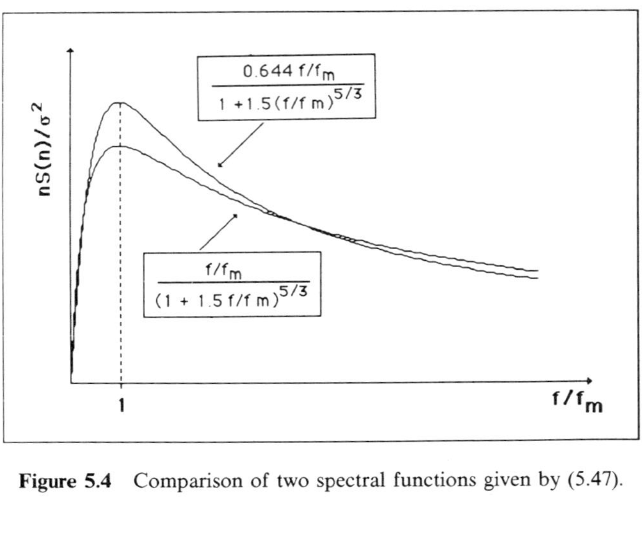

Figure 5.4 Comparison of two spectral functions given by (5.47).

As seen in Figure 5.4, the function (5.47b) has a lower maximum value, equal to 0.258, while for the function (5.47a), the value of the maximum is 0.217. The areas under both curves are the same and equal 1.

5.6 SPECTRA AND COSPECTRA IN THE SURFACE LAYER

In the surface layer, the Monin–Obukhov similarity theory proved to be a very successful tool in describing a vertical distribution of basic meteorological parameters. It is not surprising that spectral and cospectral densities follow the Monin–Obukhov similarity in the surface layer.

All spectral functions (5.37) obtained for the inertial interval can be transformed to the dimensionless form

$$\begin{aligned}\frac{nS_u(n)}{u_*^2} &= (2\pi\kappa)^{-2/3}\alpha_u\phi_\epsilon^{2/3}f^{-2/3}\\ \frac{nS_v(n)}{u_*^2} &= (2\pi\kappa)^{-2/3}\alpha_v\phi_\epsilon^{2/3}f^{-2/3}\\ \frac{nS_w(n)}{u_*^2} &= (2\pi\kappa)^{-2/3}\alpha_w\phi_\epsilon^{2/3}f^{-2/3}\\ \frac{nS_\theta(n)}{T_*^2} &= (2\pi\kappa)^{-2/3}\alpha_T\phi_\epsilon^{-1/3}\phi_N f^{-2/3}\end{aligned} \tag{5.48}$$

and

$$\begin{aligned}\frac{nC_{uw}(n)}{u_*^2} &= -(2\pi\kappa)^{-4/3}\alpha_{uw}\phi_\epsilon^{1/3}\phi_m f^{-7/3}\\ \frac{nC_{w\theta}(n)}{u_*T_*} &= -(2\pi\kappa)^{-4/3}\alpha_{w\theta}\phi_\epsilon^{1/3}\phi_h f^{-7/3}\\ \frac{nC_{u\theta}(k)}{u_*T_*} &= (2\pi\kappa)^{-3/2}\alpha_{u\theta}\phi_\epsilon^{1/4}\phi_m^{1/4}\phi_h f^{-3/2}\end{aligned} \tag{5.49}$$

where functions ϕ are similarity functions defined in Chapter 4.

Including ϕ_ϵ in the normalization of u, v, and w spectra removes the dependence on z/L from the equations. This brings all spectra, for different z/L, into coincidence in the inertial subrange. Similar procedures are valid for the spectrum of temperature and cospectral functions. Generalized spectral curves for u, v, and w obtained by Kaimal et al. (1972) are shown in Figures 5.5 to 5.7.

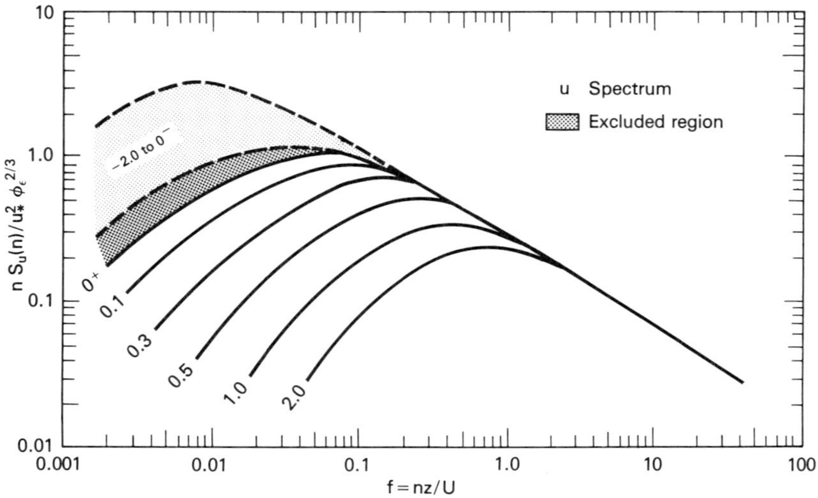

Figure 5.5 Generalized u spectrum for z/L values ranging from $+2.0$ to -2.0. $0+$ and $0-$ denote $z/L = 0$ approached from positive and negative sides, respectively. (After Kaimal et al., 1972. *Quarterly Journal of Royal Meteorological Society* published by the Royal Meteorological Society.)

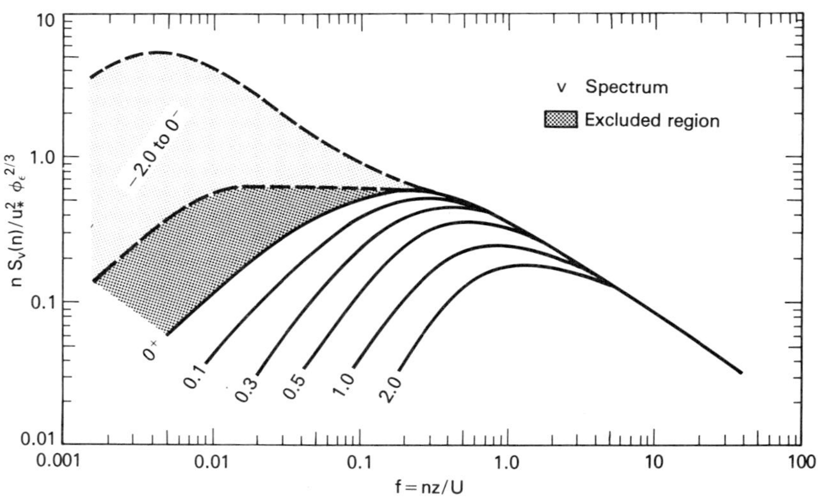

Figure 5.6 Generalized v spectrum for z/L values ranging from $+2.0$ to -2.0. (After Kaimal et al., 1972. *Quarterly Journal of Royal Meteorological Society* published by the Royal Meteorological Society.)

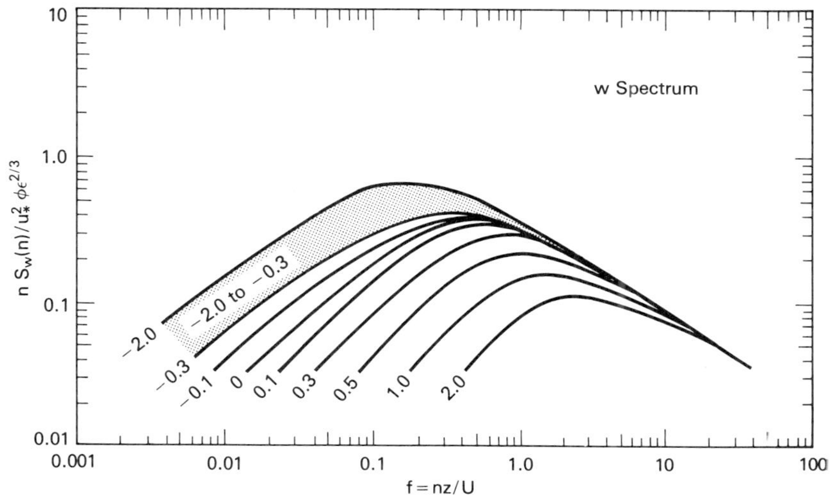

Figure 5.7 Generalized w spectrum for z/L values ranging from $+2.0$ to -2.0. Stippling indicates absence of any well-defined trend with z/L. (After Kaimal et al., 1972. *Quarterly Journal of Royal Meteorological Society* published by the Royal Meteorological Society.)

For $z/L > 0$, u and v spectral curves exhibit the same shape and behavior as w. Order of spectral densities for positive values of z/L breaks down as z/L changes sign. The unstable spectra for $z/L < 0$ are confined to the stippled area, with no particular regard to z/L. The u and v spectra are excluded from the region between stable and unstable conditions. The excluded region of spectral intensities is less spread in the u spectral plot than in the v plot.

The heavy curves designated by 0^+ in the figures depicit the neutral limits for spectra on the stable side. The corresponding limits 0^- on the unstable side are not uniquely defined and do not converge with either of the two envelopes indicated by dashed curves. The unstable v spectrum has two distinct regimes. The first one, in the range $f \geq 0.2$, has an identical shape as the neutral spectrum, including the curvature near its maximum. Another in the range $f < 0.2$ is dominated by a larger peak in the range $0.005 < f < 0.02$. These two regimes are present in the unstable u spectrum. However, they are not as clearly distinguishable as in v. A change in the spectral slope at $f \sim 0.2$ indicates the point for the generalized u plot where both regimes separate.

The stable spectra of temperature, like those of velocity, dissolve into distinct categories of z/L (Figure 5.8). The unstable spectra are found in the relatively narrow band indicated by the stippled area. The progression from stable to unstable spectra is not a smooth one. The spectrum shifts from 0^+ to the outer edge of the stippled area as z/L changes sign. With increasingly

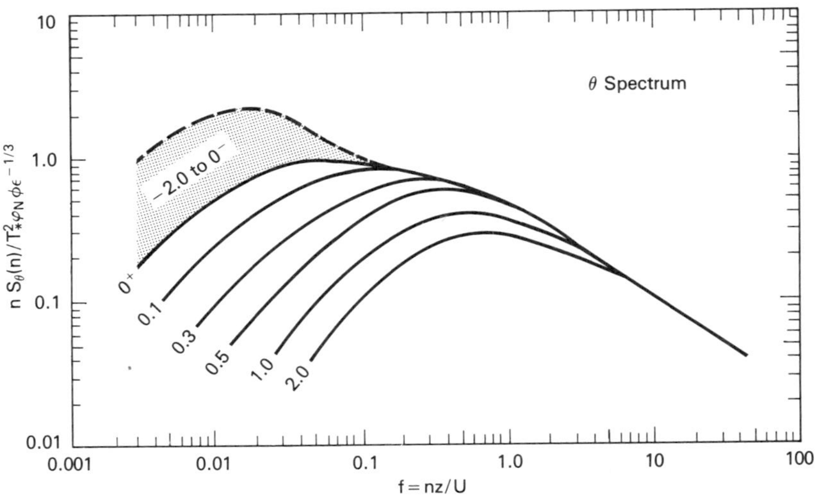

Figure 5.8 Generalized θ spectrum for z/L values ranging from $+2.0$ to -2.0. (After Kaimal et al., 1972. *Quarterly Journal of Royal Meteorological Society* published by the Royal Meteorological Society.)

negative z/L the spectrum moves back toward 0^+. The trend is reversed when z/L changes from negative to positive.

Generalized cospectra of uw, $u\theta$, and $w\theta$ are shown in Figures 5.9 to 5.11. The cospectra uw and $w\theta$ show a $-7/3$ power law in the inertial subrange. Cospectra $u\theta$ show an average slope of $-5/2$ in the range $1 < f < 10$ (compare with 5.33-5.34).

The position of stable uw cospectra varies as a function of z/L. All unstable cospectra belong in the narrow band indicated in Figure 5.9 by the stippled area. The dashed curves represent the upper and lower limits of scatter in the unstable cospectral densities. There is an overlap between the stable and unstable regions, as 0^+ is located within the stippled area. The logarithmic cospectral peak extends over only two decades of f and is separated between 0.01 and 0.1 for $z/L < 0$ and between 0.1 and 1.0 for $z/L > 0$.

The normalized $u\theta$ curves in Figure 5.10 are spread over a much wider range of cospectral intensities then either uw or $w\theta$. The stable cospectra show the same systematic advancement with z/L as seen in earlier figures. The overlap between the stable and unstable regions is larger than those for uw and $w\theta$. There is a rough separation according to z/L for unstable spectra. The top third of the stippled area is occupied mainly by cospectral curves in the z/L range 0 to -0.3, the middle third by the z/L range -0.3 to -1.0, and the lower third by the z/L range -1.0 to -2.0.

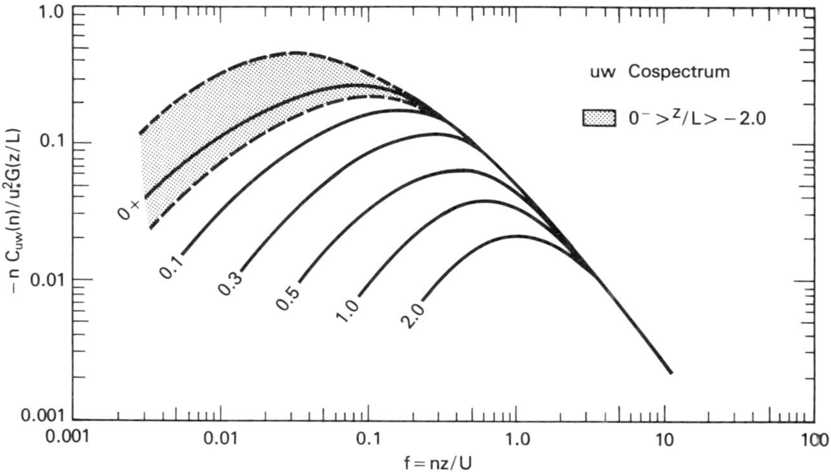

Figure 5.9 Generalized *uw* cospectrum for z/L values ranging from $+2.0$ to -2.0. Function $G = 1$ for $-2 \le z/L \le 0$ and $G = 1 + 7.9z/L$ otherwise. (After Kaimal et al., 1972. *Quarterly Journal of Royal Meteorological Society* published by the Royal Meteorological Society.)

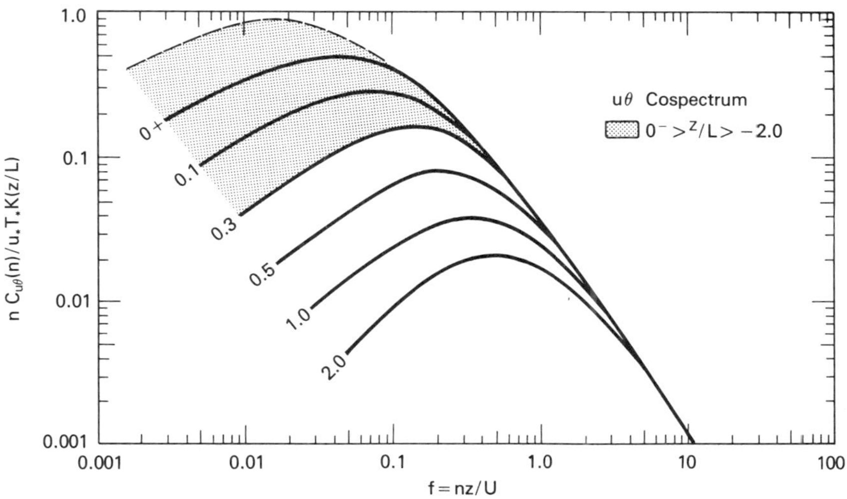

Figure 5.10 Generalized $u\theta$ cospectrum for z/L values ranging from $+2.0$ to -2.0. Function $K = 1$ for $-2 \le z/L \le 0$ and $K = 1 + 17.4z/L$ otherwise. (After Kaimal et al., 1972. *Quarterly Journal of Royal Meteorological Society* published by the Royal Meteorological Society.)

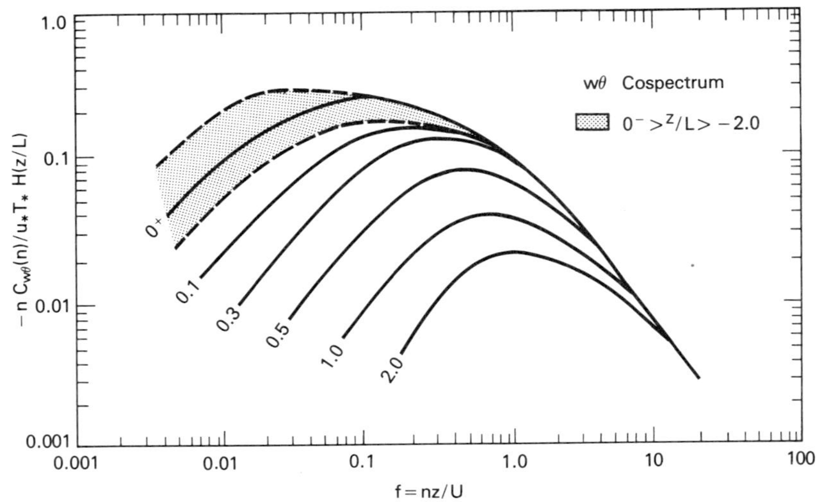

Figure 5.11 Generalized $w\theta$ cospectrum for z/L values ranging from $+2.0$ to -2.0. Function $H = 1$ for $-2 \le z/L \le 0$ and $H = 1 + 6.4z/L$ otherwise. (After Kaimal et al., 1972. *Quarterly Journal of Royal Meteorological Society* published by the Royal Meteorological Society.)

The normalized $w\theta$ cospectrum, shown in Figure 5.11, exhibits behavior very similar to the stress cospectrum in the location of its peak and the roll-off on the low-frequency side. However, on the high-frequency side, the curves are shifted at least an octave higher on the frequency scale. This extended contribution at the high-frequency end gives $w\theta$ cospectra, especially for unstable cases, a flatter appearance in the midrange. The inertial subrange cospectral level for $w\Theta$ is about three times as large as it is for uw. Smaller eddies evidently transport heat more effectively than momentum.

5.7 SPECTRA AND COSPECTRA IN THE MIXED LAYER

In the mixed layer the velocity spectrum can be normalized by w_*. For the inertial subrange, in analogy to (5.48), we can obtain

$$\frac{nS_u(n)}{w_*^2} = (2\pi\kappa)^{-2/3}\alpha_u \varphi_\epsilon^{2/3} f_i^{-2/3} \tag{5.50}$$

where $\varphi_\epsilon = \epsilon z_i/w_*^3$ and $f_i = nz_i/U$ is the dimensionless frequency.

Logarithmic spectra normalized in the described manner show systematic behavior when plotted as a function of f_i. Spectra of the wind velocity components obtained during the Minnesota experiment are shown in Figure

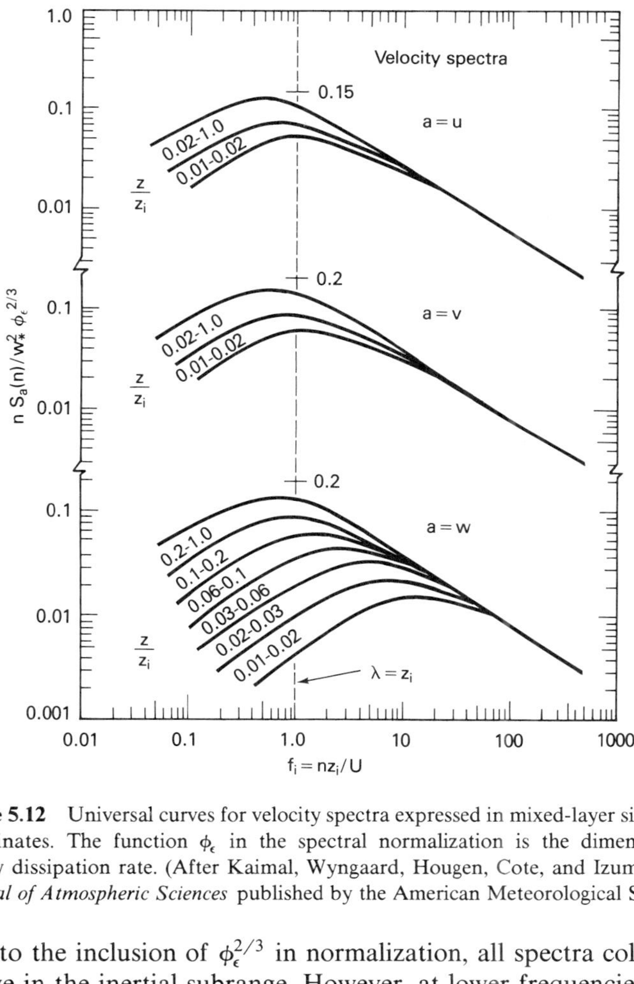

Figure 5.12 Universal curves for velocity spectra expressed in mixed-layer similarity coordinates. The function ϕ_ϵ in the spectral normalization is the dimensionless energy dissipation rate. (After Kaimal, Wyngaard, Hougen, Cote, and Izumi, 1976. *Journal of Atmospheric Sciences* published by the American Meteorological Society.)

5.12. Due to the inclusion of $\phi_\epsilon^{2/3}$ in normalization, all spectra collapse into a single curve in the inertial subrange. However, at lower frequencies the curves separate as a function of z/z_i.

For u and v spectral densities, only two categories exist: from 0.01 to 0.02 and from 0.02 to 1.0. The shift in the spectral peak for the horizontal velocity components is insignificant. As seen in the figure, of the three velocity components, spectra of the vertical velocity show the largest spread with height. The position of the spectral peak shifts to increasingly lower values of f_i as z/z_i increases.

The peak wavelengths $\lambda_m (\lambda = U/n)$ for three components of wind velocity are shown in Figure 5.13. For the u and v components, λ_m/z_i shows little, if any, variation with z/z_i.

Results obtained during the Minnesota experiment can be generalized by a model presented by Højstrup (1982). In the model, wind velocity spectra are

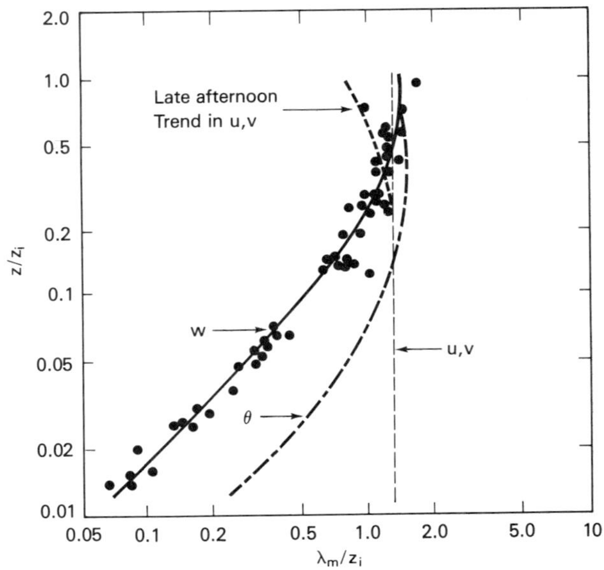

Figure 5.13 Dimensionless peak wavelength for the velocity components and temperature plotted as a function of z/z_i. Scatter in the u, v, and θ data points (not shown) is about twice as large as for w. (After Kaimal, Wyngaard, Hougen, Cote, and Izumi, 1976. *Journal of Atmospheric Sciences* published by the American Meteorological Society.)

expressed, following Kaimal's (1978) idea, as a sum of the low-wavenumber and high-wavenumber components:

$$\begin{aligned} nS_u(n) &= A_u(f_i) + B_u(f) \\ nS_v(n) &= A_v(f_i) + B_v(f) \\ nS_w(n) &= A_w(f_i) + B_w(f) \end{aligned} \tag{5.51}$$

where $A(f_i)$ is "buoyancy-produced," the low-wavenumber part, and $B(f)$ is "shear-produced," the high-wavenumber portion. Since little interaction is assumed between both portions, they can be modeled separately. The model exhibits a continuous transition from the unstable spectrum to a neutral limit (Figure 5.14).

Part A for horizontal components of the wind velocity is modeled as a surface layer spectral function which, when scaled by $w_*^2 = u_*^2(z_i/\kappa L)^{2/3}$, is dependent only on f_i. For the w spectra a slightly different approach is used. Since the presence of a solid boundary exerts a mechanical impedance upon the vertical fluctuations which increases for decreasing height or wavenumber, part A for the vertical component of the wind velocity is modeled as a lateral component function, on which a high-pass filter is imposed. Part B is modeled

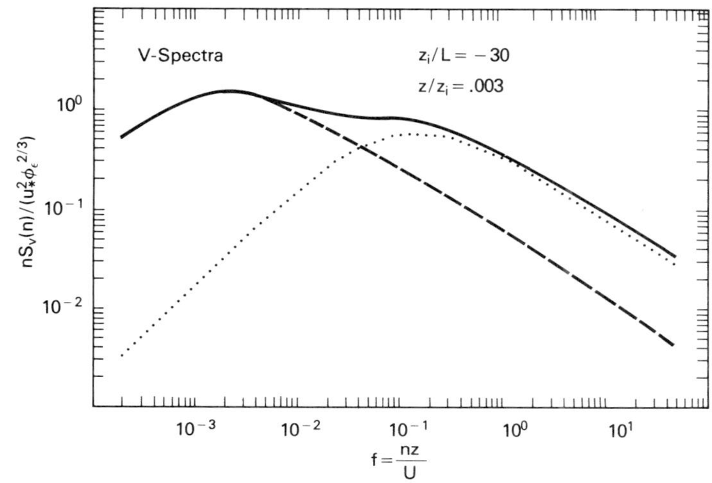

Figure 5.14 Model v spectrum. The dashed line is a plot of the function $A(f_i)$, the dotted line is $B(f)$. The solid line is the resultant spectrum. (After Højstrup, 1982. *Journal of Atmospheric Sciences* published by the American Meteorological Society.)

as a neutral surface layer spectral function and, when scaled by $\mathbf{U}_*^2$, is dependent only on $f = nz/u$.

Assuming that $\mathbf{U}_* = u_*(1 - z/z_i)$, Højstrup (1982) obtained a spectrum for the u component of the wind velocity in the form

$$\frac{nS_u(n)}{u_*^2} = \frac{0.5f_i}{1 + 2.2f_i^{5/3}}\left(\frac{z_i}{-L}\right)^{2/3} + \frac{105f_{ru}}{(1 + 33f_{ru})^{5/3}}\frac{(1 - z/z_i)^2}{[1 + 15(z/z_i)]^{2/3}} \tag{5.52}$$

where

$$f_{ru} = \frac{f}{1 + 15(z/z_i)}$$

and a spectrum for the v component of the wind velocity in the form

$$\frac{nS_v(n)}{u_*^2} = \frac{0.95f_i}{1 + 2.0f_i^{5/3}}\left(\frac{z_i}{-L}\right)^{2/3} + \frac{17f_{rv}}{(1 + 9.5f_{rv})^{5/3}}\frac{(1 - z/z_i)^2}{[1 + 2.8(z/z_i)]^{2/3}} \tag{5.53}$$

where

$$f_{rv} = \frac{f}{1 + 2.8z/z_i}$$

The peak wavelength λ_B of the surface layer function B is proportional to z. The Minnesota data show that for horizontal components of the wind velocity, the peak wavelength is independent of height. Therefore, λ_B in the model has to be limited by the substitution of $f_r = f/(1 + bz/z_i)$ for f in B. In order to maintain the same spectral level in the inertial subrange, the spectrum is divided by $(1 + bz/z_i)^{2/3}$. The values of b are chosen such that the peak wavelength of the A and B functions coincide at $z/z_i = 0.5$. This gives $b = 15$ for u-spectrum and $b = 2.8$, for v-spectrum.

The spectrum of the w component of the wind velocity is modeled in the following way:

$$\frac{nS_w(n)}{u_*^2} = F\left(f, \frac{z}{z_i}\right)\frac{0.95f_i}{1 + 2.0f_i^{5/3}}\left(\frac{z_i}{-L}\right)^{2/3} + \frac{2f}{(1 + 5.3f)^{5/3}}\left(1 - \frac{z}{z_i}\right)^2 \tag{5.54}$$

where the filter function is defined as

$$F\left(f, \frac{z}{z_i}\right) = \left\{\frac{f^2 + \left[0.3(z/z_i)^2\right]}{f^2 + (0.15)^2}\right\}^{1/2} \tag{5.55}$$

In the surface layer, the model shows a rapid variation in shape when z/L decreases, exhibiting the same kind of behavior that was found in the

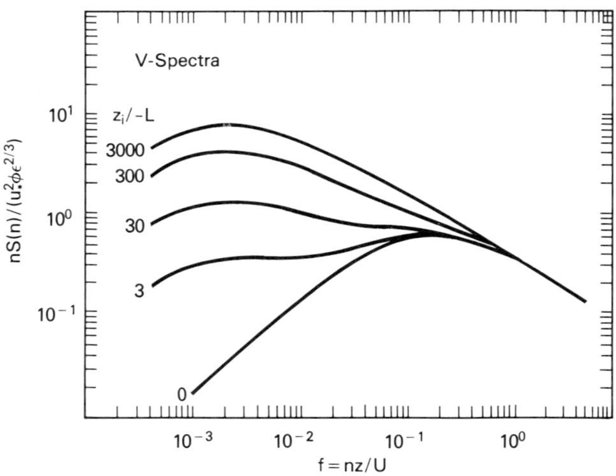

Figure 5.15 Model v spectrum plotted for different stabilities and for $z/(-L) = 0.009, 0.09, 9$. Note the rapid increase in spectral level at low frequencies for only slightly unstable conditions. (After Højstrup, 1982. *Journal of Atmospheric Sciences* published by the American Meteorological Society.)

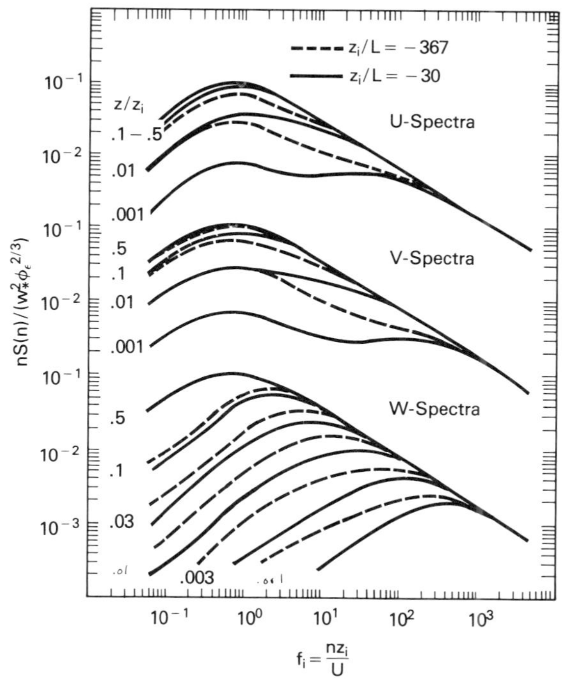

Figure 5.16 Model spectra obtained by Højstrup (1982). Two curves are plotted for each height, one for conditions as the most unstable found in Minnesota (dashed line) and one for the most neutral conditions (solid line). (After Højstrup, 1982. *Journal of Atmospheric Sciences* published by the American Meteorological Society.)

Kansas data (Figure 5.15). This provides an explanation for a jump between stable and unstable regions in Figures 5.5-5.6.

The mixed-layer spectra obtained from the model are shown in Figure 5.16. For each height, two curves are plotted: one for the $z_i/L = -30$ and a second for $z_i/L = -367$. These two curves represent an envelope in the same range as observed during the Minnesota experiment. At low levels, the spectra have double peaks.

In the low-frequency range the spectrum of temperature exhibits run-to-run variations introduced by entrainment effects in the upper half of the boundary layer. Some broad generalizations lead to the idealized spectral curves, shown in Figure 5.17. At the low-frequency end, all curves converge into a single one which extends upward in response to the diurnal trend in temperature. At inertial subrange frequencies, the spectral intensity decreases steadily with height to $0.5z_i$, stays at a low value between 0.5 and $0.7z_i$, and starts to rise again above $0.7z_i$. The arrows in the figure illustrate this trend.

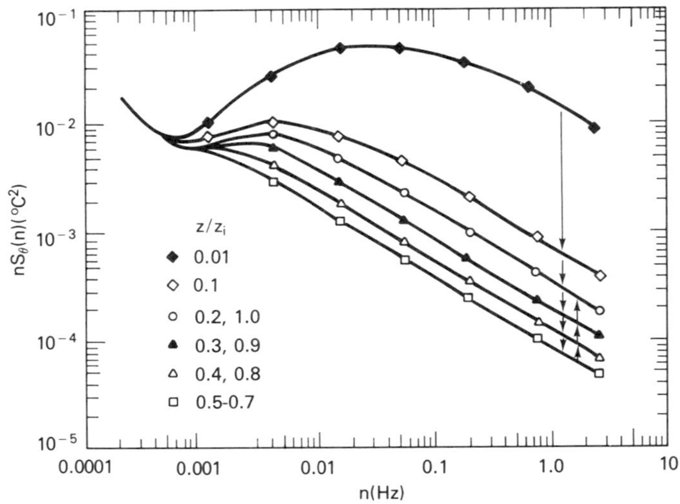

Figure 5.17 Idealized temperature spectra in dimensional coordinates showing variation in spectral behavior as a function of z/z_i. (After Kaimal, Wyngaard, Hougen, Cote, and Izumi, 1976. *Journal of Atmospheric Sciences* published by the American Meteorological Society.)

The behavior of temperature spectra can be explained in terms of local similarity theory. Following the local similarity approach, the logarithmic temperature spectrum can be split into two components:

$$nS_\theta(n) = nS_{\theta b}(n) + nS_{\theta t}(n) \tag{5.56}$$

Each component nondimensionalized by local scale should then be a function of the reduced frequency only:

$$\begin{aligned} \frac{nS_{\theta b}(n)}{\Theta_b^2} &= c_{\theta b} F_b(f) \\ \frac{nS_{\theta t}(n)}{\Theta_t^2} &= c_{\theta t} F_t(f_t) \end{aligned} \tag{5.57}$$

where f and f_t are the reduced frequencies $f = nz/U$, $f_t = n(z_i - z + \Delta)/U$, and $c_{\theta b}$ and $c_{\theta t}$ are universal constants describing the distribution of temperature variance (see Chapter 4). Functions F_b and F_t integrate over the entire spectral range to 1. Moreover, in the inertial subrange, their values are

$$\begin{aligned} F_b(f) &= nS_{\theta b}(n)/\Theta_b^2 = (2\pi)^{-2/3} \alpha_T c_{\epsilon Nb} f^{-2/3} \\ F_t(f_t) &= nS_{\theta t}(n)/\Theta_t^2 = (2\pi)^{-2/3} \alpha_T c_{\epsilon Nt} f_t^{-2/3} \end{aligned} \tag{5.58}$$

The coefficients $c_{\epsilon Nb}$, $c_{\epsilon Nt}$ describe the distribution of the product $\epsilon^{-1/3}N$ in terms of local scales. It is easy to see that in the inertial range, nS_θ defined by (5.56)–(5.57), changes with height as:

$$\begin{aligned} nS_\theta(n) &= \left[c_{\theta b}c_{\epsilon Nb}\Theta_b^2(f)^{2/3} + c_{\theta t}c_{\epsilon Nt}\Theta_t^2(f_t)^{2/3}\right] \\ &= \Theta_*^2(f_i)^{-2/3}\left[C_b(1 - z/z_i)^{4/3}(z/z_i)^{-4/3}\right. \\ &\qquad \left. + C_t R^{4/3}(z/z_i)^{4/3}(1 - z/z_i + D)^{-4/3}\right] \end{aligned} \tag{5.59}$$

where $f_i = nz_i/U$, $C_b = (2\pi)^{-2/3}\alpha_T c_{\epsilon Nb}$ and $C_b = (2\pi)^{-2/3}\alpha_T c_{\epsilon Nt}$. Values of the function nS_θ initially decrease with increasing height and then increase when $z \to z_i$.

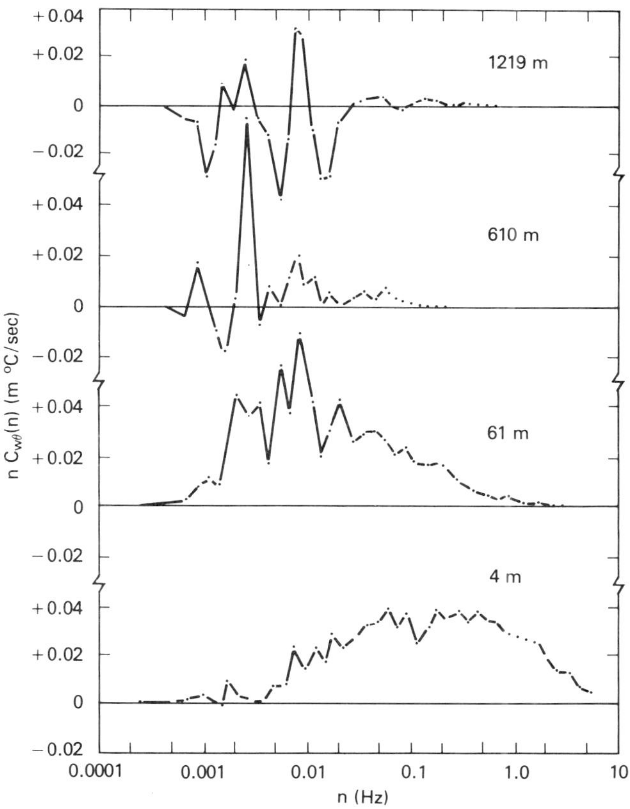

Figure 5.18 Cospectra of heat flux at four heights in the convective boundary layer. (After Kaimal, Wyngaard, Hougen, Cote, and Izumi, 1976. *Journal of Atmospheric Sciences* published by the American Meteorological Society.)

Cospectral curves of heat flux and stress in the mixed layer depart significantly from the surface layer universal form as z exceeds $0.1z_i$. Large excursions in positive and negative directions are a characteristic feature. These emerging patterns are not consistent with the development of universal forms.

The cospectra of heat flux for run 2A1 ($z_i = 1250$ m) during the Minnesota experiment are shown in Figure 5.18. It can be seen that the range of frequencies contributing to the upward transport of heat narrows with height. Near the inversion base, while there is still some upward flux at 0.003 Hz, the net flux is distinctly downward, suggesting a substantial entrainment into the boundary layer from the layer above.

5.8 SPECTRA AND COSPECTRA IN THE STABLE OUTER LAYER

In Chapter 4 it was shown that variances and covariances of the wind velocity and temperature in the stable ABL obey the local similarity theory. As a consequence, it can also be expected that stable spectra and cospectra should also obey the local scaling. From this argument, the logarithmic spectral and cospectral densities, scaled by local scales $\mathbf{U}_*$ and $\mathbf{T}_*$, should be functions only of the reduced frequency f and nondimensionalized parameter $Z = z/\Lambda$:

$$
\begin{aligned}
\frac{nS_i(n)}{\mathbf{U}_*^2} &= F_i(Z, f) \\
\frac{nS_\theta(n)}{\mathbf{T}_*^2} &= F_\theta(Z, f) \\
\frac{nC_{uv}(n)}{\mathbf{U}_*^2} &= F_{uv}(Z, f) \\
\frac{nC_i(n)}{\mathbf{U}_*\mathbf{T}_*} &= F_i(Z, f) \qquad \text{for } i = u\theta \text{ or } w\theta
\end{aligned}
\tag{5.60}
$$

where S_i and C_i are spectral and cospectral densities, n is a cyclic frequency, f is a reduced frequency, $f = nz/u$, and u is the wind velocity. In the surface layer, where $Z = z/L$, $\mathbf{U}_* = u_*$, and $\mathbf{T}_* = T_*$, the equations (5.60) take the form of the Monin–Obukhov spectral hypothesis (Monin, 1959).

The spectral "power laws," which were proven to be valid in the surface layer, can be generalized for the layer above and expressed in terms of the

local similarity functions (Sorbjan, 1986c):

$$\frac{nS_u(n)}{\mathbf{U}_*^2} = (2\pi\kappa)^{-2/3}\alpha_u\Phi_\epsilon^{2/3}f^{-2/3}$$

$$\frac{nS_v(n)}{\mathbf{U}_*^2} = (2\pi\kappa)^{-2/3}\alpha_v\Phi_\epsilon^{2/3}f^{-2/3}$$

$$\frac{nS_w(n)}{\mathbf{U}_*^2} = (2\pi\kappa)^{-2/3}\alpha_w\Phi_\epsilon^{2/3}f^{-2/3}$$

$$\frac{nS_\theta(n)}{\mathbf{T}_*^2} = (2\pi\kappa)^{-2/3}\alpha_T\Phi_\epsilon^{-1/3}\Phi_N f^{-2/3} \tag{5.61}$$

$$\frac{nC_{uw}(n)}{\mathbf{U}_*^2} = -(2\pi\kappa)^{-4/3}\alpha_{uw}\Phi_\epsilon^{1/3}\Phi_m f^{-7/3}$$

$$\frac{nC_{w\theta}(n)}{\mathbf{U}_*\mathbf{T}_*} = -(2\pi\kappa)^{-4/3}\alpha_{w\theta}\Phi_\epsilon^{1/3}\Phi_h f^{-7/3}$$

$$\frac{nC_{u\theta}(k)}{\mathbf{U}_*\mathbf{T}_*} = (2\pi\kappa)^{-3/2}\alpha_{u\theta}\Phi_\epsilon^{1/4}\Phi_m^{1/4}\Phi_h f^{-3}$$

where $\alpha_u \cdots \alpha_{u\theta}$ are universal constants and universal functions Φ are defined by (4.45) and (4.48).

Only the cospectra of wind velocity and temperature are considered in (5.61). However, there is no reason to expect a significant difference between cospectra of wind and various scalars (e.g., Schmidt et al., 1979).

The form (5.60), strictly speaking, is justified only in the universal-equilibrium range, in which it takes the form (5.61). Observations show, however, that (5.60) can be extended to the lower wavenumbers. Based on this fact, it can be assumed that (5.60) is valid for all f and that every function F_i can be expressed as a product of two functions: a *slope function* q_i depending only on f and a *magnitude function* p_i depending only on Z:

$$F_i(Z, f) = p_i(Z)q_i(f) \tag{5.62}$$

where $i = u, v, w, \theta, uw, w\theta$, and $u\theta$.

The integration of (5.62) with respect to $\ln f$ should result in the appropriate dimensionless variance or covariance function ($\Phi_{w\theta} = -1$, $\Phi_{uw} = 1$):

$$\int_0^\infty q_i(f)\, d \ln f = 1 \tag{5.63}$$

$$p_i(Z) = \Phi_i \tag{5.64}$$

Applying the arguments which led to (5.47), the following form of the spectral and cospectral densities can be deduced:

$$q_i(f) = \begin{cases} \dfrac{0.644(f/f_m)}{1 + \frac{3}{2}(f/f_m)^{5/3}} & \text{for } i = u, v, w, \text{ and } \theta \\ \dfrac{f/f_m}{\left[1 + \frac{3}{4}(f/f_m)\right]^{7/3}} & \text{for } i = uw \text{ and } w\theta \\ \dfrac{f/f_m}{1 + \frac{2}{3}(f/f_m)^{5/2}} & \text{for } i = u\theta \end{cases} \tag{5.65}$$

where f_m is a reduced frequency for which q_i has a maximum. Constants in (5.65) are chosen so that q_i integrates to unity and has a maximum at $f = f_m$. The high-frequency slopes in (5.65) coincide with (5.61). The low-frequency slope of all functions in (5.65) is $+1$, which agrees with observations in the f/f_m range about 0.05 to 0.5.

Comparing the high-frequency limits of (5.65) with (5.61), it is found that

$$(f_m) = \begin{cases} a_i \Phi_\epsilon \Phi_i^{-3/2} & \text{for } i = u, v, w \\ a_\theta \Phi_N^{3/2} \Phi_\epsilon^{-1/2} \Phi_\theta^{-3/2} & \text{for } i = \theta \\ a_{uw} \Phi_m^{3/4} \Phi_\epsilon^{1/4} \Phi_{uw}^{3/4} & \text{for } i = uw \\ a_{w\theta} \Phi_\epsilon^{1/4} \Phi_h^{3/4} & \text{for } i = w\theta \\ a_{u\theta} \Phi_\epsilon^{1/6} \Phi_m^{1/6} \Phi_h^{2/3} \Phi_{u\theta}^{-2/3} & \text{for } i = u\theta \end{cases} \tag{5.66}$$

where $a_u = 0.57$, $a_v = 0.88$, $a_w = 0.88$, $a_\theta = 1.16$, $a_{uw} = 0.092$, and $a_{w\theta} = 0.155$.

For the particular case of the Minnesota data, we can obtain using (4.62)

$$(f_m) = \begin{cases} b_i\left(1 + \dfrac{3.7z}{L}\right) & \text{for } i = u, v, w \\ b_\theta\left(0.74 + \dfrac{4.7z}{L}\right)^{3/2}\left(1 + \dfrac{3.7z}{L}\right)^{-1/2} & \text{for } i = \theta \\ b_{uw}\left(1 + \dfrac{4.7z}{L}\right)^{3/4}\left(1 + \dfrac{3.7z}{L}\right)^{1/4} & \text{for } i = uw \\ b_{w\theta}\left(0.74 + \dfrac{4.7z}{L}\right)^{3/4}\left(1 + \dfrac{3.7z}{L}\right)^{1/4} & \text{for } i = w\theta \\ b_{u\theta}\left(1 + \dfrac{3.7z}{L}\right)^{1/6}\left(1 + \dfrac{4.7z}{L}\right)^{1/6}\left(0.74 + \dfrac{4.7z}{L}\right)^{2/3} & \text{for } i = u\theta \end{cases} \tag{5.67}$$

where $b_u = 0.058$, $b_v = 0.33$, $b_w = 0.12$, $b_\theta = 0.12$, $b_{uw} = 0.097$, $b_{w\theta} = 0.163$, $b_{u\theta} = 0.05$. The peak wavelength $\lambda_m = z/f_m$ can be obtained from (5.67) in the form

$$\frac{\lambda_m}{h} = \begin{cases} c_i \frac{z}{h}\left(1 + \frac{3.7z}{L}\right) & \text{for } i = u, v, w \\ c_\theta \frac{z}{h}\left(0.74 + \frac{4.7z}{L}\right)^{3/2}\left(1 + \frac{3.7z}{L}\right)^{-1/2} & \text{for } i = \theta \\ c_{uw} \frac{z}{h}\left(1 + \frac{4.7z}{L}\right)^{3/4}\left(1 + \frac{3.7z}{L}\right)^{1/4} & \text{for } i = uw \\ c_{w\theta} \frac{z}{h}\left(0.74 + \frac{4.7z}{L}\right)^{3/4}\left(1 + \frac{3.7z}{L}\right)^{1/4} & \text{for } i = w\theta \\ c_{u\theta} \frac{z}{h}\left(1 + \frac{3.7z}{L}\right)^{1/6}\left(1 + \frac{4.7z}{L}\right)^{1/6}\left(0.74 + \frac{4.7z}{L}\right)^{2/3} & \text{for } i = u\theta \end{cases} \tag{5.68}$$

where $c_u = 17.2$, $c_v = 4.5$, $c_w = 3.0$, $c_\theta = 8.3$, $c_{uw} = 10.3$, $c_{w\theta} = 5.1$, and $c_{u\theta} = 20.0$.

Spectral and cospectral densities obtained during the Minnesota experiment and presented by Caughey et al. (1979) are shown in Figures 5.19 and 5.20. Equations (5.65) are also plotted in the same figures.

Analyzing the spectral data, it can be noticed that the curves underestimate the w spectrum for $f/f_m < 0.1$. However, the Minnesota w spectra tend to have more energy at low frequencies for all heights and all stabilities in comparison with the Kansas spectra. The cospectral data are more scattered than the spectral data, as they are inherently more difficult to measure and particularly sensitive to instrumental errors. The Minnesota normalized spectral and cospectral densities have maximum values of about 0.25 to 0.26 for u, v, w, and θ, about 0.27 for uw and $w\theta$ and between 0.28 and 0.29 for $u\theta$. For comparison, the same spectral values obtained from the Kansas data are 0.25 for u, v, w, and θ. Equations (5.65) give maximum values 0.26 for u, v, w, and θ; 0.27 for uw and $w\theta$, and 0.28 for $u\theta$.

In Figures 5.21 and 5.22, the vertical profiles of the characteristic wavelength λ_m for all spectra and cospectra are plotted. The theoretical curves coincide with the empirical points in the surface layer. The discrepancy increases with height. Above the surface layer, the empirical values are greater than predicted by the theoretical curves. The best agreement is obtained for u, v, and uw. The differences between theoretical and empirical results are due to the rather flat slopes of the empirical spectra and cospectra and to the scatter resulting from measurement of low-intensity turbulence. Caughey (1977) reported that in more stable conditions, the turbulent intensities are so low that

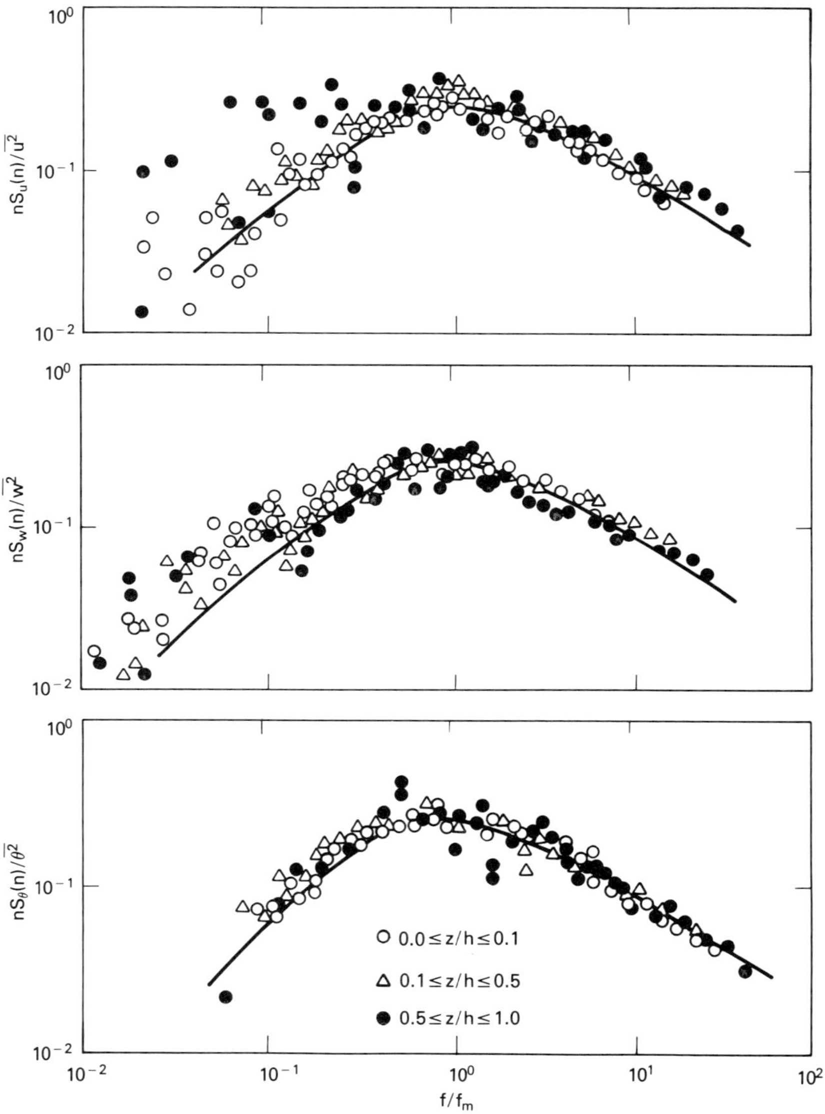

Figure 5.19 Normalized logarithmic spectra of u, w, and θ plotted against f/f_m. Data obtained from the Minnesota experiment. (After Caughey, Wyngaard, and Kaimal, 1979. *Journal of Atmospheric Sciences* published by the American Meteorological Society.)

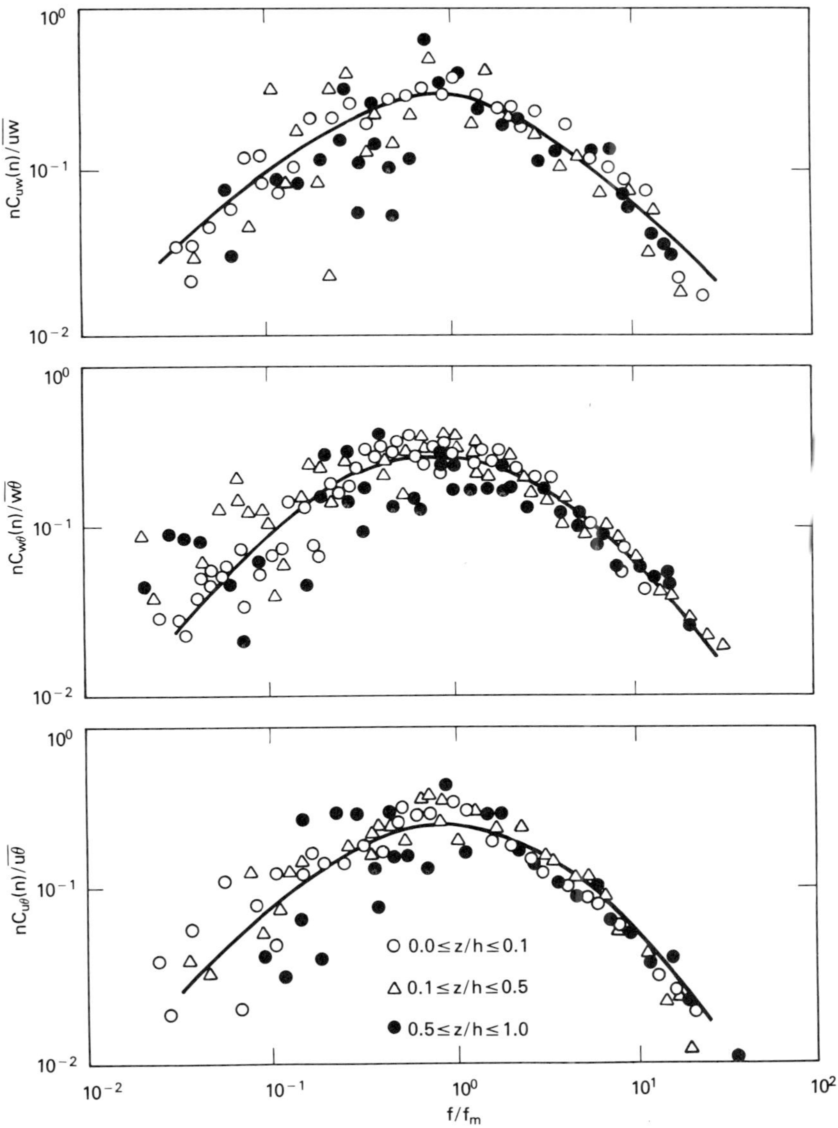

Figure 5.20 Normalized logarithmic cospectra of uw, $w\theta$, and $u\theta$ plotted against f/f_m. Data obtained from the Minnesota experiment. (After Caughey, Wyngaard, and Kaimal, 1979. *Journal of Atmospheric Sciences* published by the American Meteorological Society.)

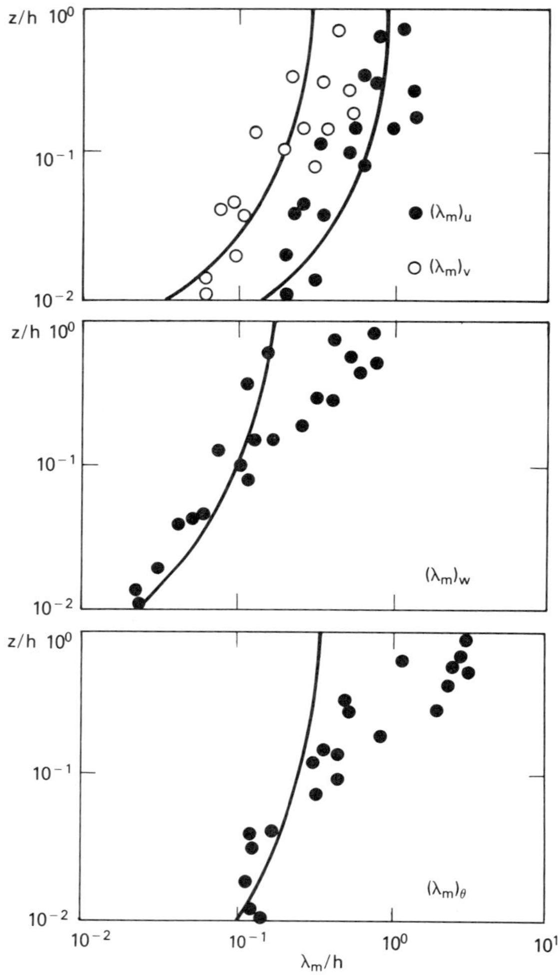

Figure 5.21 Vertical profiles of λ_m for u, v, w, and θ. The curves are obtained for $h/L = 5$. The points obtained from the Minnesota experiment. (After Caughey, Wyngaard, and Kaimal, 1979. *Journal of Atmospheric Sciences* published by the American Meteorological Society.)

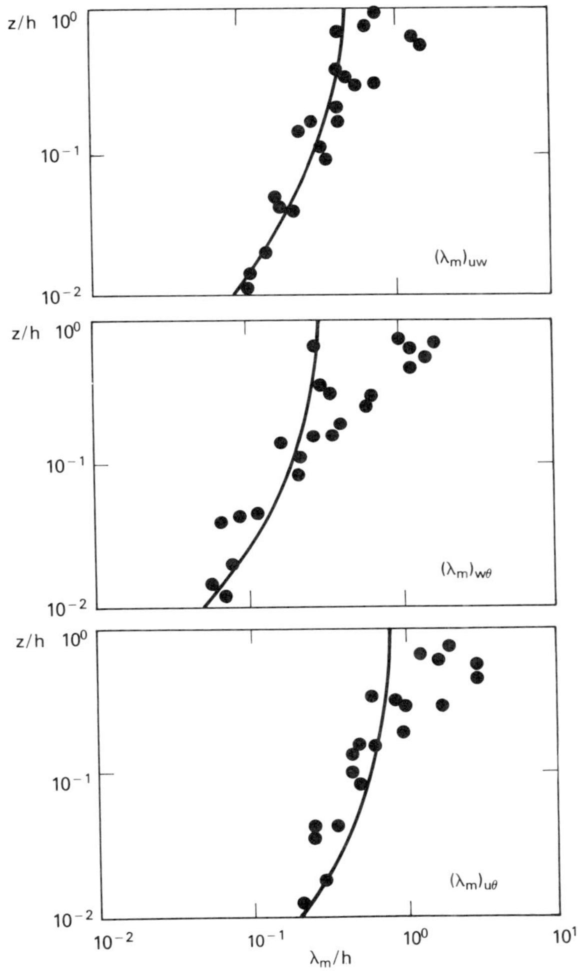

Figure 5.22 Vertical profiles of λm for *uw*, $w\theta$, $u\theta$. The curves are obtained for $h/L = 5$. The points obtained from the Minnesota experiment. (After Caughey, Wyngaard, and Kaimal, 1979. *Journal of Atmospheric Sciences* published by the American Meteorological Society.)

spectral estimates fall to near-noise limits. Similarly, Okamoto and Webb (1970) found that for Richardson number Ri ~ 0.2, some runs produced a near-zero spectral slope.

PROBLEMS

5.1. Assuming that $R(t) = A\exp(-|t|/\tau_0)$, find the energy spectrum $\Phi(\omega)$. Plot both functions for $\tau_0 = 100$ s, $A = 1$ m^2 s^{-2}.

5.2. Assuming that $R(t) = \exp(-at^2)$, find the energy spectrum $\Phi(\omega)$. Plot both functions for $a = 0.001$ s^{-2}, $A = 1$ m^2 s^{-2}. *Hint:* Use the Fourier transform $\mathscr{F}[\exp(-at^2)] = 1/\sqrt{\pi a}\exp[-(\omega^2/4a)^2]$.

5.3. Assuming that $R_{uv}(t) = A\exp(-at^2)$, $R_u(t) = B\exp(-a_u t^2)$, and $R_v(t) = C\exp(-a_v t^2)$, find:

(a) The quadrature spectrum $Q(\omega)$

(b) The cospectrum $C(\omega)$

(c) Amplitude

(d) Phase angle

(e) Coherence

5.4. Assuming $R_{11}(r_1) = \overline{u'^2}\exp(-|r_1|/L)$ find $k\mathscr{H}_{11}/\overline{u'^2}$, $k\mathscr{H}_{22}/\overline{u'^2}$, and $k\psi_E/\overline{u'^2}$ in the isotropic case. Make a plot of the obtained functions versus kL.

Notice that at the origin the three-dimensional energy spectra vanish because there is no energy at zero wavelength. Notice also that at the origin the one-dimensional spectra do not vanish. This is a result of the fact that a mode of wave-number k, traveling nearly normal to the direction defining the one-dimensional spectra appears to be of a very low wave-number.

5.5. Assuming isotropy, find relationships between

(a) $\mathscr{H}_{ii}(k_i)$, for $i = 1, 2, 3$.

(b) $\mathscr{H}_{ii}(k_j)$, for $j \neq i$ and $i = 1, 2, 3$.

CHAPTER SIX

MODELING

6.1 GENERAL REMARKS

In this chapter we review mathematical models of the atmospheric boundary layer. Mathematical modeling provides a theoretical basis for the interpretation of measured characteristics and allows their prediction. It consists of two stages: selection of an appropriate closure scheme and solution of the closed set of the averaged governing equations.

The closure problem is fundamental in mathematical modeling of the boundary layer processes. As mentioned in Chapter 2, the averaged equations are filtered out of a certain range of eddies. The *ensemble averaging operator* completely removes randomness and filters out eddies of all sizes, while the *volume averaging operator* filters out eddies smaller than its spatial scale. The impact of filtered eddies is retained in additional terms (Reynolds terms) which appear in averaged equations. To close the system of governing equations, these additional terms have to be related with other variables or parameters of the system.

A variety of the ABL models is depicted in the following diagram:

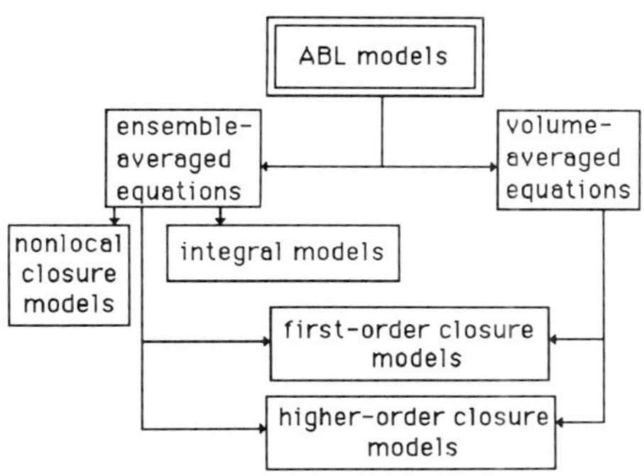

Ensemble-averaged equations must contain parametrization covering the entire eddy range. Obviously, such a parametrization cannot be uniformly valid in a wide range of flows. Turbulent flows differ from one to another principally in their large-eddy structure, and only small-scale eddies in all turbulent flows seem to be statistically similar.

The *volume-averaged equations* are averaged over a small spatial region. The main object is to remove the small eddies from the flow field so that equations for the large eddies may be derived. The effects of the small eddies on the large ones is then parametrized (*subgrid parametrizations*). Thus these parametrizations might be expected to have some universal properties and be independent of the large-scale differences in flow. From these arguments, it is clear that *large eddy simulation* (LES) based on volume-averaged equations is the best available method of calculating the three-dimensional, time-dependent structure of large Reynolds number turbulent flow. Currently, computations of the LES are limited to scientists who have access to very large, fast supercomputers. Therefore, the major utility of this method is the improvement in understanding the physics of turbulent flows, thereby helping to develop, test, evaluate, and improve the ABL models. The boundary layer models, based on ensemble averaged equations, are cheaper and therefore in common use.

Integral models employ ensemble-averaged equations which are integrated over at least one coordinate direction. This decreases the number of independent variables and greatly simplifies the mathematical problem. One of their principal drawbacks is that they need to be reworked when a new type of flow is to be computed. Integral models predict only the mean properties and, by their nature, provide no information on the distributions within the PBL.

Nonlocal closure models assume that the flux of a quantity at a given level is not related to the properties at this level, but depends on properties in all domains of the flow.

The *first-order closure models* evaluate the fluxes at a given level through the mean quantities at this particular level.

The *second-order (or higher) closure models* employ equations for the second moments which are closed on the third moments level. Generally, an n-order closure employs equations for the n moments and all equations are closed on the level of $n + 1$ moments.

One method used in numerical studies of small-scale turbulence is called *full simulation.* The method has no application in ABL modeling and therefore was not included in the earlier diagram. Full simulation provides the numerical solution of the exact Navier–Stokes equations. The only errors made are numerical ones, which, with care, can be kept as small as desired. To solve the equations, Fourier transform is first applied, with the calculations performed in wavenumber space (spectral method). In this way, spatial derivatives are evaluated to the Nth order, where N is the number of lattice points in a given direction. By its nature, this approach is limited to low Reynolds numbers. It means that the ratio of the largest scale to the smallest is typically about 100. In the atmospheric boundary layer this ratio is about 10^6.

Further characteristics of the models depicted in the diagram are presented next.

6.2 INTEGRAL MODELS

Integral (or slab) models are useful and convenient tools for predicting the averaged properties of the atmospheric boundary layer. This approach is particularly useful in the convective entraining planetary boundary layer (mixed layer), with a uniformly distributed mean temperature (Ball, 1960; Lilly, 1968; Carson, 1973; Stull, 1973). Integral models eliminate the need for a detailed knowledge of eddy fluxes. The closure problem reduces to the assumptions concerning mean profiles, surface-flux parametrizations, and the rate of change of the boundary layer height.

A typical distribution of temperature and heat flux in the mixed layer is shown in Figure 6.1. The potential temperature decreases sharply in the surface layer and then stays approximately constant above it. On the top of the mixed layer the sudden jump of the temperature marks the interfacial sublayer between the mixed layer and a capping inversion layer. This sublayer is a result of the continuous bombardment (called *penetrative convection*) of the inversion layer by convective elements generated by surface heating and causes an irregular inflow of warm air downward into the mixed layer—called *entrainment.*

Changes of the mean temperature Θ_m, in the mixed layer, can be determined from the vertically integrated first law of thermodynamics:

$$\frac{d\Theta_m}{dt} = \frac{1}{z_i}\left(\overline{w'\theta'}\big|_0 - \overline{w'\theta'}\big|_i\right) \tag{6.1}$$

where the subscripts 0 and i denote the turbulent fluxes on the bottom and the top of the ABL. Notice that in the case when entrainment makes the heat flux at the inversion base negative, the boundary layer heats up more rapidly.

The magnitude of the temperature jump $\Delta\Theta$ on the top of the ABL changes in two ways: it increases as the height of the mixed layer increases, and decreases if the boundary layer warms up:

$$\frac{d\,\Delta\Theta}{dt} = \frac{d\Theta}{dz}\frac{dz_i}{dt} - \frac{d\Theta_m}{dt} \tag{6.2}$$

where $d\Theta/dz$ refers to the lapse rate above the inversion base. Notice that when a subsidence generated by large scale processes occurs on the top of the mixed layer, the growth expressed by the term dz_i/dt will be reduced. In this

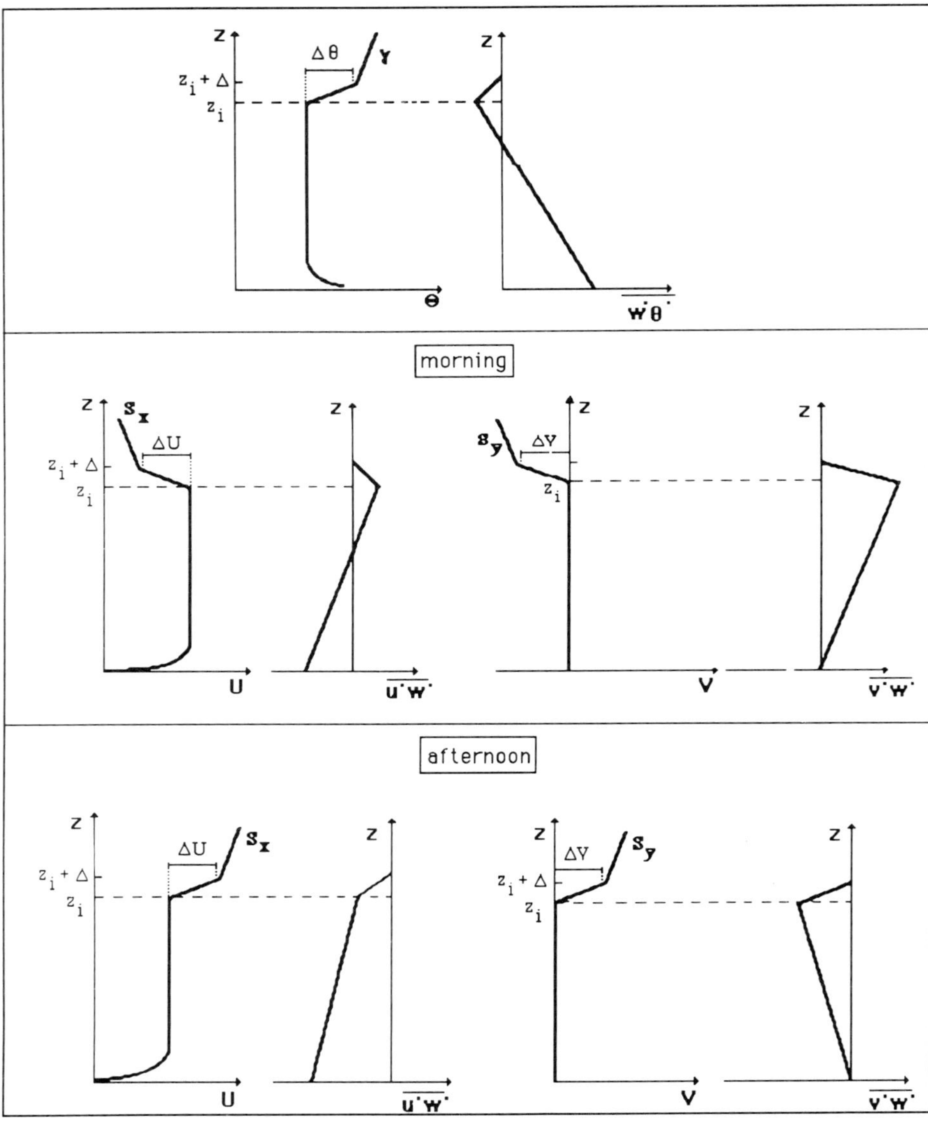

Figure 6.1 Vertical distribution of the potential temperature, wind velocity, and the turbulent fluxes in the mixed layer.

case, the term dz_i/dt in (6.2) has to be substituted by the *entrainment velocity*, defined as $w_e = dz_i/dt - w_h$, where w_h is the mean vertical velocity of the air at the top of the mixed layer. We will limit our discussion assuming here and after that $w_h = 0$.

Setting $d\Theta/dz = \gamma$ and substituting for $d\Theta_m/dt$ from (6.1) yields

$$\frac{d\,\Delta\Theta}{dt} = \gamma\frac{dz_i}{dt} - \frac{1}{z_i}\left(\overline{w'\theta'}\big|_0 - \overline{w'\theta'}\big|_i\right) \tag{6.3}$$

An equation for wind components can be obtained in a similar way. If the geostrophic wind in the boundary layer has average components U_g and V_g, the net rates of change of the mixed-layer wind velocity components, obtained by integration of the equation of motion, are

$$\begin{aligned}\frac{dU_m}{dt} &= f(V_m - V_g) + \frac{1}{z_i}\left(\overline{u'w'}\big|_0 - \overline{u'w'}\big|_i\right)\\ \frac{dV_m}{dt} &= -f(U_m - U_g) + \frac{1}{z_i}\left(\overline{v'w'}\big|_0 - \overline{v'w'}\big|_i\right)\end{aligned} \tag{6.4}$$

The equations for the rates of change of wind velocity jumps ΔU and ΔV, on the top of the mixed layer, are obtained in the same way as the equation for $\Delta\Theta$ (Deardorff, 1973; Zeman and Tennekes, 1977):

$$\begin{aligned}\frac{d\,\Delta U}{dt} &= S_x\frac{dz_i}{dt} - \frac{1}{z_i}\left(\overline{u'w'}\big|_0 - \overline{u'w'}\big|_i\right) + f\Delta V\\ \frac{d\,\Delta V}{dt} &= S_y\frac{dz_i}{dt} - \frac{1}{z_i}\left(\overline{v'w'}\big|_0 - \overline{v'w'}\big|_i\right) - f\Delta U\end{aligned} \tag{6.5}$$

where $S_x = (dU_g/dz)_i$ and $S_y = (dV_g/dz)_i$ are components of the thermal wind in the inversion layer.

The obtained equations for Θ_m, U_m, V_m, $\Delta\Theta$, ΔU, ΔV, and z_i are not closed. They require specification of the turbulent fluxes on the bottom and the top of the mixed layer. The bottom fluxes can be parametrized in terms of the bulk transfer formulas (analogous to those in Problem 4.5).

The crucial closure problem is the parametrization of heat and momentum fluxes at the inversion base. These fluxes can be related to the temperature and velocity jumps over the interface and to the entrainment rate through the following relationships (Lilly, 1968):

$$\begin{aligned}-\overline{w'\theta'}\big|_i &= \Delta\Theta\frac{dz_i}{dt}\\ -\overline{w'u'}\big|_i &= \Delta U\frac{dz_i}{dt}\\ -\overline{w'v'}\big|_i &= \Delta V\frac{dz_i}{dt}\end{aligned} \tag{6.6}$$

When the large scale subsidence is present, the entrainment velocity w_e has to replace the term dz_i/dt in (6.6).

A simplified case without entrainment (and without subsidence) can also be considered. In this situation, the heat flux on the top is zero. Convection is nonpenetrative and mixed-layer growth is due to *encroachment*. From (6.6) it follows that as long as $\Delta\Theta$ is positive, z_i does not change with time. Equation (6.6) also implies that z_i changes with time only when $\Delta\Theta = 0$. In the last case, it can be derived from (6.3) that $dz_i/dt = \overline{w'\theta'}|_0/(\gamma z_i)$. Equation (6.1) indicates that when $\Delta\Theta = 0$, the mean temperature in the layer increases at a rate given by the time integral of $(\overline{w'\theta'}|_0/z_i)$.

Vertical profiles of temperature, wind velocity, and turbulent fluxes are sketched schematically in Figure 6.1. Notice that the heat fluxes on the top and the bottom of the mixed layer have opposite signs. If there is no subsidence, entrainment implies that $dz_i/dt > 0$. Moreover, $\Delta\theta$ is positive because of the presence of the inversion layer above the mixed layer. As the result, the heat flux on the top is negative, confirming the first expression in (6.6).

Equations (6.6) indicate that in the morning, when ΔU is negative (nocturnal low-level jet), the term $\overline{u'w'}|_i$ is positive. Contrastingly, in the afternoon, ΔU is usually positive, due to an increase of wind speed aloft. This, with a substantial entrainment, can produce a large negative value of $\overline{u'w'}|_i$ at the top of the mixed layer. If $U_m > 0$, surface friction opposes the momentum from the mixed layer and $\overline{u'w'}|_0$ must be negative. If the x-axis of the coordinate system is oriented along the surface wind, $\overline{v'w'}|_0$ is zero.

Many other models have been proposed to parametrize the process of entrainment across the upper interface, driven by buoyancy. The simplest one assumes heat flux at the inversion base to be a constant fraction of the known surface heat flux (Tennekes, 1973).

A more complex model of Zilitinkevich (1975b) utilizes the turbulence kinetic energy budget and finds dz_i/dt inversely proportional to the overall Richardson number defined as:

$$R_b = \beta \Delta\Theta \, z_i/V_s^2 \tag{6.7}$$

where V_s is the velocity scale. In the convective case $V_s = w_*$. In the case of wind-driven mixed layer: $V_s = u_*$. In cases where both convective and mechanical production of turbulence is present V_s is a combination of u_* and w_* (Tennekes, 1973).

According to Artaz and André (1980), the basic parameters that govern the structure and evolution of the entrainment at the top of the convectively driven mixed layer are: the upward buoyancy flux at the ground $\overline{w'\theta'}|_0$, the lapse rate γ in the overlying stable layer, the depth of the mixed layer z_i, and the buoyancy parameter $\beta = g/T_0$. With such assumptions, the dimensionless properties and characteristics of the entrainment would be functions only of the dimensionless number x that can be constructed from the governing parameters, namely (Artaz and André, 1980),

$$x = \left(\frac{\overline{w'\theta'}\big|_0^2}{\beta \gamma^3 z_i^4} \right)^{1/3}$$

$$w_n(x) = \frac{1}{w_*} \frac{dz_i}{dt} \tag{5.8}$$

$$R(x) = \frac{\overline{w'\theta'}\big|_i}{\overline{w'\theta'}\big|_0}$$

The power ($\frac{1}{3}$) involved in the definition of x is for mathematical convenience. Notice that x is inversely proportional to the overall Richardson number R_b, $x = R_b^{-1}$.

Most parametrizations reduce to the same expression for the rate of growth of the mixed layer, but with different values of numerical parameters

TABLE 6.1 Parameters in Equations (6.9)

(a) Constants

	Tennekes (1973)	Deardorff (1974b)	Zilitinkevich (1975b)	Tennekes (1975)	Tennekes revised
a	0	0	0	0	0
b	1.4	1.8	1.4	3.6	3.4
c	0	9	7	18.5	31.2
A	0.2	0.4	0.2	0.51	0.49
B	0	−0.45	0	−2.64	−4.46
C	0	9	7	5.28	8.93

(b) Functions of x

	Zeman (1975)	Zeman revised	Mahrt and Lenschow (1976)
a	$-0.16[x(1+x)]^{1/2}$	$-0.41[x(1+x)]^{1/2}$	0
b	3.5	6	1.4
c	24.85	40	$x^{-0.5}$
A	$0.5 - 0.024[(1+x)/x]^{1/2}$	$0.86 - 0.06[(1+x)/x]^{1/2}$	0.2
B	−3.55	−5.77	0
C	7.1	11.4	$x^{-0.5}$

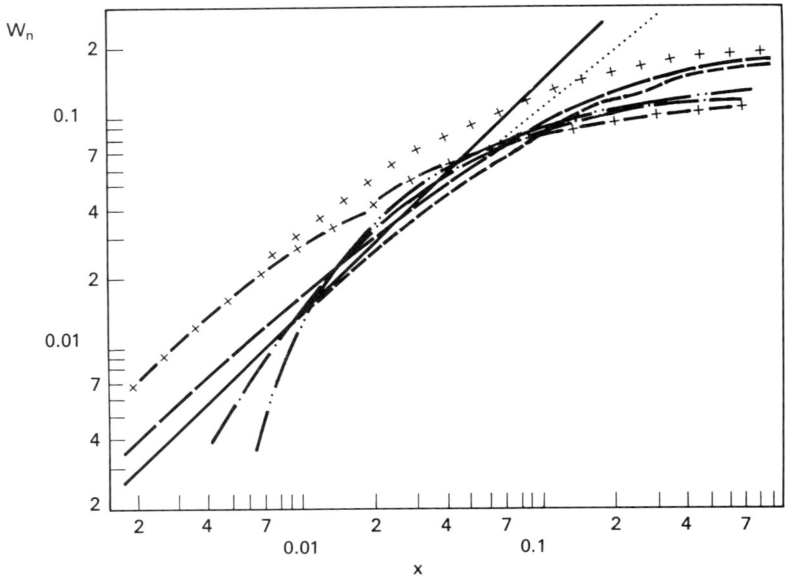

Figure 6.2 Dimensionless variation of the mixed-layer growth rate in the models of Zilitinkevich (1975b) (– – – –), Tennekes (1975) (+ + + +), Zeman (1975) (-·-·-) and (-··-··-), Deardorff (1974b) (---), and Mahrt and Lenschow (1976) (....). The solid line represents the constant ratio model. (After Artaz and André, 1980.)

(Artaz and André, 1980):

$$
\begin{aligned}
w_n(x) &= \frac{a + bx}{1 + cx} \\
R(x) &= -\frac{A + Bx}{1 + Cx}
\end{aligned}
\qquad (6.9)
$$

The parameters a, b, c, A, B, and C are listed in Table 6.1. Models assuming these parameters to be constants are listed in Table 6.1a. The other group of models uses parametrizations that can be expressed as in (6.9), but with the difference that parameters a and A are no longer constants but are functions of x (Table 6.1b). For some models, the diagnostic expression $\Delta\Theta = \gamma z_i/7$ proposed by Tennekes (1973) was used to obtain equations in the form given by (6.9).

A comparison of results given by (6.9) with laboratory and field experiments (Figures 6.2 to 6.5) shows that mixed-layer depth is predicted quite accurately by the simplest, constant-flux ratio model of Tennekes (1973). In this model $a = c = 0$, $b = 1.4$, and $R(x) = 0.2$. This comparison also shows that the ratio R between the heat fluxes at the inversion level and at the ground is not constant and shows a tendency to decrease with increasing

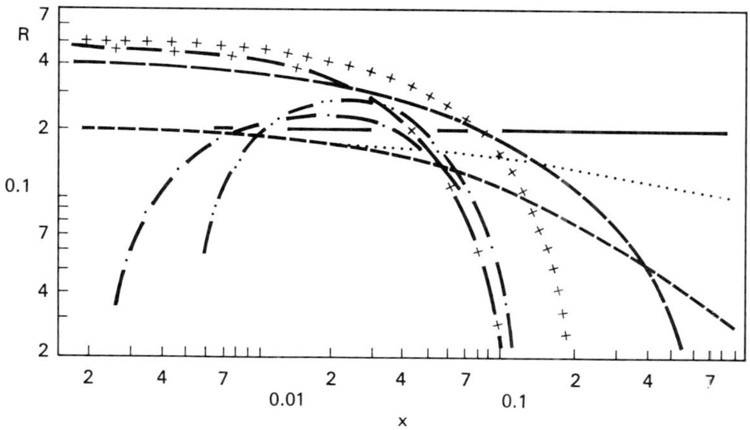

Figure 6.3 Variation of the heat flux ratio in the models of Zilitinkevich (1975b) (— — — —), Tennekes (1975) (+ + + +) and (+ — + —), Zeman (1975) (-·-·-) and (-·-·-·-), Deardorff (1974b) (---), and Mahrt and Lenschow (1976) (....). The solid line represents the constant ratio model. (After Artaz and André, 1980.)

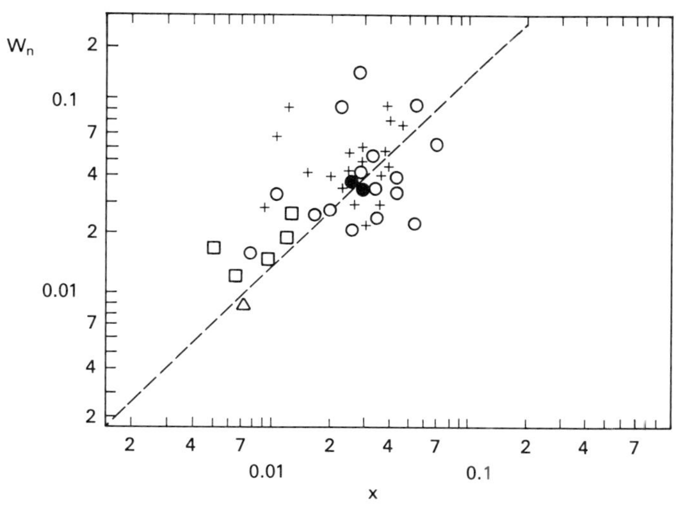

Figure 6.4 Dimensionless variation of the mixed-layer growth rate measured during the atmospheric experiments of Wangara (□), AMTEX (△), and Haswell (●); Limagne and Phitiviers (○) and Voves (+). The dashed line represents the constant heat flux ratio model ($w_n = 1.4x$). (After Artaz and André, 1980.)

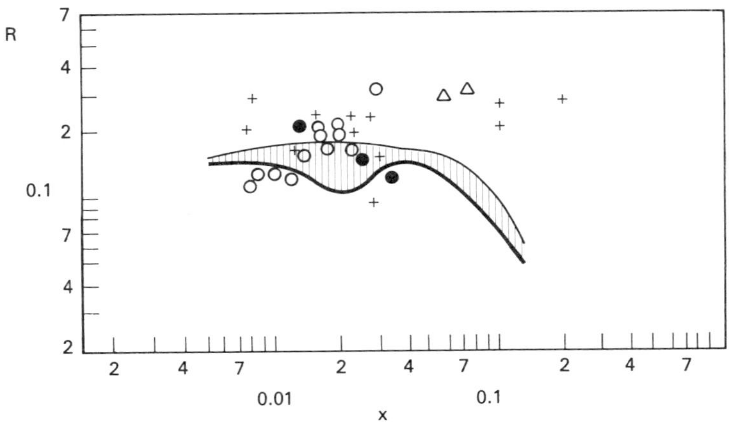

Figure 6.5 A variation of the heat flux ratio measured by Deardorff et al. (1969) (○), Willis and Deardorff (1974) (□), Cattle and Weston (1975) (△), Yamamoto et al. (1977) (●), during the Limagne experiment (+) and from numerical model of André et al. (1978) (| | |). (After Artaz and André, 1980.)

convective intensity x. This indicates the influence of other factors, among which shear-induced entrainment is probably the most important.

The considered integral modes describe the evolution of the mixed layer free of clouds. The mixed layer can contain a stratocumulus cloud deck in its upper portion, where specific humidity reaches its saturation value (e.g., Stage and Businger, 1981). The dynamics of the mixed layer in this case are considerably more complicated and depend on two new factors influencing turbulence: radiative cooling at the top of the cloud and absorption of shortwave radiation in the cloud's interior. So far, no satisfactory entrainment equations have been obtained for this case. However, the behavior of the mixed layer is more sensitive to "external" factors, such as the correct estimate of surface heat flux, than to actual entrainment formulation. The entrainment rate is usually small, especially for stratocumulus layers.

6.3 FIRST-ORDER CLOSURES

K-models are based on the hypothesis of Boussinesq, who, in 1877, suggested that turbulent shearing stress in analogy to viscous stress (2.18) can be related to the rate of mean strain as follows:

$$-\overline{u_i'u_j'} = k_m\left[\left(\frac{\partial U_i}{\partial x_j} + \frac{\partial U_j}{\partial x_i}\right) - \frac{2}{3}\delta_{ij}\frac{\partial U_\alpha}{\partial x_\alpha}\right] - \frac{2}{3}\frac{\overline{u_\alpha' u_\alpha'}}{2}\delta_{ij} \qquad (6.10)$$

where k_m is the eddy viscosity. The last term in (6.10) contains the turbulent kinetic energy. Its presence in (6.10) can be easily explained in the case of incompressible flow if a summation is performed for all $i = j$.

Equation (6.10) is substantially simplified in the case of incompressible, horizontally homogeneous and horizontal flow:

$$\overline{u'w'} = -k_m \frac{\partial U}{\partial z}$$
$$\overline{v'w'} = -k_m \frac{\partial U}{\partial z} \tag{6.11}$$

Similarly, the turbulent heat flux can be related to the mean temperature gradient:

$$\overline{u_i'\theta'} = -k_h \frac{\partial \Theta}{\partial x_i} \tag{6.12}$$

The coefficient k_h is related to k_m by $k_h = k_m/\alpha_\theta$, where α_θ is a function of height and stability. Observations show α_θ to be 0.74 in the neutral and stable surface layer, and decreasing with thermal instability.

By analogy with molecular diffusion, K-closures replace unknown turbulent fluxes with the product of an eddy viscosity and the appropriate mean gradient. Thus the closure problem is shifted to one of prescribing an eddy viscosity. There are a great variety of K-models, differing in the evaluation of eddy viscosity k_m. In many of them, the eddy viscosity is prescribed explicitly and somewhat arbitrarily. Early ABL models set k_m equal to a constant or a simple function of height. This permitted ABL equations to be solved analytically. Prandtl's mixing-length method, by analogy with kinetic theory of gases, sets k_m equal to the product of a velocity scale and a length scale:

$$k_m = u_t l \tag{6.13}$$

where u_t and l are characteristic velocity and length scales of turbulence, respectively. The characteristic length scale l is also called a *mixing length.* Later focus switched to specification of the length scale l. The logarithmic wind profile in the neutral surface layer can be obtained from (6.11) and (6.13) by assuming that $l = \kappa z$, $u_t = u_* =$ constant, and kinematic stress is equal to u_*^2.

More general formulations can be obtained by assuming that the characteristic velocity scale is proportional to the turbulent kinetic energy. This yields

$$k_m = clE^{1/2} \tag{6.14}$$

where c is a numerical constant.

The turbulent kinetic energy can be obtained from its simplified equation, keeping only the dynamic production and dissipation rate. The dissipation rate can be evaluated from a dimensional analysis as $\epsilon = c_\epsilon E^{3/2}/l$. Consideration of surface values of ϵ, k_m, and l gives $c_\epsilon = c^3$. All these expressions, together with (6.14), yield the formula first obtained by Prandtl in 1932:

$$k_m = l^2 \left[\left(\frac{\partial U}{\partial z} \right)^2 + \left(\frac{\partial V}{\partial z} \right)^2 \right]^{1/2} \tag{6.15}$$

Taking the stability effect into consideration, (6.15) can be modified to

$$k_m = l^2 \left[\left(\frac{\partial U}{\partial z} \right)^2 + \left(\frac{\partial V}{\partial z} \right)^2 \right]^{1/2} f(\mathrm{Ri}) \tag{6.16}$$

where f is an empirical function of the Richardson number Ri (defined in Problem 3.6):

$$f(\mathrm{Ri}) = \begin{cases} 1 - \dfrac{a \cdot \mathrm{Ri}}{1 + b|\mathrm{Ri}|^{1/2}} & \text{for unstable stratification} \\ \dfrac{1}{(1 + c \cdot \mathrm{Ri})^2} & \text{for stable stratification} \end{cases} \tag{6.17}$$

where a, b, and c are empirical constants. The function f increases when $\mathrm{Ri} \to -\infty$ (convective case), equals 1 when $\mathrm{Ri} = 0$, and decreases with increasing Ri.

In the formulations above, the main question that remains to be solved concerns the choice of the mixing length l. Blackadar (1962) assumed that the mixing length is linear in the surface layer and is limited by a constant value l_∞ above:

$$l = \frac{\kappa z}{1 + \kappa z / l_\infty} \tag{6.18}$$

The mixing length l_∞ can be parametrized in several different ways:

$$l_\infty = \begin{cases} 2.7 \times 10^{-4} \dfrac{G}{f} & \text{(Blackadar, 1962)} \\ 0.009 \dfrac{u_*}{f} & \text{(Blackadar, 1962)} \\ -2\kappa \dfrac{\Psi}{\partial \Psi / \partial z} & \Psi^2 = \left(\dfrac{\partial U}{\partial z} \right)^2 + \left(\dfrac{\partial V}{\partial z} \right)^2 - \alpha_\theta \dfrac{g}{T} \left(\dfrac{\partial \Theta}{\partial z} \right) \\ & \text{(Zilitinkevich and Laikhtman, 1965)} \\ 0.1 \dfrac{\int_0^\infty \sqrt{E}\, z \, dz}{\int_0^\infty \sqrt{E} \, dz} & \text{(Mellor and Yamada, 1974)} \end{cases} \tag{6.19}$$

Another approach directly specifies the shape of the vertical k_m profile throughout the ABL by similarity expressions for k_m in the surface layer

(O'Brien, 1970). This technique has been used extensively by Pielke and his colleagues (Segal et al., 1982; McNider and Pielke, 1981; Pielke, 1984) in dynamic PBL modeling in the mesoscale.

Eddy-viscosity models appear quite satisfactory in neutral or stable planetary boundary layers; however, they fail in convective situations. The concept of eddy viscosity is then physically meaningless, resulting in negative values of k_m. This failure may be traced to the importance of third moments, which contribute significantly to the levels of Reynolds stresses and fluxes in a convective planetary boundary layer (Zeman, 1981).

Some simple models of the atmospheric boundary layer are reviewed in the following sections.

6.3.1 Flow in the surface layer

The equations of the surface layer can be written in the form

$$\begin{aligned} k_m \frac{dU}{dz} &= u_*^2 \\ \frac{1}{\alpha_\theta} k_m \frac{d\Theta}{dz} &= -Q_0 \\ k_m\left(\frac{dU}{dz}\right)^2 - \beta Q_0 - c_0 \frac{E^2}{k_m} &= 0 \\ k_m &= c_0^{1/4}\sqrt{E}\, l \end{aligned} \tag{6.20}$$

The first two equations were obtained from (2.76) using first-order closure (6.11)–(6.12). The third equation is the surface-layer form of the turbulent kinetic energy equation (3.36), with dissipation parametrization discussed in the preceding section. Also, the last formula was obtained in the preceding section (c_0 is a new constant, $c_0 = c^4$). In the system above the values of Q_0, u_*, and c_0 are known. The values of U, k, Θ, and E can be obtained as a solution only when the system is closed by specifying the mixing length.

It will be convenient to express the system above in nondimensional form:

$$\begin{aligned} k_n \frac{dU_n}{d\zeta} &= 1 \\ \frac{1}{\alpha_\theta} k_n \frac{d\Theta_n}{d\zeta} &= 1 \\ \frac{1}{k_n} - \frac{1}{\kappa} - \frac{E_n^2}{k_n} &= 0 \\ k_n &= \sqrt{E_n}\, l_n \end{aligned} \tag{6.21}$$

where quantities subscripted with n are dimensionless. The wind velocity was scaled by u_*, the turbulent kinetic energy by $u_*^2/c_0^{1/2}$, temperature by $-Q_0/u_*$, and the eddy diffusivity by u_*L. The dimensionless height is defined as $\zeta = z/L$, where L is the Monin–Obukhov length [see equation (3.38)]. From the last two equations in the system above we can obtain

$$\left(\frac{l_n}{k_n}\right)^4 - \frac{l_n}{\kappa}\left(\frac{l_n}{k_n}\right)^3 - 1 = 0 \tag{6.22}$$

Obukhov (1946) assumed that $l_n = \kappa\zeta$. With this closure it is easy to see that the quantities $\kappa\zeta/k_n$, $E_n^{-1/2}$, and $\kappa\zeta\, dU/d\zeta$, $\kappa\zeta/\alpha_\theta\,(d\Theta/dz)$ are the roots of the same equation:

$$\phi^4 - \zeta\phi^3 - 1 = 0 \tag{6.23}$$

Equation (6.23) is called KEYPS, which stands for the initials of Kazansky, Ellison, Yamamoto, Panofsky, and Sellers, who derived it independently. It seems that it should be called O'KEYPS since Obukhov was the first person to derive it. Consider now the asymptotic properties of (6.23).

6.3.1.1 Free convection regime In the case when there is no vertical shear due to intensive convective mixing, then $\phi = \kappa\zeta\, dU/d\zeta \to 0$ and $\zeta \to -\infty$. Moreover, $|\zeta\phi^3| \gg \phi^4$ and (6.23) takes the form

$$-\zeta\phi^3 - 1 = 0 \tag{6.24}$$

This yields

$$\phi = (-\zeta)^{-1/3} \tag{6.25}$$

which is equivalent to

$$\begin{aligned} k_n &\sim (-\zeta)^{4/3} \\ E_n &\sim (-\zeta)^{2/3} \\ \frac{dU_n}{d\zeta} &\sim \zeta^{-4/3} \\ \frac{d\Theta_n}{d\zeta} &\sim \zeta^{-4/3} \end{aligned} \tag{6.26}$$

Integration of (6.26) gives the exponential ("$-\frac{1}{3}$ law") wind and temperature profiles in the convective surface layer.

6.3.1.2 Extremely stable case In the case when the turbulent energy is small, $\phi = E^{-1/2} \to \infty$, $\zeta \to \infty$, and (6.23) takes the form

$$\phi^4 - \zeta\phi^3 = 0 \tag{6.27}$$

The solution of (6.27) is

$$\phi = \zeta \tag{6.28}$$

From this, we can obtain

$$
\begin{aligned}
k_n &= \kappa \\
E_n &\sim \zeta^{-2} \\
\kappa \frac{dU_n}{d\zeta} &= 1 \\
\frac{\kappa}{\alpha_\theta} \frac{d\Theta_n}{d\zeta} &= 1
\end{aligned}
\tag{6.29}
$$

Integration of (6.29) gives the linear wind and temperature profiles in the stable surface layer.

6.3.1.3 Neutral stratification In the case of *neutral stratification*, when $\zeta \to 0$, (6.23) takes the form

$$\phi^4 - 1 = 0 \tag{6.30}$$

The solution to this equation is $\phi = 1$, which yields

$$
\begin{aligned}
k_n &= \zeta\kappa \\
E_n &= 1 \\
\frac{dU_n}{d\zeta} &\sim \zeta^{-1} \\
\frac{d\Theta_n}{d\zeta} &\sim \zeta^{-1}
\end{aligned}
\tag{6.31}
$$

Integration of (6.26) gives the logarithmic wind and temperature profiles in the neutral surface layer.

The foregoing asymptotic solutions ϕ are shown in Figure 6.6. The model properly describes the vertical distribution of flow in the surface layer and coincides with the Monin–Obukhov similarity predictions.

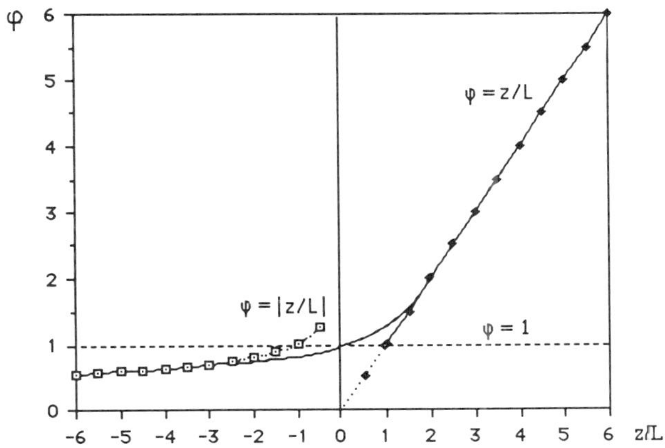

Figure 6.6 Distribution of the function ϕ.

6.3.2 The Ekman model

The dynamics of flow above the surface layer is strongly influenced by the Coriolis force. Action of the Coriolis force results in a change of wind direction with height. Applying the simple K-parametrization of the Reynolds stress components to the steady-state form of (2.69) yields the equation of the atmospheric boundary layer in the form

$$\begin{aligned} \frac{d}{dz}k_m\frac{dU}{dz} + f\left(V - V_g\right) = 0 \\ \frac{d}{dz}k_m\frac{dV}{dz} - f\left(U - U_g\right) = 0 \end{aligned} \tag{6.32}$$

System (6.32) expresses the balance of pressure gradient, Coriolis and Reynolds stress forces. Its solution was found in 1905 by Ekman, with the assumption that k_m is constant. To solve the system (6.32), it is convenient to multiply the second equation by $i = \sqrt{-1}$. This gives for U_g, V_g = constant:

$$\frac{d^2}{dz^2}\Phi - 2a^2 i\Phi = 0 \tag{6.33}$$

where $\Phi = (U - U_g) + i(V - V_g)$ and $a = \sqrt{f/2k_m}$.

The solution of (6.33) has the form

$$\Phi = c_1 e^{a(1+i)z} + c_2 e^{-a(1+i)z} \tag{6.34}$$

From the boundary conditions,

$$\Phi = \begin{cases} -U_g - iV_g & \text{for } z = 0 \\ 0 & \text{for } z = \infty \end{cases} \tag{6.35}$$

it can be determined that $c_1 = 0$ and $c_2 = -U_g - iV_g$. Thus the solution of the problem is

$$U + iV = \left(U_g + iV_g\right)\left(1 - e^{-a(1+i)z}\right)$$

or

$$\begin{aligned} U = U_g - e^{-az}\left(U_g \cos az + V_g \sin az\right) \\ V = V_g - e^{-az}\left(V_g \cos az - U_g \sin az\right) \end{aligned} \tag{6.36}$$

Rotating the coordinate system so that the geostrophic wind is along the x-axis yields ($U_g = G, V_g = 0$):

$$\begin{aligned} U = G\left(1 - e^{-az}\cos az\right) \\ V = Ge^{-az}\sin az \end{aligned} \tag{6.37}$$

The hodograph of the wind velocity vector described by (6.37), called the *Ekman spiral*, is shown in Figure 6.7.

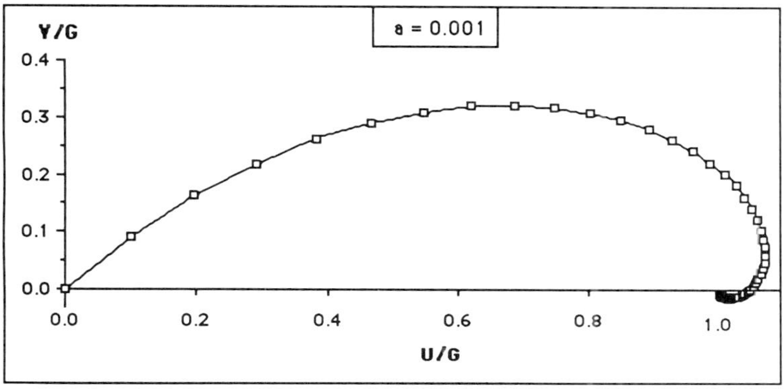

Figure 6.7 Ekman spiral obtained for $a = 0.001$. Points are plotted every 100 m, starting on the surface.

The angle between the wind vector and the geostrophic wind vector can be found from

$$\tan \alpha = \frac{V}{U} = \frac{e^{-az} \sin az}{1 - e^{-az} \cos az} \tag{6.38}$$

From the l'Hospital rule, and for $z \to 0$, we obtain $\alpha \to 45°$. Moreover, on the top of the boundary layer, where the wind coincides with the geostrophic wind vector, $\alpha = 0$, which is equivalent to $\sin ah = 0$ or $ah = \pi$. This allows estimating the height of the boundary layer h as

$$h = \pi \sqrt{\frac{2k_m}{f}} \tag{6.39}$$

For $k_m = 10\ \text{m}^2/\text{s}$ and $f = 10^{-4}\ \text{s}^{-1}$, $h = 1405$ m. For the case in Figure 6.7, obtained with $a = 0.001$, we find that $h = 3142$ m.

In Figure 6.8, the U-component of the wind vector is depicted, showing that, above the 1500-m level, this component exceeds values of the geostrophic wind. To explain this result, notice that the first momentum equation, integrated over the entire boundary layer, gives

$$\int_0^h U\, dz = U_g \tag{6.40}$$

where boundary conditions assuming the vanishing of the y component of the Reynolds stress on the top and the bottom of the ABL were adopted. If the integral of U is equal to U_g and U is 0 on the surface, apparently, somewhere within the ABL, $U > U_g$.

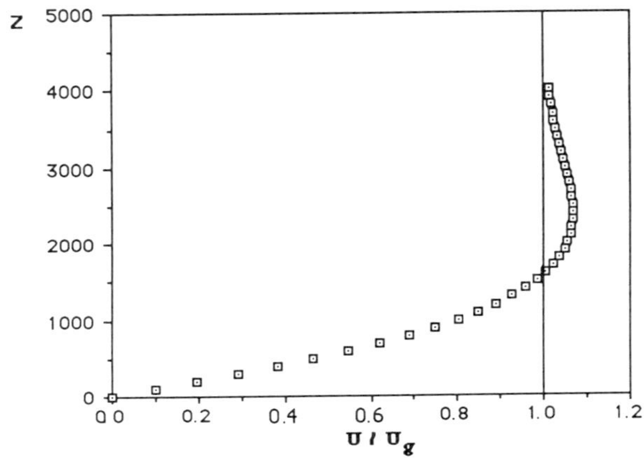

Figure 6.8 The U component of the wind vector. Points are plotted every 100 m.

The momentum equations (6.32) can be expressed in the vector form

$$\underbrace{\frac{d}{dz} k_m \frac{d\mathbf{V}}{dz}}_{\mathbf{T}} = \underbrace{f\mathbf{k} \times (\mathbf{V} - \mathbf{G})}_{\mathbf{C} + \mathbf{P}} \tag{6.41}$$

where **V** and **G** are actual and geostrophic wind vectors and **k** is the unit vector of the z-axis. The term on the left-hand side (denoted by **T**) is the Reynolds stress force acceleration. The right-hand side reflects the sum of the Coriolis force **C** and the pressure gradient force **P** accelerations. The Coriolis force is perpendicular to the wind vector **V**. The pressure gradient force **P** is perpendicular to the geostrophic wind vector **G**.

On the surface, where the wind velocity is zero, the Coriolis force vanishes and the pressure gradient force is balanced by the Reynolds stress force: $\mathbf{P} = \mathbf{T}$. On the top of the boundary layer, the Reynolds stress force vanishes and the pressure gradient force is balanced by the Coriolis force only: $\mathbf{P} = \mathbf{C}$.

The shift of the wind direction in the ABL is the result of the decrease of Reynolds stress force and the presence of Coriolis force. This is illustrated in Figure 6.9. In the figure two levels z_1 and z_2 are considered, in which $z_1 < z_2$. Since vector $\mathbf{T}_2$ is smaller and is shifted to the left in order to balance the pressure gradient force **P**, the Coriolis force $\mathbf{C}_2$ also has to be shifted to the left and become bigger. As a result, the wind vector $\mathbf{V}_2$ turns to the right with height and increases.

Using the continuity equation

$$\frac{\partial w}{\partial z} = -\nabla_h \cdot \mathbf{V} \tag{6.42}$$

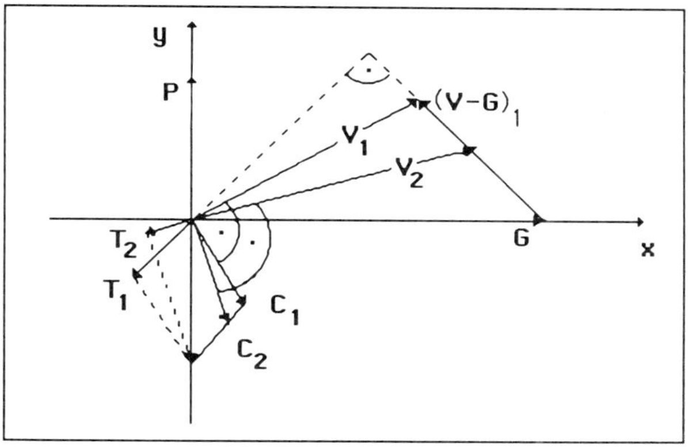

Figure 6.9 The balance of forces on two levels z_1 and z_2 ($z_1 < z_2$) in the atmospheric boundary layer. V, wind vector; G, geostrophic wind vector; T, Reynolds stress vector; C, Coriolis force; P, pressure gradient force.

where $\nabla_h = \partial/\partial x + \partial/\partial y$, it can be obtained that

$$\nabla_h \cdot \mathbf{V} = -e^{-az} \sin az \left(\frac{\partial V_g}{\partial x} - \frac{\partial U_g}{\partial y} \right) \tag{6.43}$$

Substituting this result in the integrated continuity equation (6.42) yields

$$\begin{aligned} w(x, y, z) - w(x, y, 0) &= -\int_0^z \nabla_h \cdot \mathbf{V}\, dz \\ &= \frac{[1 - \exp(-az)[\sin az + \cos az]}{2a\rho f} \nabla_h^2 p \end{aligned} \tag{6.44}$$

where $w(x, y, 0)$ can be assumed zero. At the top of the Ekman layer, where $z = h = \pi/a$,

$$w(h) = \frac{\nabla_h^2 p [1 - \exp(-\pi)]}{2a\rho f} \tag{6.45}$$

In a cyclone, $\nabla_h^2 p > 0$ and therefore the frictional vertical velocity $w(h)$ is positive. In an anticyclone $\nabla_h^2 p < 0$, and consequently $w(h)$ is negative. Equation (6.45) describes *Ekman pumping*, which is interpreted as a secondary circulation imposed on large-scale vertical motions.

6.3.3 Effects of baroclinicity

Observed hodographs of the wind vector in the atmospheric boundary layer (Figure 6.10) often have a different shape from that of the Ekman spiral. One of the reasons for this discrepancy is the geostrophic wind changing with height. Differentiating the equations of the geostrophic wind components [see

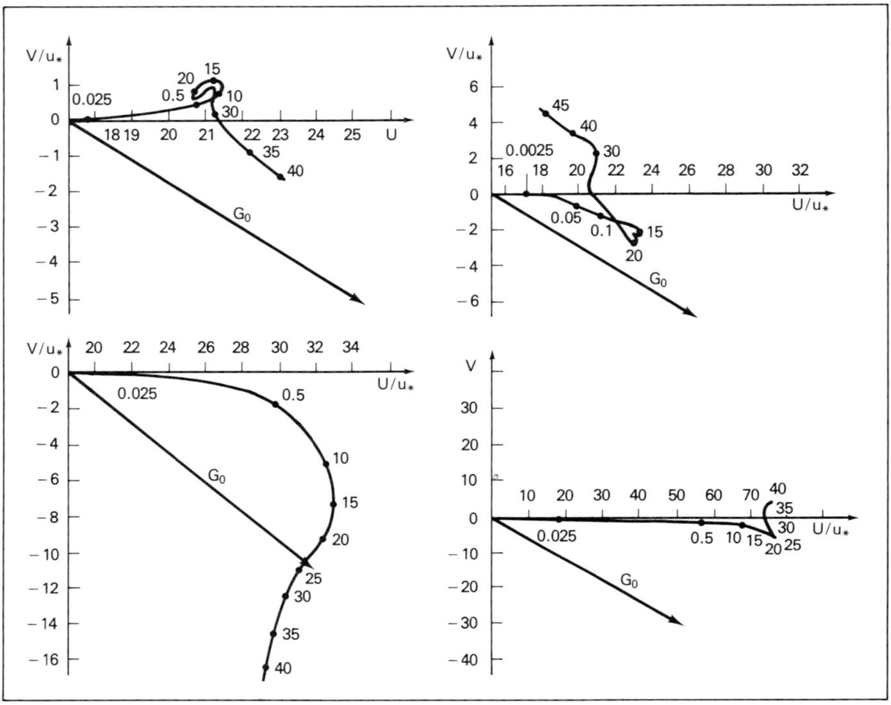

Figure 6.10 Empirical hodographs of the wind vector in the baroclinic boundary layer. G_0, surface geostrophic wind. In the figure velocities are scaled by U_* and height by U_*/f. (After Clarke, 1970. *Quarterly Journal of Royal Meteorological Society* published by the Royal Meteorological Society.)

(2.45)] and making use of the hydrostatic equation yields equations for the vertical change of the geostrophic wind (called the *thermal wind*):

$$\begin{aligned}\frac{\partial U_g}{\partial z} &= -\frac{R}{f}\left(\frac{\partial T}{\partial z}\frac{\partial \ln p}{\partial y} - \frac{\partial T}{\partial y}\frac{\partial \ln p}{\partial z}\right)\\ \frac{\partial V_g}{\partial z} &= \frac{R}{f}\left(\frac{\partial T}{\partial z}\frac{\partial \ln p}{\partial x} - \frac{\partial T}{\partial x}\frac{\partial \ln p}{\partial z}\right)\end{aligned} \tag{6.46}$$

When the atmospheric boundary layer is *barotropic*, the surfaces of constant pressure and constant temperature are parallel ($p \sim T$) and it follows from (6.46) that the thermal wind is zero. When the atmospheric boundary layer is *baroclinic*, the surfaces of constant pressure and constant temperature intersect and the thermal wind is not equal to zero.

Notice that for the typical values of $\partial T/\partial x \sim \partial T/\partial y \sim 0.1°\text{C/km}$, $\partial T/\partial z \sim 10°\text{C/km}$, $\partial p/\partial x \sim \partial p/\partial y \sim 0.01$ mbar/km, $\partial p/\partial z \sim$

TABLE 6.2 Vertical Gradient of the Geostrophic Wind, as a Function of the Horizontal Temperature Gradient

Horizontal Temperature Gradient (°C/100 km)	Vertical Gradient of the Geostrophic Wind (m/s/km)
0.50	1.51
1.0	3.03
1.25	3.79
1.50	4.54
1.75	5.30
2.00	6.06

100 mbar/km, and $p \sim 1000$ mbar, the first term in parentheses is of the order 10^{-4}°C/km², while the second one is of order 10^{-2}°C/km². For this reason the first term in parentheses can be neglected as being smaller and

$$\begin{aligned}\frac{\partial U_g}{\partial z} &\approx -\frac{g}{fT}\frac{\partial T}{\partial y} = -m \\ \frac{\partial V_g}{\partial z} &\approx \frac{g}{fT}\frac{\partial T}{\partial x} = n\end{aligned} \tag{6.47}$$

The numerical values obtained from (6.47) are shown in Table 6.2.

Assuming baroclinicity and substituting the geostrophic wind relations

$$\begin{aligned}U_g &= U_{g0} - mz \\ V_g &= V_{g0} + nz\end{aligned} \tag{6.48}$$

into (6.36) yields

$$\begin{aligned}U + iV &= \left[\left(U_{g0} + iV_{g0}\right) + (-m + in)z\right]\left(1 - e^{-a(1+i)z}\right) \\ &\approx \left(U_{g0} + iV_{g0}\right)\left(1 - e^{-a(1+i)z}\right) + (-m + in)z\end{aligned}$$

or

$$\begin{aligned}U &\approx U_{g0}\left(1 - e^{-az}\cos az\right) - V_{g0}e^{-az}\sin az - mz \\ V &\approx U_{g0}e^{-az}\sin az + V_{g0}\left(1 - e^{-az}\cos az\right) + nz\end{aligned} \tag{6.49}$$

where U_{g0} and V_{g0} are components of the geostrophic wind on the earth's surface.

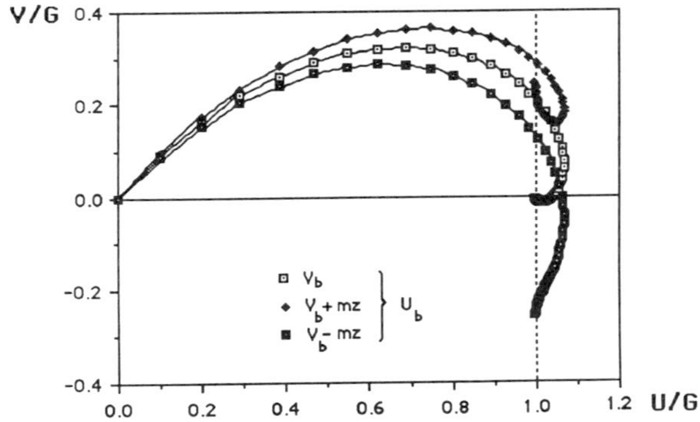

Figure 6.11 Ekman spirals obtained for the baroclinic correction of the V component of the wind velocity, $a = 0.001$, $m = 0.0001$. Points are plotted every 100 m, starting on the surface. U_b, V_b-barotropic components of the wind vector. Dotted line shows directions of the thermal wind vectors.

If the x-axis of the coordinates system is oriented along the geostrophic wind, then

$$\begin{aligned} U &\approx G_0(1 - e^{-az}\cos az) - mz = U_{\text{barotropic}} - mz \\ V &\approx G_0 e^{-az}\sin az + nz = V_{\text{barotropic}} + nz \end{aligned} \tag{6.50}$$

The hodograph of the wind velocity vector described by (6.50) is shown in Figures 6.11 to 6.13. From these figures it can be seen that the baroclinicity substantially modifies the Ekman spiral.

Baroclinicity can also strongly affect the stress profiles. This is easy to show in the mixed layer, where vertical gradients of the wind velocity compo-

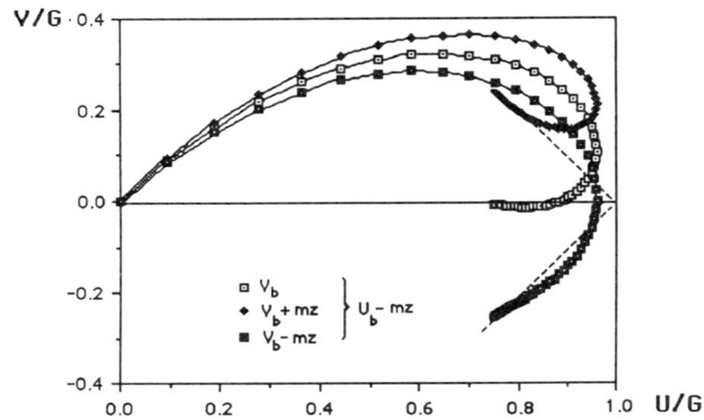

Figure 6.12 Ekman spirals obtained for baroclinic correction of both components of the wind velocity, $a = 0.001$, $m = 0.0001$. Points are plotted every 100 m, starting on the surface. U_b, V_b-barotropic components of the wind vector. Dotted line shows directions of the thermal wind vectors.

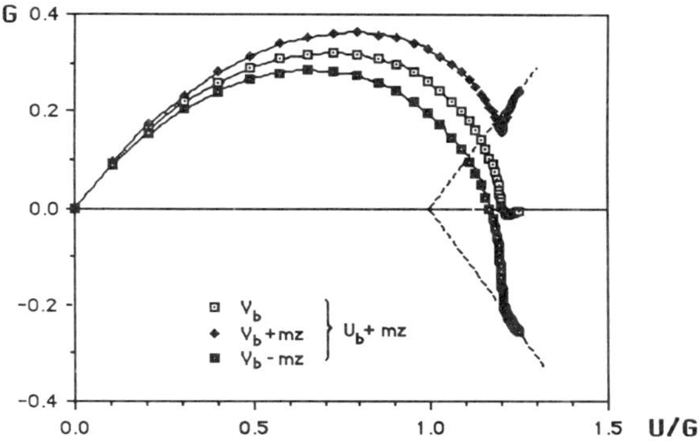

Figure 6.13 Ekman spirals obtained for baroclinic correction of both components of the wind velocity, $a = 0.001$, $m = 0.0001$. Points are plotted every 100 m, starting on the surface. U_b, V_b-barotropic components of the wind vector. Dotted line shows directions of the thermal wind vectors.

nents and potential temperature vanish. In this case, differentiation of the momentum equations gives (Wyngaard et al., 1974b)

$$\begin{aligned} \frac{\partial^2 \overline{u'w'}}{\partial z^2} &= -f\frac{\partial V_g}{\partial z} = -\frac{g}{T_0}\frac{\partial T}{\partial x} \\ \frac{\partial^2 \overline{v'w'}}{\partial z^2} &= f\frac{\partial U_g}{\partial z} = -\frac{g}{T_0}\frac{\partial T}{\partial y} \end{aligned} \tag{6.51}$$

In the instance of warm air advection ($\partial T/\partial x$, $\partial T/\partial y < 0$), both fluxes $\overline{u'w'}$ and $\overline{v'w'}$ have a minimum somewhere within the ABL (since their second derivatives are positive) with values several times those at the surface. The opposite happens in the case of cold air advection.

6.3.4 Effects of thermal stability

Observations show that the direction of the surface wind with respect to the surface geostrophic wind (or with respect to isobars) depends strongly on thermal stability. The Ekman model of the barotropic and baroclinic ABL fails to describe this feature properly since it does not include any information about temperature stratification. In this section a model of the thermally stratified, steady ABL in the stable case will be considered. A model of the unstable, nonsteady ABL is presented in Section 6.3.6.

A steady-state stable ABL is established through a balance of cooling by turbulent heat flux divergence and warming by mean advection. It can occur, for example, over Antarctica, where observations show slight diurnal variations of the temperature, especially during the polar night (Wendler and Kodama, 1984).

Let us again consider the steady-state version of the system (2.69)–(2.70). It is convenient to express this system in the nondimensional form, introducing new nondimensional variables (formalism of Rossby number similarity):

$$X = -\frac{\overline{u'w'}}{u_*^2}$$
$$Y = -\frac{\overline{v'w'}}{u_*^2} \tag{6.52}$$
$$R = -\frac{\overline{\theta'w'}}{T_* u_*}$$
$$K = \frac{k_m f}{u_*^2}$$

and also

$$Z = \frac{z}{L_E} \qquad L_E = \frac{\kappa u_*}{f}$$

From (6.11), (6.12), and (6.52) it follows that

$$X = \kappa K \frac{dU}{dZ}$$
$$Y = \kappa K \frac{dV}{dZ} \tag{6.53}$$
$$R = K \frac{d}{dZ} \frac{\Theta}{T_*}$$

The equation of motion (2.69)–(2.70) can be expressed in the form

$$\kappa \frac{u''}{u_*} = \frac{dY}{dZ} + \eta_z Z$$
$$\kappa \frac{v''}{u_*} = -\frac{dX}{dZ} + \eta_y Z \tag{6.54}$$

where $u'' = U - U_g$, $v'' = V - V_g$, and

$$\eta_x = \frac{\kappa^2}{f} \frac{dU_g}{dZ}$$
$$\eta_y = \frac{\kappa^2}{f} \frac{dV_g}{dZ} \tag{6.55}$$

are the components of the thermal wind.

Assuming that the atmosphere is barotropic, and differentiating (6.54) gives

$$K\frac{d^2X}{dz^2} + Y = 0$$
$$K\frac{d^2Y}{dz^2} - X = 0 \tag{6.56}$$

The system above can be supplied with the following boundary conditions:

$$\text{At } Z = Z_0: \quad X = 1 \quad Y = 0$$
$$\text{At } Z = Z_t: \quad X = 0 \quad Y = 0 \tag{6.57}$$

The turbulent structure is represented in the model by the dimensionless eddy diffusivity K, the turbulent heat flux R, and the height of the boundary layer Z_T (see Section 4.4.1):

$$K = \frac{Z}{1 + \beta_0\mu_0 Z}\left(1 - \frac{Z}{Z_T}\right)^q$$
$$R = R_0\left(1 - \frac{Z}{Z_T}\right)^p \tag{6.58}$$
$$Z_T = c\mu_0^r \qquad \mu_0 \gg 1$$

where $\mu_0 = L_E/L$ and β_0 is a constant. The vertical distribution of R in the model is expressed by an exponential formula. The height of the stable boundary layer is described by the formula obtained from the similarity arguments of Zilitinkevich (1972, 1975a), with $r = -0.5$ and $c = 0.8$.

From (6.55) we can obtain

$$C_g = \frac{\kappa}{\sqrt{(dX/dZ)_0^2 + (dY/dZ)_0^2}}$$
$$\alpha_0 = \arctan\left[-\frac{(dX/dZ)_0}{(dY/dZ)_0}\right] \tag{6.59}$$

where $C_g = u_*/G$ is the geostrophic drag coefficient and α_0 is a cross-isobar angle, defined as an angle between surface stress and surface wind vectors.

The model calculation, performed for $q = 1$, $p = 1$ in (6.58) and two values of $\mu_0 = L_E/L = 10$ and 100, are shown in Figures 6.14 and 6.15. The stress profiles increase in their curvature with increasing stability (Figure 6.14). Both U and V components overshoot their respective geostrophic values before returning to the geostrophic equilibrium at a greater height (Figure 6.15). Both plots show that the height of the stable boundary decreases with stability. The cross isobar angle α_0 increases with stability, while the geostrophic drag coefficient C_g decreases (Figure 6.16).

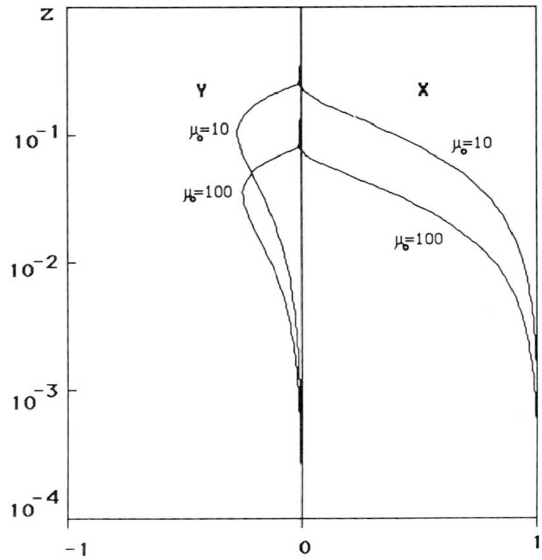

Figure 6.14 Profiles of dimensionless components of the Reynolds stress for two values of stability parameter $\mu_0 = 10$ and 100 (northern hemisphere).

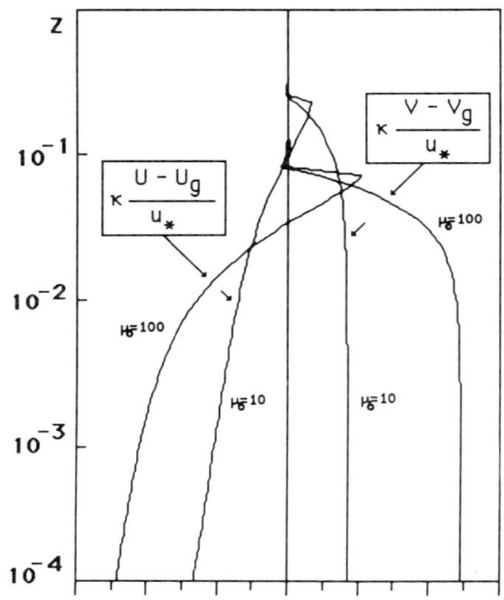

Figure 6.15 Profiles of dimensionless components of the wind velocity defects for $\mu_0 = 10$ and 100.

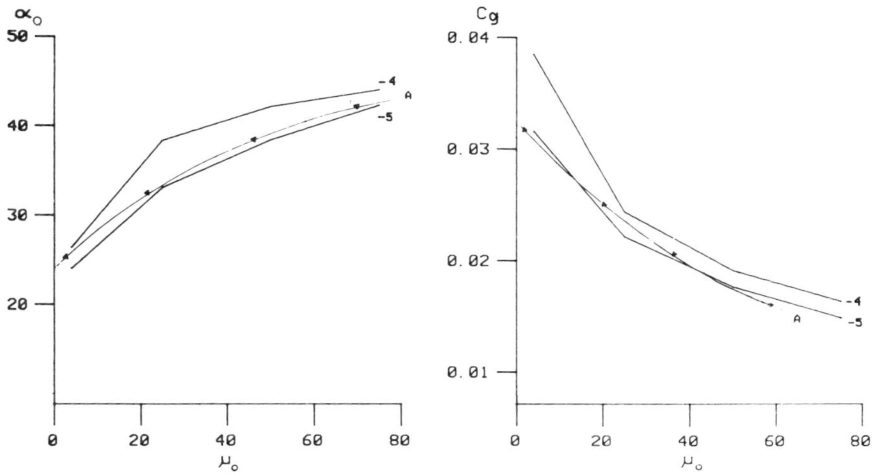

Figure 6.16 Comparison of the cross-isobar angle α_0 and geostrophic drag coefficient C_g, calculated from a model for $Z_0 = 10^{-4}$ and 10^{-5}, with those obtained by Arya (1975a) from the Wangara data (marked lines, Ro = 10^7). (After Sorbjan, 1984. *Journal of Atmospheric Sciences* published by the American Meteorological Society.)

6.3.5 Effects of slightly inclined terrain

Slightly sloping terrain extending over hundreds of kilometers generates buoyancy forces which are responsible for the deformation of the wind hodographs in the boundary layer. The buoyancy forces arise as a result of the radiational cooling (or heating) of a surface sloped at a small angle ψ to the horizontal (Figure 6.17). During this process the hydrostatic balance

$$-\frac{1}{\rho}\frac{\partial p}{\partial z} = g \tag{6.60}$$

is perturbed by a change of the air density ρ very close to the ground surface. This causes an additional acceleration of the air:

$$-\frac{1}{\mathrm{\rho}}\frac{\partial p}{\partial z} = g + a \tag{6.61}$$

where $\mathrm{\rho}$ is the density of the air close to the earth's surface and a is the acceleration of drainage force. After some simple rearrangement we obtain

$$a = -g\left(\frac{\mathrm{\rho} - \rho}{\mathrm{\rho}}\right) = -g\left(\frac{T - \mathbf{T}}{T}\right) \tag{6.62}$$

where $\mathbf{T}$ and T are temperatures at two different points on the same horizontal plane. The first point is very close to the sloped surface and its temperature is governed by the surface radiational budget.

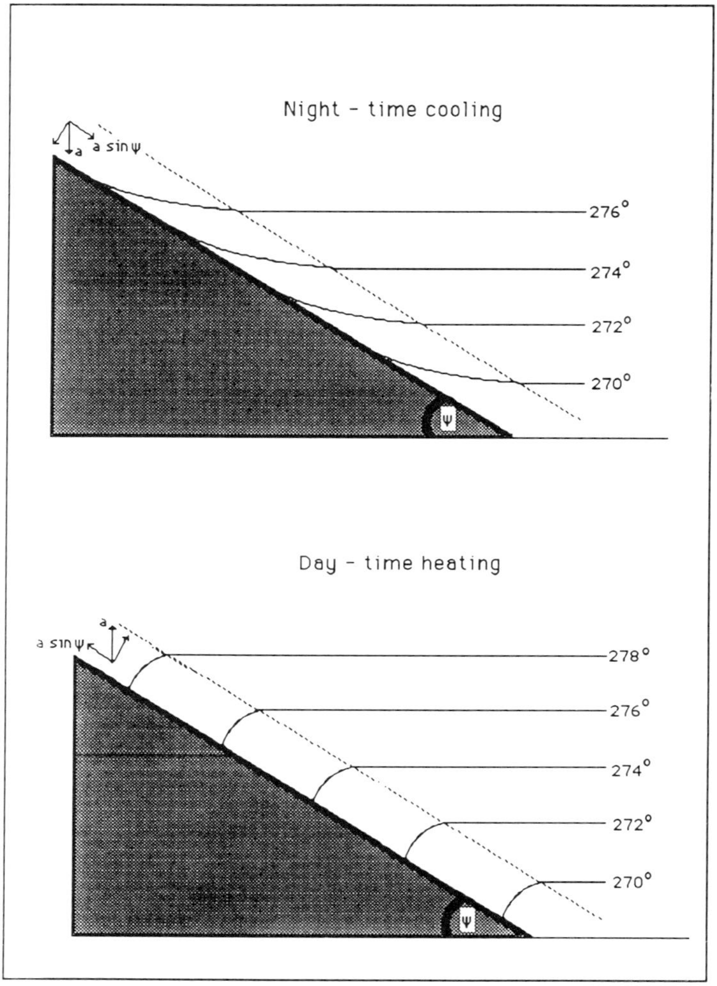

Figure 6.17 Flow over a slightly inclined terrain. The slope is largely exaggerated.

The component of the drainage acceleration, which is parallel to the surface,

$$a_s = a \sin \psi \approx \beta \psi \Theta'' \tag{6.63}$$

will force the air to move along the slope and cause katabatic (down the slope) or anabatic (up the slope) flow. In (6.63), β is the buoyancy parameter, $\beta = g/T_0$, and the temperature deficit is defined as $\Theta'' = \mathbf{\Theta} - \Theta$ (since the altitude is small, it is assumed that the actual temperature and potential temperature are equal, $T \approx \Theta$ and $\mathbf{T} \approx \mathbf{\Theta}$). Moreover, the angle ψ is assumed to be small and therefore, $\sin \psi \approx \psi$.

Equations of the flow defined by the dynamic balance between the Coriolis, frictional, and drainage forces over a slightly inclined terrain have the following form (Sorbjan, 1984):

$$\begin{aligned} \frac{d}{dz} k_m \frac{dU}{dz} + f(V - V_g) - \beta \psi \Theta'' = 0 \\ \frac{d}{dz} k_m \frac{dV}{dz} - f(U - U_g) = 0 \end{aligned} \tag{6.64}$$

The system above is written in a coordinate system following the slope, for which the x-axis is oriented along the slope and downward. The temperature deviation Θ'', which is negative in stable conditions, produces a height-dependent, additional pressure gradient that is the additional component of the thermal wind. This component, acting along the y-axis, is parallel to the contour lines of the terrain and is proportional to deficit Θ'' and to the slope of the ground.

The dimensionless form of (6.64) is

$$\begin{aligned} \kappa \frac{u''}{u_*} &= \frac{dY}{dZ} + \eta_x Z \\ \kappa \frac{v''}{u_*} &= \eta \frac{\Theta''}{T_*} - \frac{dX}{dZ} + \eta_y Z \end{aligned} \tag{6.65}$$

where η_x and η_y are dimensionless components of the thermal wind, defined by (6.55) and $\eta = \mu_0 \psi / \kappa^2$. The primes denote deviations of the boundary layer parameters from the reference values

$$u'' = U - U_{g0} \qquad v'' = V - V_{g0} \tag{6.66}$$

where U_{g0} and V_{g0} are components of the surface geostrophic wind, defined previously by (6.49).

Differentiation of (6.65) leads to the following result:

$$\begin{aligned} \frac{d^2X}{dZ^2} + \frac{Y - \eta R}{K} = \eta_x \\ \frac{d^2}{dZ^2}(Y - \eta R) - \frac{X}{K} = -\eta_y \end{aligned} \tag{6.67}$$

The lower boundary condition, for $Z = Z_0$, can be written in the form

$$X = \cos\delta \qquad Y = \sin\delta \qquad R = R_0 = \alpha_\theta \tag{6.68}$$

$$\frac{\kappa u''}{u_*} = -\frac{\kappa \cos\delta}{C_g} \qquad \frac{\kappa v''}{u_*} = -\frac{\kappa \sin\delta}{C_g} \qquad \frac{\Theta''}{T_*} = \left(\frac{\Theta''}{T_*}\right)_0 \tag{6.69}$$

where δ is the angle between the x-axis and the surface stress vector and c_g is the geostrophic drag coefficient, $C_g = u_*/G$. It can be assumed that at the top

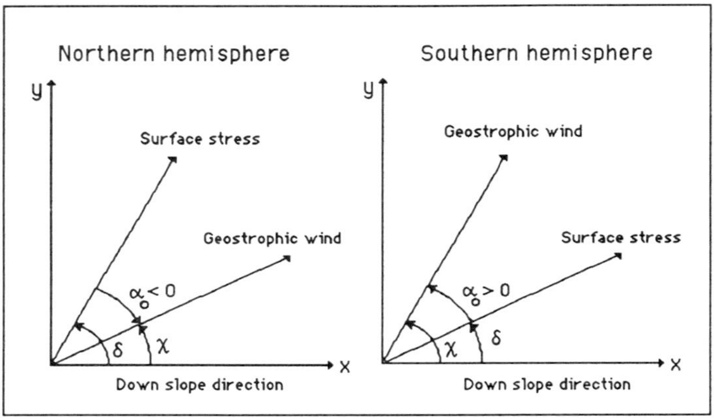

Figure 6.18 Orientation of the angles α_0, δ, and χ.

of the boundary layer ($Z = Z_T$),

$$X = 0 \qquad Y = 0 \qquad R = 0 \qquad \frac{\Theta''}{T_*} = 0 \tag{6.70}$$

If α_0 is the angle between the surface stress vector and the geostrophic wind vector, then $\alpha_0 = \chi - \delta$, independent of the hemisphere, where χ is the angle between the geostrophic wind vector and the downslope direction (Figure 6.18).

The ABL characteristics $\kappa u''/u_*$, $\kappa v''/u_*$, X, and Y can now be obtained through numerical integration of the governing equations. Each depends on the six governing parameters ψ, μ_0, δ, η_x, η_y, and η.

The surface value Θ_0''/T_* is calculated by integrating the third equation in (6.53) from the bottom to the top of the boundary layer:

$$-\frac{\Theta_0''}{T_*} = \int_{Z_0}^{Z_T} \frac{R}{K} dZ + \alpha_\theta \beta_0 \mu_0 (Z_T - Z_0) \tag{6.71}$$

From (6.67) and (6.69), it can be obtained that the geostrophic drag coefficient and the cross-isobar angle are of the form

$$\begin{aligned} C_g &= \frac{\kappa}{\sqrt{[dX/dZ - \eta(\Theta''/T_*)]_0^2 + [dY/dZ]_0^2}} \\ \alpha_0 &= \arctan\left\{ -\frac{[dX/dZ - \eta(\Theta''/T_*)]_0}{(dY/dZ)_0} \right\} - \delta \end{aligned} \tag{6.72}$$

The results of the numerical integration of governing equations, obtained for $p = q = 1$ in (6.58) and for different values of an angle δ, are shown in Figures 6.19 to 6.24. These figures show the substantial influence of slope

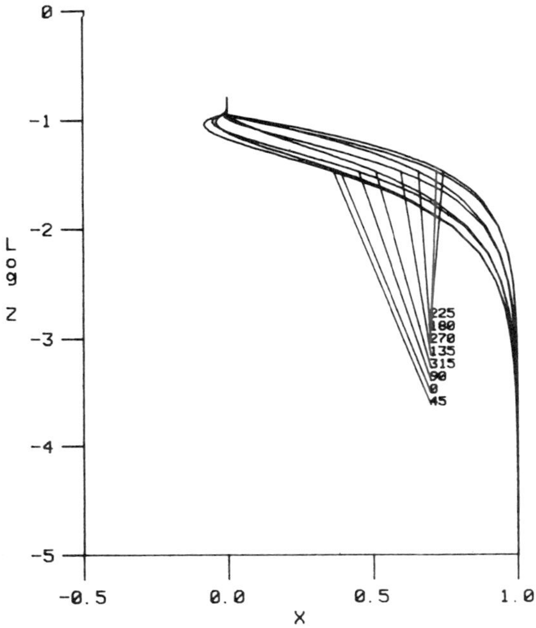

Figure 6.19 Dimensionless component X of the Reynolds stress vector for different values of the angle δ for the southern hemisphere (x-axis along the surface stress vector) $\mu_0 = 50$, $\psi = 0.003$, $Z_0 = 10^{-5}$. (After Sorbjan, 1984. *Journal of Atmospheric Sciences* published by the American Meteorological Society.)

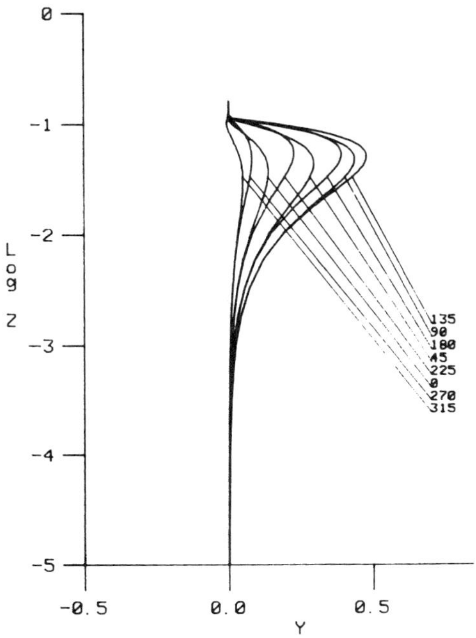

Figure 6.20 Dimensionless component Y of the Reynolds stress vector for different values of the angle δ for the southern hemisphere (x-axis along the surface stress vector) $\mu_0 = 50$, $\psi = 0.003$, $Z_0 = 10^{-5}$. (After Sorbjan, 1984. *Journal of Atmospheric Sciences* published by the American Meteorological Society.)

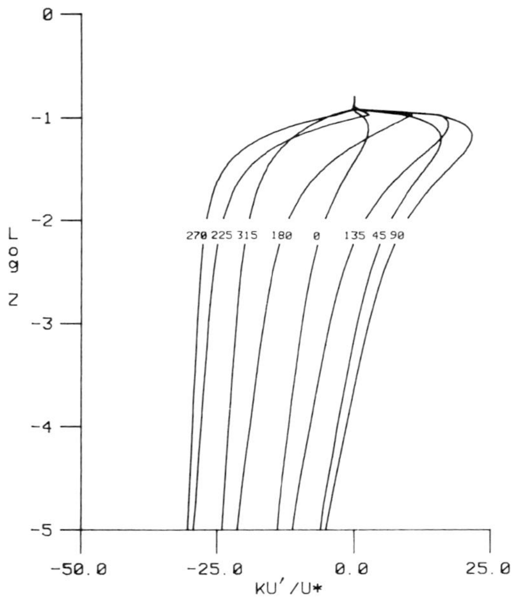

Figure 6.21 Dimensionless velocity defect for different values of the angle δ for the southern hemisphere (x-axis along the surface stress vector) $\mu_0 = 50$, $\psi = 0.003$, $Z_0 = 10^{-5}$. (After Sorbjan, 1984. *Journal of Atmospheric Sciences* published by the American Meteorological Society.)

orientation (compare these with Figures 6.14 and 6.15). Profiles of X, Y and $\kappa(U - U_g)/u_*$, $\kappa(V - V_g)/u_*$ are displayed in Figures 6.19 to 6.22. The cross-isobar angle α_0 in Figure 6.23 exceeds its horizontal case values for $45° < \delta < 225°$, but is smaller in the range $225° < \delta < 45°$. In Figures 6.23 and 6.24, the slope-induced changes of the geostrophic drag coefficient are shown. The x-axis on the plots is parallel to the surface stress vector.

6.3.6 The effects of nonstationarity

A real ABL is nonstationary and exhibits the diurnal variation of the wind vector, temperature, and other parameters. These variations are generated by the diurnal change in the radiative heat input at ground level. During the daytime, solar radiation warms the ground, which becomes warmer than air, thereby directing the turbulent heat flux upward. During the night, the ground loses heat by infrared radiation, becomes cooler than the air, and causes a downward heat flux.

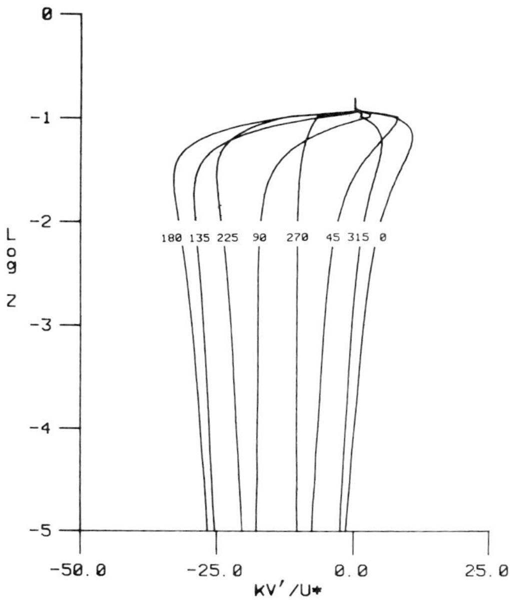

Figure 6.22 Dimensionless velocity defect for different values of the angle δ for the southern hemisphere (x-axis along the surface stress vector), $\mu_0 = 50$, $\psi = 0.003$. (After Sorbjan, 1984. *Journal of Atmospheric Sciences* published by the American Meteorological Society.)

The diurnal dynamics of the ABL has to be studied in terms of equations, including the time derivatives of wind components:

$$\begin{aligned} \frac{\partial U}{\partial t} &= \frac{\partial}{\partial z} k_m \frac{\partial U}{\partial z} + f(V - V_g) \\ \frac{\partial V}{\partial t} &= \frac{\partial}{\partial z} k_m \frac{\partial V}{\partial z} - f(U - U_g) \end{aligned} \tag{6.73}$$

After simple transformations, we can obtain

$$\frac{\partial W}{\partial t} = \frac{\partial}{\partial z} k_m \frac{\partial W}{\partial z} + if\Lambda z e^{ift} \tag{6.74}$$

where W and Λ are defined in the following way:

$$\begin{aligned} W &= \left[(U + iV) - \left(U_{g0} + iV_{g0}\right)\right] e^{ift} \\ \Lambda &= \frac{\partial U_g}{\partial z} + i \frac{\partial V_g}{\partial z} \end{aligned} \tag{6.75}$$

and the geostrophic wind on the surface is assumed to be constant with time.

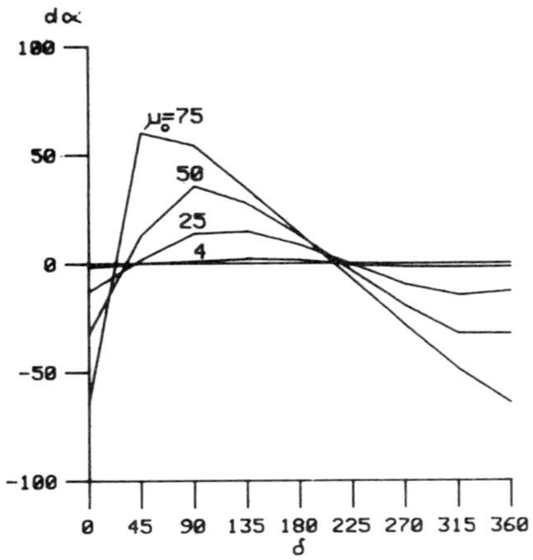

Figure 6.23 Slope-caused deviation of the cross-isobar angle from its horizontal value, for different values of the stability parameter μ_0, $\psi = 0.003$, $Z_0 = 10^{-5}$. (After Sorbjan, 1984. *Journal of Atmospheric Sciences* published by the American Meteorological Society.)

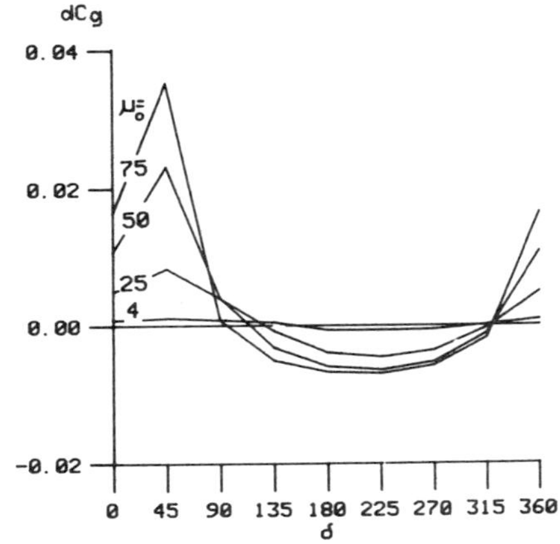

Figure 6.24 Slope-caused deviation of the geostrophic drag coefficient from its horizontal value, for different values of the stability parameter μ_0, $\psi = 0.003$, $Z_0 = 10^{-5}$. (After Sorbjan, 1984. *Journal of Atmospheric Sciences* published by the American Meteorological Society.)

Equation (6.74) can be solved for a given eddy viscosity k_m and boundary conditions (see the Appendix). In the simplest case k_m can be assumed constant, which leads to an analytical solution. However, this approach is rather formal and does not reflect the influences of the diurnal variation of the thermal regime in the atmospheric boundary layer. A more realistic model has to take thermal stability into consideration. An example of such a model is the simulation of a daytime ABL by Sun and Chang (1986a). Their model employs (6.73) with the following additional equations:

$$\frac{\partial \Theta}{\partial t} = \frac{\partial}{\partial z} k_h \frac{\partial \Theta}{\partial z}$$

$$\frac{\partial q}{\partial t} = \frac{\partial}{\partial z} k_h \frac{\partial q}{\partial z}$$

$$\frac{\partial E}{\partial t} = -\frac{g}{\Theta_0} k_h \frac{\partial \Theta}{\partial z} + k_m \left[\left(\frac{\partial U}{\partial z} \right)^2 + \left(\frac{\partial V}{\partial z} \right)^2 \right] + 0.2 \frac{\partial}{\partial z} k_m \frac{\partial E}{\partial z} + 0.41 \frac{E^{3/2}}{l}$$

$$k_m = 0.1 l E^{1/2} \tag{6.76}$$

where q is a mixing ratio, E the turbulent kinetic energy, and l a mixing length.

The system of equations (6.76) is closed by an empirical equation for the mixing length in the form $l = 0.45 z_i [1 - \exp(-4z/z_i) - 0.0003 \exp(8z/z_i)]$.

For stable stratification this equation is replaced by the equation $l = 0.76\{E/[\beta(d\Theta/dz)]\}^{1/2}$, for $l < \Delta z$, and $l = \Delta z$, for $l \geq \Delta z$, where Δz is the vertical coordinate increment.

After Deardorff (1980), it is assumed that

$$k_h = \begin{cases} k_m & \text{for stable stratification} \\ 3k_m & \text{for unstable stratification} \end{cases} \tag{6.77}$$

At the upper boundary, all turbulence fluxes and wind shear are set to zero and the temperature and moisture are fixed. On the surface, the turbulent fluxes are assumed to be known. The heat flux at the ground surface is prescribed as a function of time. Data from the Wangara experiment at 0900 local time on day 33 (August 16, 1967) are used as the initial conditions for the model (Figure 6.26a). The observed potential–temperature profiles in Figure 6.25a are well simulated during the daytime of day 33 (Figure 6.26a).

The model reproduces the shallow superadiabatic layer near the ground surface, a deep well-mixed layer in the middle, and an inversion layer on top. Also, the simulated depth of the mixed layer is comparable to that observed. The simulated temperature lapse rate $\partial\Theta/\partial z$ in the mixed layer is slightly negative. In the real mixed layer, the temperature lapse rate $\partial\Theta/\partial z$ is near zero or even slightly positive. Therefore, the real heat flux in this layer is not directed down the gradient of potential temperature (i.e., toward smaller values of Θ), but has a *countergradient* orientation. The model discussed is unable to simulate the *countergradient flux* portion of the mixed layer. This flaw can be avoided by employing, as suggested by Deardorff (1966), the following expression instead of the flux-gradient relation (6.12):

$$\overline{w'\theta'} = -k_h\left(\frac{\partial\Theta}{\partial z} - \gamma_c\right) \tag{6.78}$$

where $\gamma_c \sim 0.7 \times 10^{-3}$ °C m^{-1} [compare it with (3.42)]. Figure 6.26e shows that the calculated vertical heat flux in a convective layer is a linear function of height, positive by the surface and slightly negative near the inversion.

The calculated humidity profiles shown in Figure 6.26b indicate that the mixing ratio decreases only slightly with height in the mixed layer. This is consistent with observational data. As the mixed layer grows, the surface supply of moisture is insufficient to maintain the early morning mixing ratio near the ground, and both calculated and observed values decrease with time.

When a cool, moist air parcel penetrates the overlying stable layer from below, a warm, dry air parcel from the stable layer moves into the mixed layer to compensate. As a result, heat flux in the topping inversion is downward, but moisture flux is upward.

Calculated x and y components of the mean wind are presented in Figures 6.26c and d, respectively. The computed U is close to the observation until 1500 EST. At 1800 EST, the computed U is about -6.5 m/s in the

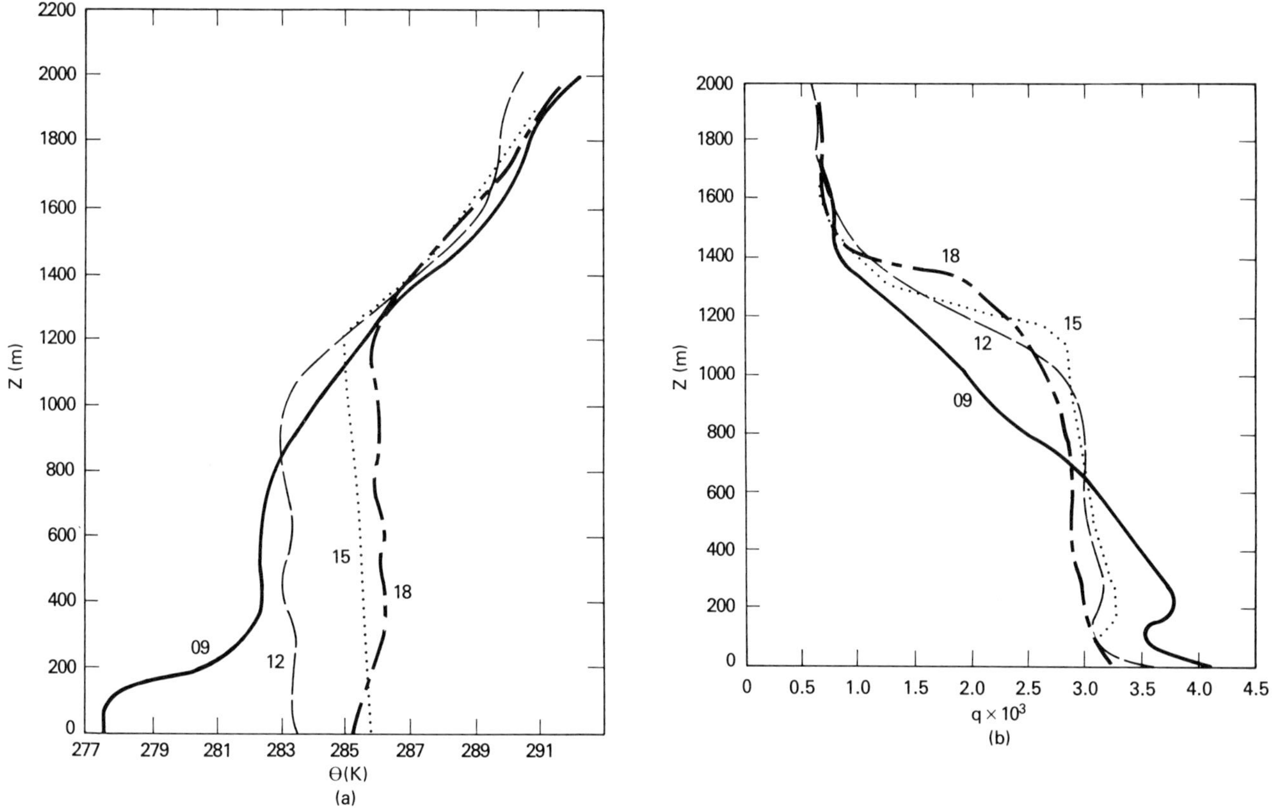

Figure 6.25 Observed profiles of: (a) potential temperature θ during day 33 of the Wangara experiment; (b) mixing ratio q; (c) U component of wind velocity; (d) V component of wind velocity. (After Sun and Chang, 1986a. *Journal of Climate and Applied Meteorology* published by the American Meteorological Society.)

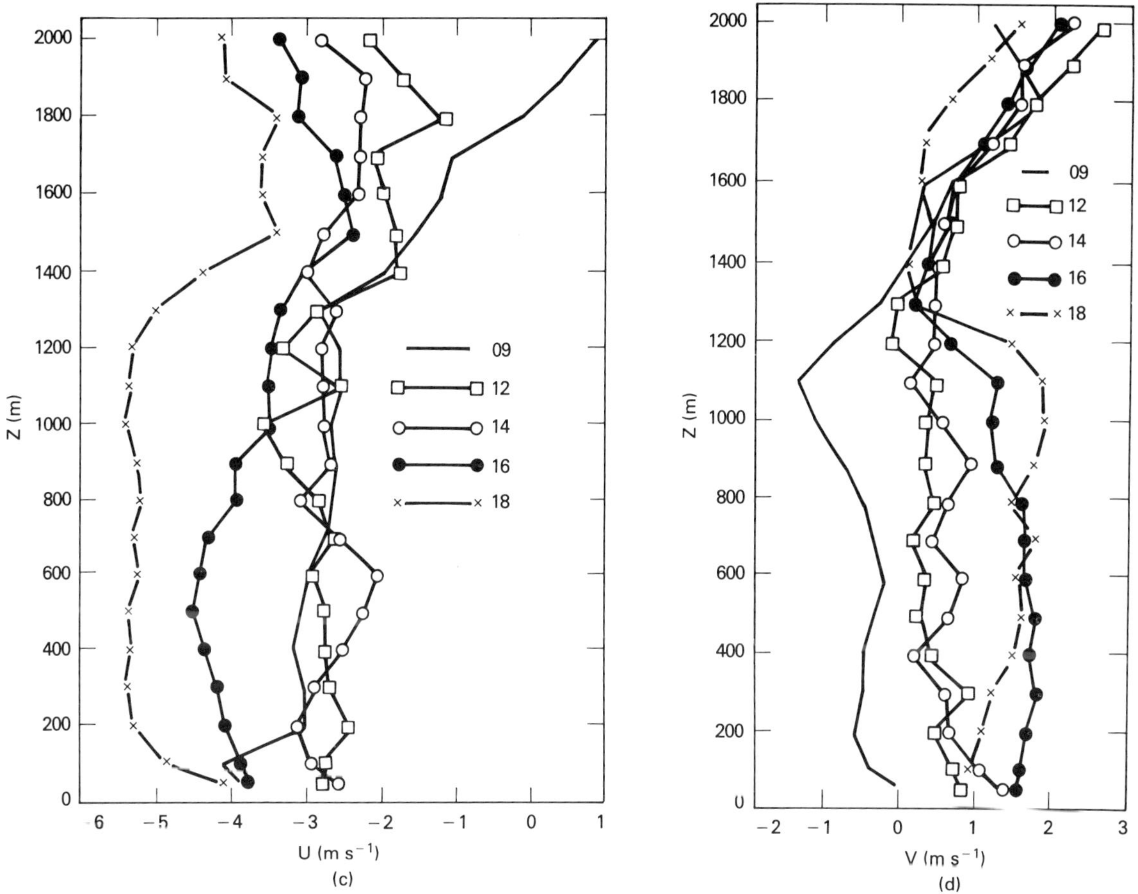

Figure 6.25 *continued*

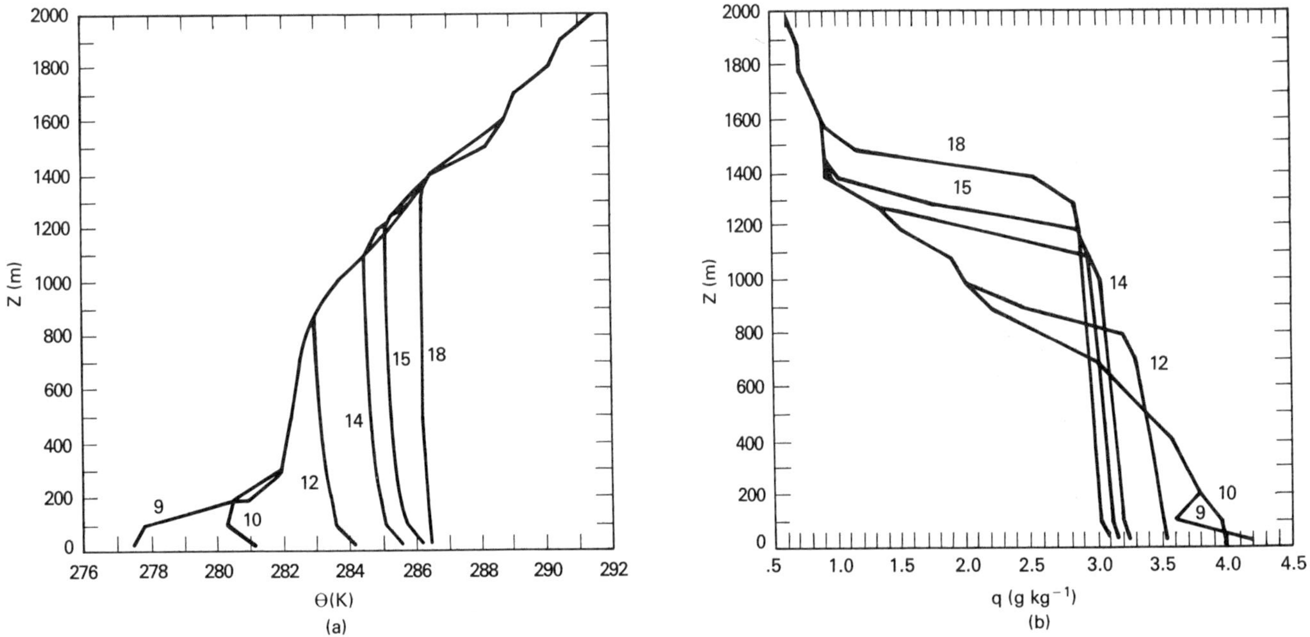

Figure 6.26 Calculated profiles of: (a) mean virtual temperature during day 33 of the Wangara experiment; (b) mean mixing ratio; (c) U component of mean wind; (d) V component of mean wind; (e) turbulent heat flux; (f) moisture flux. (After Sun and Chang, 1986a. *Journal of Climate and Applied Meteorology* published by the American Meteorological Society.)

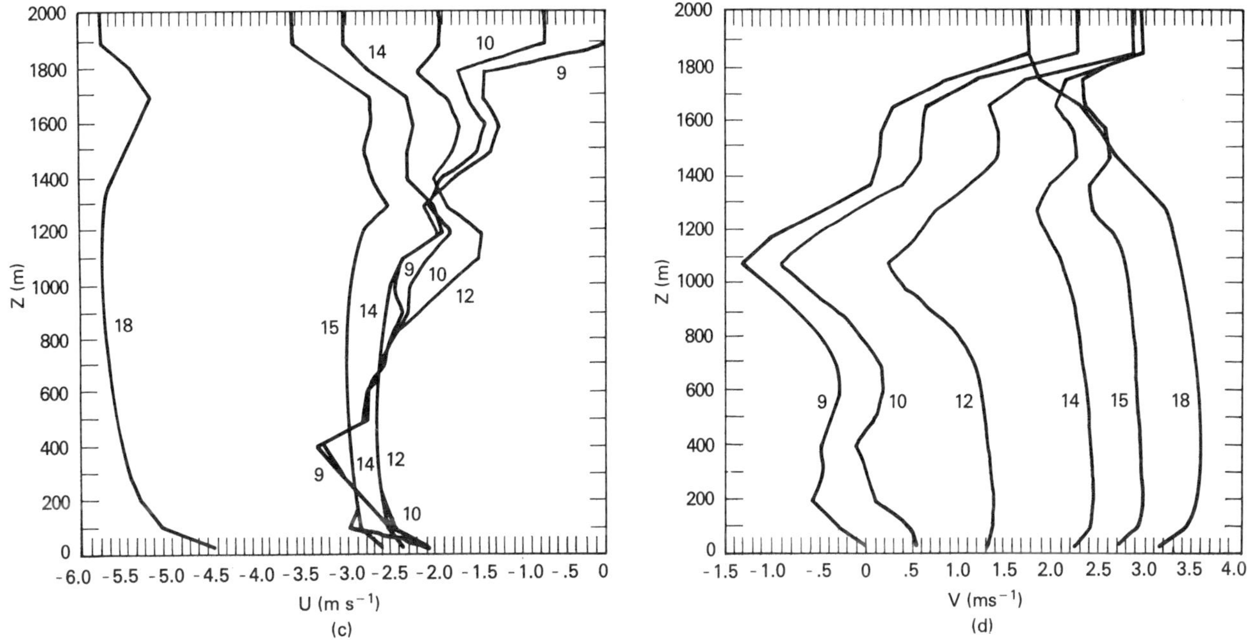

Figure 6.26 *continued*

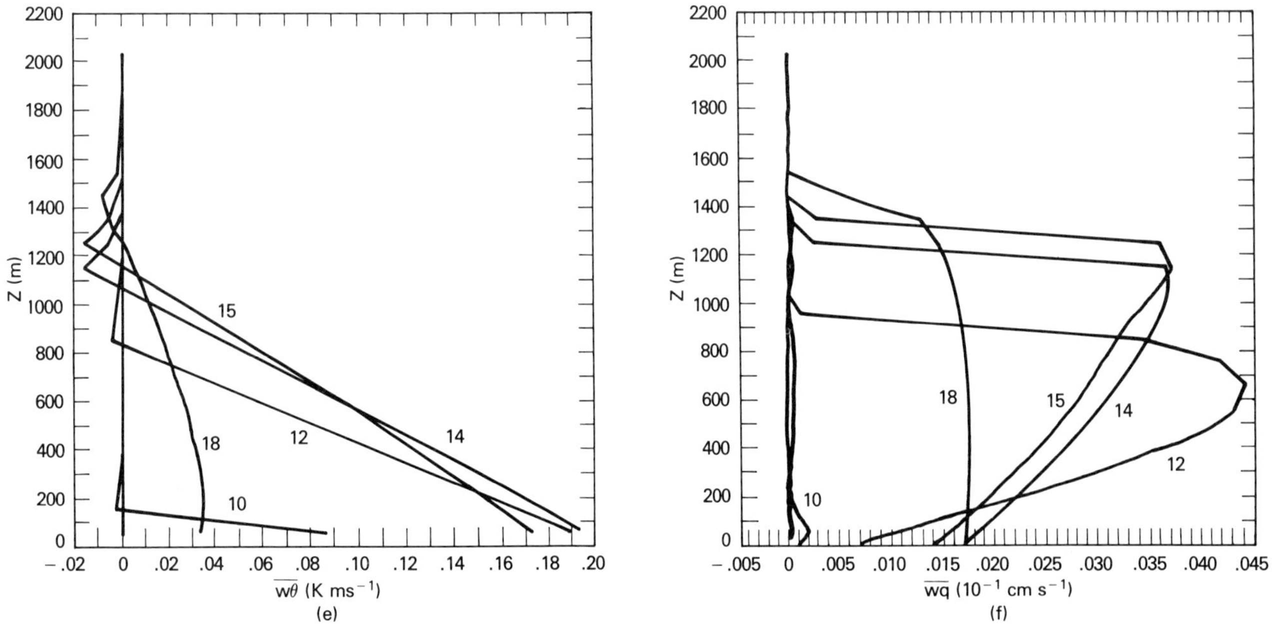

Figure 6.26 *continued*

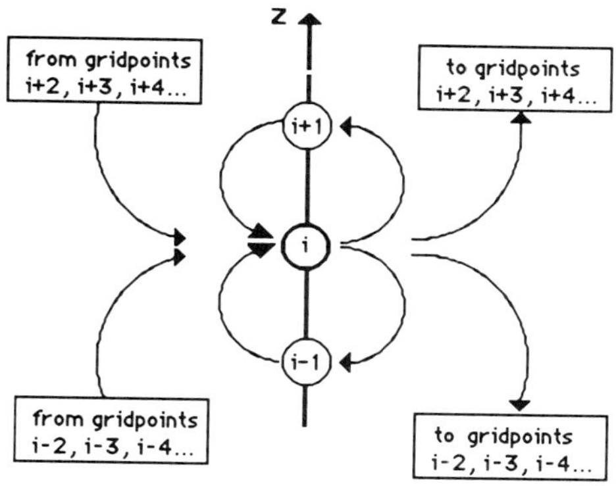

Figure 6.27 Schematic of the turbulent mixing between air in grid point i and all other grid points.

mixed layer, also comparable to the observation. However, the model minimum value of $U = -5$ m/s occurs at the height of 1700 m, while the observational value is only about -3.5 m/s at this height. The observed wind in Figure 6.25c, d is not as well mixed as the computed wind. The magnitude of the computed V is also larger than that observed in the late afternoon. At 1800 EST, the calculated y-component wind speed is about 3.5 m/s in the mixed layer, in contrast to the observed value of 1.6 m/s.

6.4 NONLOCAL CLOSURES

In contrast to local diffusive methods such as the K-theory closure, nonlocal closure is based on the concept that all turbulent eddies can transport fluid elements across finite distances. Nonlocal closure models were applied to model puff and plume dispersion (e.g., Berkowicz and Prahm, 1980), mixing in the convective boundary layer (e.g., Fiedler, 1984), and atmospheric and oceanic turbulence (e.g., Stull, 1984; Stull and Kraus, 1987).

The idea of nonlocal closure models can be explained by considering the mixing of conservative quantity S, between grid points i and j (Figure 6.27). The forecast of the conservative quantity S, one time step in advance, at any grid point i can be obtained (notice the analogy with Markov's process, e.g., Moore and Yackel, 1974):

$$S_i(t + \Delta t) = \sum_{j=1}^{n} c_{ij} S_j(t) \tag{6.79}$$

where coefficients c_{ij} describe mixing between all possible pairs of grid points (Stull, 1984). Notice that if c_{ij} is the fraction of air entering grid point i from grid point j, then, by definition, the conservation of air mass requires that the sum of all mixing fractions be unity:

$$\sum_{j=1}^{n} c_{ij} = 1 \tag{6.80}$$

Also, if the amount of the quantity initially in grid point j is conserved as it mixes out of j and into the other grid points i, the conservation of tracer amount requires that

$$\sum_{i=1}^{n} c_{ij} = 1 \tag{6.81}$$

For a one-dimensional, continuous-in-space domain of turbulence between $z = a$ and $z = b$, (6.79) can be rewritten as

$$\begin{aligned} S(z, t + \Delta t) &= \int_a^b C(Z, z) S(Z, t)\, dZ \\ &= \int_{a-z}^{b-z} C(z, \zeta, \Delta t) S(z + \zeta, t)\, d\zeta \end{aligned} \tag{6.82}$$

where $C(z, \zeta, \Delta t)$ is a "transilient" coefficient and $\zeta = Z - z$ is a separation distance.

Substracting $S(z, t)$ from both sides and applying an integral analog of (6.80) in the form

$$\int_{a-z}^{b-z} C(z, \zeta, \Delta t)\, d\zeta = 1 \tag{6.83}$$

yields

$$S(z, t + \Delta t) - S(z, t) = \int_{a-z}^{b-z} C(z, \zeta, \Delta t)[S(z + \zeta, t) - S(z, t)]\, d\zeta \tag{6.84}$$

Both sides of the equation can be divided by Δt. As Δt approaches zero, the left side of this equation defines the time derivative dS/dt, while the right side defines a new coefficient γ:

$$\frac{dS}{dt}(z, t) = \int_{a-z}^{b-z} \gamma(z, \zeta, t)[S(z + \zeta, t) - S(z, t)]\, d\zeta \tag{6.85}$$

The coefficient γ is a function of separation distance, ζ. For nonhomogeneous turbulence, it is also a function of z; for nonstationary turbulence, it is a function of time. The right-hand side of (6.85) can be interpreted as the flux divergence:

$$\frac{dF}{dz} = \int_{a-z}^{b-z} \gamma(z, \zeta, t)[S(z + \zeta, t) - S(z, t)]\, d\zeta \tag{6.86}$$

Parametrization (6.86) can be applied in the left-hand sides in (2.69). In the case of the momentum equations, S in (6.86) has to be replaced by wind components U and V, while for Equation (2.70) S has to be replaced by temperature Θ.

Assuming that the turbulence flux F is zero outside the turbulent domain and integrating (6.86) from one edge of the domain ($z = a$, for example) to any height $z = z_1$ gives

$$F(z_1) = -\int_a^{z_1} \int_{a-z}^{b-z} \gamma(z, \zeta, t)[S(z + \zeta, t) - S(z, t)]\, d\zeta\, dz \tag{6.87}$$

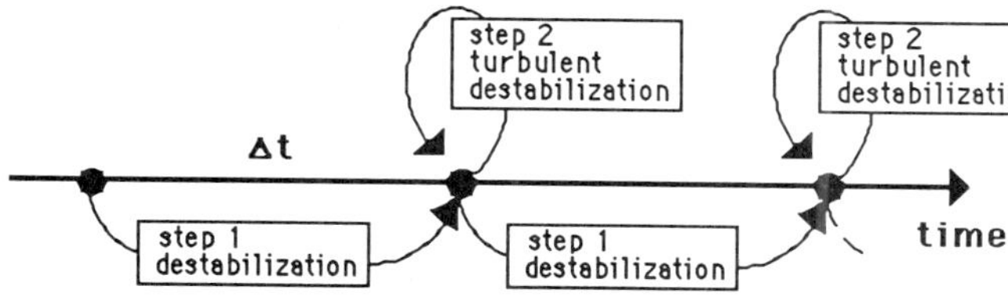

Figure 6.28 Time-splitting algorithm.

Equation (6.87) is equivalent to

$$F(z_1) = -\int_{a-z_1}^{b-z_1} K(\zeta, z_1) \frac{dS}{d\zeta}(\zeta, t)\, d\zeta \tag{6.88}$$

This yields a nonlocal generalization of (6.11)–(6.12). The spectral diffusivity K has to be parametrized. It can be done in various ways (e.g., Berkowicz and Prahm, 1980; Fiedler, 1984; Berkowicz, 1984; Stull and Driedonks, 1987). Within the boundary layer model, nonlocal closure can be implemented by splitting each time step into two parts, destabilization and mixing (Stull, 1984). During the first part (Figure 6.28), turbulence is excluded, while the remaining external forcings and internal processes that apply to the flow are computed. The flow may become destabilized as a result of all of these forcings. In the next part of the time step, the necessary mixing of all conserved variables is performed, employing (6.85).

6.5 HIGHER-ORDER CLOSURES

The flaws of the K-closures led to the construction of a variety of second-order closure models (Mellor and Yamada, 1974; Yamada and Mellor, 1975; Wyngaard et al., 1974a; Wyngaard, 1975; André et al., 1978; Rao and Snodgrass, 1978). Second-order closure models consist of equations of mean flow (2.56) which are supplemented by additional equations of second moments (see Section 3.4). The second-moment equations are closed on the level of third or higher moments. The proper choice of closure formulas, simplifying assumptions, and empirical constants creates a very difficult task that requires knowledge of boundary layer processes, working experience, access to suitable test data, and creativity.

Perhaps the most widely used second-order model in meteorological applications was developed by Mellor and Yamada (1974). Their approach represents a hierarchy of different closure model formulations, differing in the number of approximations made in the second-moment equations. In the models of level 4, all second moments are described by 10 time-dependent differential equations. The models of level 3 employ only two differential equations, for temperature variance and turbulent kinetic energy. The remain-

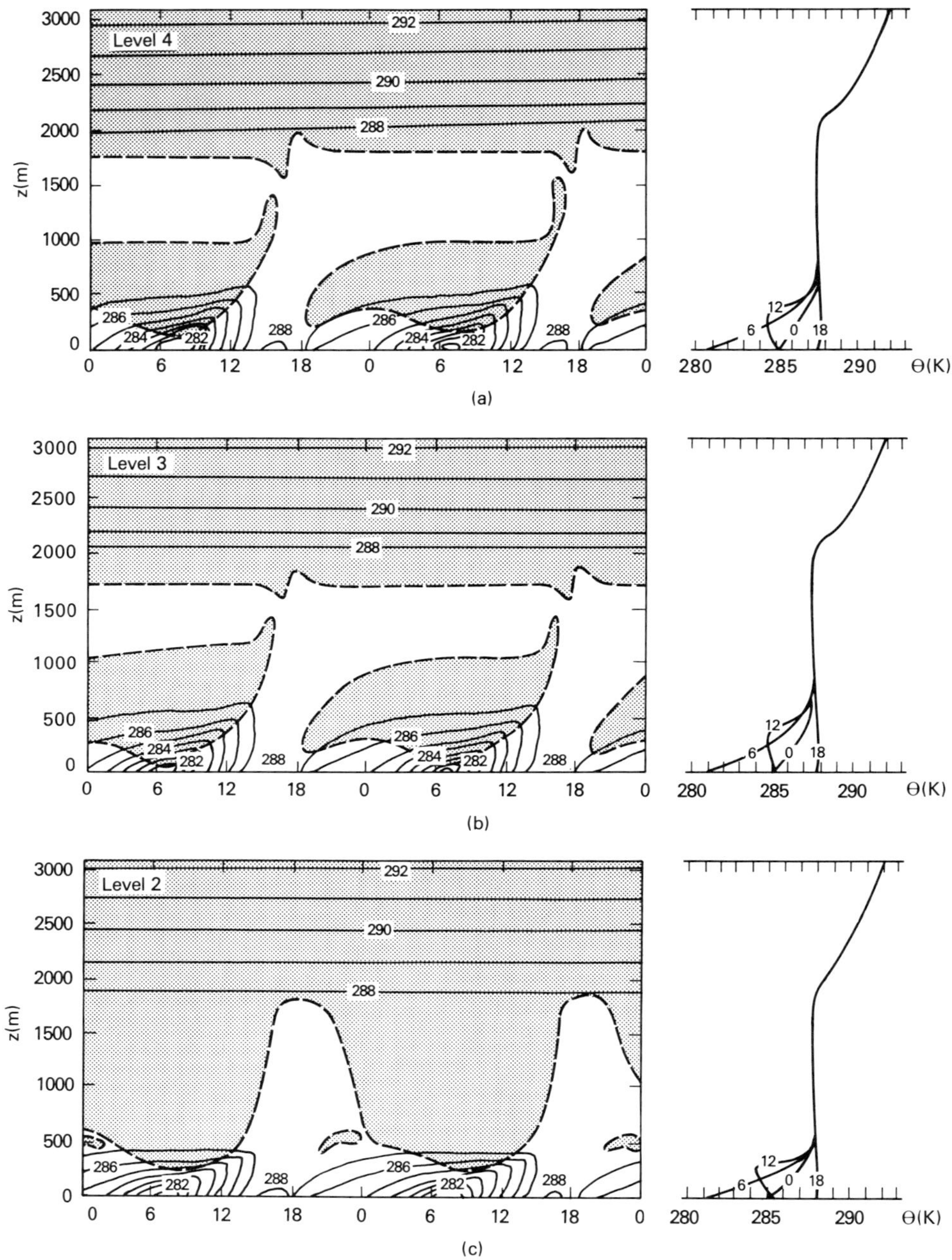

Figure 6.29 Potential temperature simulated by models of the (a) second, (b) third, and (c) fourth levels of closure. The shaded region corresponds to Ri ≥ 0.21. For level 2, this is identically a region of zero turbulence intensity. (After Mellor and Yamada, 1974. *Journal of Atmospheric Sciences* published by the American Meteorological Society.)

ing moments are described by algebraic equations obtained by simplification of full variance and covariance differential equations. The models of level 2 use only algebraic equations. The models of level 1 are equivalent to K-models. Comparison results given by different-level models is shown in Figure 6.29 (some other results of second-order closure models are presented in figures in Section 3.5).

It is not our intention to review second-order closure models in this book, but rather, to focus on major conceptual ideas developed over past years. This will be done by discussing a few closure schemes applied to second-moment equations in Section 3.4.

6.5.1 The momentum flux equation

Equation (3.25) has to be closed with respect to the following unknown quantities:

$$\epsilon_{ij} = \nu\left(\overline{u_j'\frac{\partial^2 u_i'}{\partial x_\alpha^2}} + \overline{u_i'\frac{\partial^2 u_j'}{\partial x_\alpha^2}}\right)$$

$$\Pi_{ij} = \frac{1}{\rho_0}\left(\overline{u_i'\frac{\partial p'}{\partial x_j} + u_j'\frac{\partial p'}{\partial x_i}}\right) \tag{6.89}$$

$$S_{ijk} = \overline{u_i' u_j' u_k'}$$

6.5.1.1 The dissipation term ϵ_{ij} in (6.89) can be expressed in the form

$$\epsilon_{ij} = \nu\overline{\frac{\partial u_i'}{\partial x_\alpha}\frac{\partial u_j'}{\partial x_\alpha}} \tag{6.90}$$

A similar transformation was performed in (3.30).

Dissipation is generated by small-scale turbulence, for which the assumption of isotropy (identical properties in all directions) is valid. This assumption leads to

$$\epsilon_{ij} = \tfrac{1}{3}\epsilon\,\delta_{ij} \tag{6.91}$$

where δ_{ij} represents Kronecker's delta and ϵ is the total turbulent energy dissipation. The equation of total turbulent energy dissipation ϵ is discussed in Section 6.5.4.

6.5.1.2 The second term in (6.89), Π_{ij}, is often expressed in the form

$$\Pi_{ij} = \Pi_{ij}^{(1)} - \Pi_{ij}^{(2)} \tag{6.92}$$

where

$$\Pi_{ij}^{(1)} = \frac{1}{\rho_0}\left(\overline{\frac{\partial p'u_i'}{\partial x_j} + \frac{\partial p'u_j'}{\partial x_i}}\right) \tag{6.93}$$

$$\Pi_{ij}^{(2)} = \frac{1}{\rho_0}\overline{p'\left(\frac{\partial u_i'}{\partial x_j} + \frac{\partial u_j'}{\partial x_i}\right)} \tag{6.94}$$

In the incompressible case, $\Pi_{\alpha\alpha}^{(2)}$ vanishes in the turbulent energy equation. Therefore, its role is to redistribute energy between turbulent energy components. For this reason, $\Pi_{ij}^{(2)}$ is called the *energy redistribution term.* If there is no mean shear and buoyancy influence, this term will tend to make the turbulence more isotropic.

Unfortunately, there is little available information regarding values of $\Pi_{ij}^{(1)}$ in the atmosphere. In many models, this term has vanished. Direct and indirect (imbalance in the budget equation) measurements indicate that in the unstable surface layer the pressure term $\partial\overline{(p'u_3')}/\partial x_3$ is a significant source term in the turbulent kinetic energy budget. In unstable conditions, the $\overline{p'u_3'}$ profile [compare with (4.83)] is zero near the surface and then becomes increasingly negative with increasing height (Wyngaard, 1981).

The pressure field p' can be obtained by applying the divergence operator to the first equation in (3.24) (Rotta, 1951):

$$\frac{1}{\rho_0}\nabla^2 p' = -\left(\frac{\partial u_\alpha'}{\partial x_\beta}\frac{\partial u_\beta'}{\partial x_\alpha} - \overline{\frac{\partial u_\alpha'}{\partial x_\beta}\frac{\partial u_\beta'}{\partial x_\alpha}}\right) - 2\frac{\partial U_\alpha}{\partial x_\beta}\frac{\partial u_\beta'}{\partial x_\alpha} + g_\alpha \vartheta \frac{\partial \theta'}{\partial x_\alpha} - \epsilon_{\alpha\beta 3} f \frac{\partial u_\beta'}{\partial x_\alpha} \tag{6.95}$$

where $\vartheta = 1/\Theta_0$ and $g_\alpha = g\delta_{\alpha 3}$.

In the equation above, there are no terms containing time derivatives. These terms vanish as a result of incompressibility. Equation (6.95) contains four terms indicating four different factors influencing pressure fluctuations. The first term on the right-hand side is nonlinear with respect to fluctuations of wind velocity and represents *turbulent-turbulent* interactions (also called the *return-to-isotropy term*). The remaining terms, called *rapid terms*, are linear and represent interactions with the mean shear, gravity, and rotation of the earth. The rapid terms are so-named because they involve external influences that can be changed rapidly (e.g., in laboratory situations). On the basis of scaling arguments for the terms existing on the right side of (6.95), the Coriolis term is found to be appreciably smaller than the mean shear term and can be omitted.

Equation (6.95) is a second-order elliptic differential (Poisson) equation. Its solution can be differentiated and correlated with velocity fluctuations to demonstrate that the expression for $\Pi_{ij}^{(2)}$ should contain three groups of terms: turbulent-turbulent terms $\Pi_{ij}^{(TT)}$, mean shear term $\Pi_{ij}^{(S)}$, and buoyancy term $\Pi_{ij}^{(B)}$. The contributions $\Pi_{ij}^{(TT)}$, $\Pi_{ij}^{(S)}$, and $\Pi_{ij}^{(B)}$ are usually modeled separately.

For $\Pi_{ij}^{(TT)}$, Rotta (1951) proposed

$$\Pi_{ij}^{(TT)} = c_{ij\alpha\beta}\overline{u'_\alpha u'_\beta} \tag{6.96}$$

where $c_{ij\alpha\beta}$ is a coefficient tensor.

For $\Pi_{ij}^{(S)}$ it is suggested that (Launder et al., 1975)

$$\Pi_{ij}^{(S)} = c'_{ij\alpha\beta}P_{\alpha\beta} \tag{6.97}$$

where $c'_{ij\alpha\beta}$ is a coefficient tensor, and (Launder, 1976)

$$P_{ij} = \overline{u'_i u'_\alpha}\frac{\partial U_j}{\partial x_\alpha} + \overline{u'_j u'_\alpha}\frac{\partial U_i}{\partial x_\alpha} \tag{6.98}$$

Often P_{ij} is also expressed in a simpler form (Rotta, 1951):

$$P_{ij} = q^2\left(\frac{\partial U_j}{\partial x_i} + \frac{\partial U_i}{\partial x_j}\right) \tag{6.99}$$

where $q^2 = \overline{u'_\alpha u'_\alpha} = 2E$ (doubled turbulent kinetic energy).

For $\Pi_{ij}^{(B)}$ it is accepted that

$$\Pi_{ij}^{(B)} = c''_{ij\alpha\beta}g_\alpha\vartheta\overline{u'_\beta\theta'} \tag{6.100}$$

where $c''_{ij\alpha\beta}$ is a coefficient tensor.

Approximating the fourth-order tensors in (6.96)–(6.100) as isotropic and proportional to $\delta_{ij}\delta_{\alpha\beta}$ and its permutations, while also considering the continuity equation and symmetry in i and j, yields (Zeman, 1981):

$$\begin{aligned}\Pi_{ij} &= \Pi_{ij}^{(TT)} + \Pi_{ij}^{(S)} + \Pi_{ij}^{(B)} \\ &= C_1\left(\overline{u'_i u'_j} - \tfrac{1}{3}\delta_{ij}q^2\right) + C_2\left(P_{ij} - \tfrac{1}{3}P\delta_{ij}\right) \\ &\quad - C_3 g_\beta\vartheta\left(\overline{u'_i\theta'}\delta_{\beta j} + \overline{u'_j\theta'}\delta_{i\beta} - \tfrac{2}{3}\overline{u'_\beta\theta'}\delta_{ij}\right)\end{aligned} \tag{6.101}$$

where $P = P_{\alpha\alpha}$.

On the basis of dimensional analysis, we obtain $C_1 \sim C'_1/\tau$, where τ is a turbulent time scale. This scale must be related to the mean properties of the flow. The simplest relation can be obtained by assuming that

$$\frac{1}{\tau} = \frac{\epsilon}{q^2} \tag{6.102}$$

which leads to

$$\begin{aligned}\Pi_{ij} = C'_1\frac{\epsilon}{q^2}\left(\overline{u'_i u'_j} - \tfrac{1}{3}\delta_{ij}q^2\right) + C_2\left(P_{ij} - \tfrac{1}{3}P\delta_{ij}\right) \\ - C_3 g_\beta\vartheta\left(\overline{u'_i\theta'}\delta_{\beta j} + \overline{u'_j\theta'}\delta_{i\beta} - \tfrac{2}{3}\overline{u'_\beta\theta'}\delta_{ij}\right)\end{aligned} \tag{6.103}$$

Unfortunately, (6.103) is inconsistent with the majority of boundary layer data due to departures from isotropy, lack of homogeneity, and the buoyancy effects in the atmosphere. To achieve better agreement with the data, various authors arbitrarily alter the invariant constants, which destroys their universality (Zeman and Lumley, 1976; Wyngaard, 1975).

6.5.1.3 The term $S_{ijk} = \overline{u_i'u_j'u_k'}$ is called the *diffusive* or *transport term.* Transport terms are not significant in neutral or stably stratified planetary boundary layers. However, in convective boundary layers, these terms often govern the flow dynamics and drive the entrainment processes (Tennekes, 1973; Zeman, 1975).

Employing the equation for velocity fluctuations (3.24), the following equation for S_{ijk} can be obtained (Davydov, 1959):

$$\begin{aligned}\frac{\partial S_{ijk}}{\partial t} &+ U_\alpha \frac{\partial S_{ijk}}{\partial x_\alpha}\\ &= -\left(S_{ij\alpha}\frac{\partial U_k}{\partial x_\alpha} + S_{\alpha jk}\frac{\partial U_i}{\partial x_\alpha} + S_{i\alpha k}\frac{\partial U_j}{\partial x_\alpha}\right) - \frac{\partial}{\partial x_\alpha}\overline{u_i'u_j'u_k'u_\alpha'}\\ &+ \left(\overline{u_i'u_\alpha'}\frac{\partial}{\partial x_\alpha}\overline{u_j'u_k'} + \overline{u_j'u_\alpha'}\frac{\partial}{\partial x_\alpha}\overline{u_i'u_k'} + \overline{u_k'u_\alpha'}\frac{\partial}{\partial x_\alpha}\overline{u_i'u_j'}\right)\\ &+ \frac{1}{\rho_0}\left(\frac{\partial}{\partial x_i}\overline{p'u_k'u_j'} + \frac{\partial}{\partial x_j}\overline{p'u_k'u_i'} + \frac{\partial}{\partial x_k}\overline{p'u_i'u_j'}\right)\\ &+ \vartheta\left(g_i\overline{\theta'u_k'u_j'} + g_j\overline{\theta'u_k'u_i'} + g_k\overline{\theta'u_i'u_j'}\right)\end{aligned} \tag{6.104}$$

The terms on the right side of the above equation are, respectively, advection, shear production, turbulent diffusion, turbulent production, pressure interaction, and buoyancy. Here, the Coriolis and molecular terms were omitted as being small. The equation for the third-order moment S_{ijk} is unclosed and must be completed by an additional hypothesis.

Assuming Gaussian distribution of the velocity fluctuations in energy containing range, the following expression can be obtained (for the Gaussian distribution, the fourth moment is equal to the tripled square of the second moment):

$$\overline{u_i'u_j'u_k'u_l'} = \overline{u_i'u_j'}\,\overline{u_k'u_l'} + \overline{u_i'u_k'}\,\overline{u_j'u_l'} + \overline{u_i'u_l'}\,\overline{u_j'u_k'} \tag{6.105}$$

Homogeneous turbulence is observed to be Gaussian, and departure from Gaussian behavior is associated with inhomogeneity. Therefore, (6.105) is correct only in homogeneous situations.

Moments containing pressure gradients can be treated as being equal to

$$\frac{1}{\rho_0}\left(\frac{\partial}{\partial x_i}\overline{p'u_k'u_j'} + \frac{\partial}{\partial x_j}\overline{p'u_k'u_i'} + \frac{\partial}{\partial x_k}\overline{p'u_i'u_j'}\right) = C_6\frac{\epsilon}{q^2}\overline{u_i'u_j'u_k'} \tag{6.106}$$

Notice that the term before the third moment on the right-hand side of (6.106) is a reciprocal of the time scale defined by (6.102).

Assuming in (6.104) stationarity, horizontal homogeneity, and mean motion in the horizontal plane, and taking into consideration the hypothesis (6.105)–(6.106), yields

$$\overline{u_i'u_j'u_k'} = -\frac{q^2}{C_6\epsilon}\left(\overline{u_i'u_\alpha'}\frac{\partial}{\partial x_\alpha}\overline{u_k'u_j'} + \overline{u_\alpha'u_j'}\frac{\partial}{\partial x_\alpha}\overline{u_i'u_j'} + \overline{u_k'u_\alpha'}\frac{\partial}{\partial x_\alpha}\overline{u_i'u_k'}\right) + \frac{\vartheta q^2}{C_6\epsilon}\left(g_i\overline{\theta'u_j'u_k'} + g_j\overline{\theta'u_k'u_i'} + g_k\overline{\theta'u_i'u_j'}\right) \tag{6.107}$$

If buoyancy terms are omitted, (6.107) is equivalent to the relation proposed by Hanjalic and Launder (1972). Many authors prefer to use a simpler form:

$$\overline{u_i'u_j'u_k'} = -\frac{q^4}{3C_6\epsilon}\left(\frac{\partial\overline{u_j'u_k'}}{\partial x_i} + \frac{\partial\overline{u_i'u_k'}}{\partial x_j} + \frac{\partial\overline{u_i'u_j'}}{\partial x_k}\right) \tag{6.108}$$

The expression obtained is in flux-gradient form and the term preceding the parentheses can be treated as eddy viscosity. A minus ensures that diffusion is down the gradient or from regions with high turbulence intensity to regions with low intensity. The parametrization above has been found to work well for laboratory shear flow simulations. In geophysical flows, however, buoyancy forces often play a very important role in turbulent transport. According to Wyngaard (1983), in the covective surface layer, down-gradient diffusion is a particularly poor approximation for vertical velocity variance, which has been shown empirically to be transported up-gradient. The vertical flux of turbulent kinetic energy is everywhere positive in the convective ABL, but nevertheless, down-gradient diffusion models predict negative values of this moment near the surface.

Another obvious flaw in this model can be shown by letting one component in (6.108) (say, u_i') approach zero. Since $(u_j'u_k')$ need not be zero, the condition $S_{ijk} = 0$ is violated. In this way the *realizability requirement* is not satisfied.

Realizability is the requirement that any term modeled must obey certain mathematical inequalities. Schumann (1977) was the first one to notice that second-order models should generate second-moment fields which are "realisable," that is, have nonnegative variances and correlation coefficients not greater than 1. Realizability implied by the Schwartz inequality requires that

$$\begin{aligned} \overline{u_i'u_j'} &\le \sqrt{\overline{u_i'^2}}\sqrt{\overline{u_j'^2}} \\ \overline{u_\alpha'^2} &\ge 0 \\ \overline{\theta'^2} &\ge 0 \\ \overline{u_i'^2\theta'^2} &\le \sqrt{\overline{u_i'^2}}\sqrt{\overline{\theta'^2}} \end{aligned} \tag{6.109}$$

Generally, it is not clear how to formulate parametrizations that fulfill the realizability criteria.

6.5.2 The heat flux equation

The heat flux equation (3.26) contains the following unknown terms:

$$
\begin{array}{|c|}
\hline
Q_{ij} = \overline{u_i' u_j' \theta'} \\
\hline
\nu_\theta \overline{u_i' \dfrac{\partial^2 \theta'}{\partial x_j^2}} \\
\hline
\nu \overline{\theta' \dfrac{\partial^2 u_i'}{\partial x_j^2}} \\
\hline
\overline{\theta' \dfrac{\partial p'}{\partial x_j}} \\
\hline
(\nu + \nu_\theta) \overline{\dfrac{\partial u_i'}{\partial x_\alpha} \dfrac{\partial \theta'}{\partial x_\alpha}} \\
\hline
\end{array}
\tag{6.110}
$$

where the last term in (6.110) was derived as in (3.30).

6.5.2.1 The equation for Q_{ij} has the form

$$
\begin{aligned}
\frac{\partial Q_{ij}}{\partial t} + U_\alpha \frac{\partial Q_{ij}}{\partial x_\alpha} = & -\left(Q_{i\alpha} \frac{\partial U_j}{\partial x_\alpha} + Q_{\alpha j} \frac{\partial U_i}{\partial x_\alpha} + \overline{u_i' u_j' u_\alpha'} \frac{\partial \Theta}{\partial x_\alpha} \right) - \frac{\partial}{\partial x_\alpha} \overline{u_i' u_j' u_\alpha' \theta'} \\
& + \left(\overline{u_i' u_\alpha'} \frac{\partial}{\partial x_\alpha} \overline{u_j' \theta'} + \overline{u_j' u_\alpha'} \frac{\partial}{\partial x_\alpha} \overline{u_i' \theta'} + \overline{\theta' u_\alpha'} \frac{\partial}{\partial x_\alpha} \overline{u_i' u_j'} \right) \\
& - \frac{1}{\rho_0} \left(\overline{u_j' \theta' \frac{\partial p'}{\partial x_i}} + \overline{u_i' \theta' \frac{\partial p'}{\partial x_j}} \right) + \vartheta \left(g_j \overline{u_i' \theta'^2} + g_i \overline{u_j' \theta'^2} \right)
\end{aligned}
\tag{6.111}
$$

The terms on the right-hand side of the above equation are, respectively, shear production, turbulent diffusion, turbulent production, pressure interaction, and buoyancy. Molecular terms were omitted because they were small. As before [see (6.105)],

$$\overline{u_i'u_j'u_k'\theta'} = \overline{u_i'u_j'}\,\overline{u_k'\theta'} + \overline{u_i'u_k'}\,\overline{u_j'\theta'} + \overline{u_i'\theta'}\,\overline{u_j'u_k'} \tag{6.112}$$

Usually, the following hypothesis is made:

$$\frac{1}{\rho_0}\left(\overline{u_i'\theta'\frac{\partial p'}{\partial x_j}} + \overline{u_j'\theta'\frac{\partial p'}{\partial x_j}}\right) = \frac{C_3\epsilon}{q^2}\overline{u_i'u_j'\theta'} \tag{6.113}$$

The term before the third moment on the right-hand side of (6.106) is a reciprocal of the time scale defined by (6.102).

Assuming in (6.111) stationarity, horizontal homogeneity, and horizontal mean flow, and also applying (6.112)–(6.113) yields

$$\begin{aligned} Q_{ij} &= \overline{u_i'u_j'\theta'} \\ &= -\frac{q^2}{C_3\epsilon}\left(\overline{u_i'u_\alpha'}\frac{\partial}{\partial x_\alpha}\overline{u_j'\theta'} + \overline{u_j'u_\alpha'}\frac{\partial}{\partial x_\alpha}\overline{u_i'\theta'} + \overline{u_\alpha'\theta'}\frac{\partial}{\partial x_\alpha}\overline{u_i'u_j'}\right) \\ &\quad - \frac{\vartheta q^2}{C_3\epsilon}\left(g_j\overline{u_i'\theta'^2} + g_i\overline{u_j'\theta'^2}\right) \end{aligned} \tag{6.114}$$

Equation (6.114) can be simplified in the following way:

$$\begin{aligned} \overline{u_i'u_j'\theta'} &= -\frac{q^4}{3C_3\epsilon}\left(\delta_{i\alpha}\frac{\partial}{\partial x_\alpha}\overline{u_j'\theta'} + \delta_{j\alpha}\frac{\partial}{\partial x_\alpha}\overline{u_i'\theta'}\right) \\ &= -\frac{q^4}{3C_3\epsilon}\left(\frac{\partial}{\partial x_i}\overline{u_j'\theta'} + \frac{\partial}{\partial x_j}\overline{u_i'\theta'}\right) \end{aligned} \tag{6.115}$$

Equation (6.115) is equivalent to the flux-gradient formula. A negative sign ensures a transfer of the quantity down the gradient.

6.5.2.2 The second and third terms in (6.110) are usually small.

6.5.2.3 The last term listed in (6.110) vanishes under the assumption of isotropy. Isotropy implies that tensor quantities are independent of rotation, translation, or reflection of the coordinate axes. The term

$$\overline{\frac{\partial u_i'}{\partial x_j}\frac{\partial \theta'}{\partial x_k}} \tag{6.116}$$

having an odd number of indices, changes sign under a coordinate reflection. Thus it must be zero, which in turn implies that each of its components is zero, so that

$$\overline{\frac{\partial u_i'}{\partial x_\alpha}\frac{\partial \theta'}{\partial x_\alpha}} = 0 \tag{6.117}$$

for all i.

6.5.2.4 To define the pressure–temperature correlation, (6.95) is used again. Denoting the shear portion of the fluctuating pressure field by $p^{(s)}$ and using (6.95) gives

$$\frac{1}{\rho_0}\nabla^2 p'^{(s)} = -2\frac{\partial U_\alpha}{\partial x_\beta}\frac{\partial u'_\beta}{\partial x_\alpha} \tag{6.118}$$

and

$$\frac{1}{\rho_0}\nabla^2\frac{\partial p'^{(s)}}{\partial x_k} = -2\frac{\partial^2 U_\alpha}{\partial x_\beta\,\partial x_k}\frac{\partial u'_\beta}{\partial x_\alpha} - 2\frac{\partial U_\alpha}{\partial x_\beta}\frac{\partial^2 u'_\beta}{\partial x_\alpha\,\partial x_k} \tag{6.119}$$

The form of the analytical solution of (6.119) allows the assumption that (e.g., Wyngaard, 1982)

$$\frac{1}{\rho_0}\overline{\frac{\partial p'}{\partial x_k}\theta'}^{(s)} = 2\frac{\partial^2 U_\alpha}{\partial x_\beta\,\partial x_k}F_{\alpha\beta} + 2\frac{\partial U_\alpha}{\partial x_\beta}G_{\alpha\beta k} \tag{6.120}$$

where $F_{\alpha\beta}$ and $G_{\alpha\beta k}$ are isotropic tensor functions of the vector $\overline{u'_\alpha\theta'}$ (Batchelor, 1960; Lumley, 1979):

$$\begin{aligned} F_{\alpha\beta} &= A\overline{u'_\alpha\theta'}\,\overline{u'_\beta\theta'} + B\delta_{\alpha\beta} \\ G_{\alpha\beta k} &= C\overline{u'_\beta\theta'}\delta_{\alpha k} + D\overline{u'_\alpha\theta'}\delta_{k\beta} + E\overline{u'_k\theta'}\delta_{\alpha\beta} \end{aligned} \tag{6.121}$$

It can be argued (Wyngaard, 1983) that $A = B = 0$, $C = 0.4$, and $D = E = -0.1$, which yields

$$\frac{1}{\rho_0}\overline{\frac{\partial p'}{\partial x_k}\theta'}^{(s)} = \frac{4}{5}\frac{\partial U_\alpha}{\partial x_k}\overline{u'_\alpha\theta'} - \frac{1}{5}\frac{\partial U_k}{\partial x_\alpha}\overline{u'_\alpha\theta'} \tag{6.122}$$

The buoyant contribution to the pressure field can be obtained in a similar way:

$$\frac{1}{\rho_0}\nabla^2 p'^{(B)} = g_\alpha\vartheta\frac{\partial\theta'}{\partial x_\alpha} \tag{6.123}$$

and

$$\frac{1}{\rho_0}\nabla^2\frac{\partial p'^{(B)}}{\partial x_k} = g_\alpha\vartheta\frac{\partial^2\theta'}{\partial x_\alpha\,\partial x_k} \tag{6.124}$$

From this it can be assumed that

$$\frac{1}{\rho_0}\overline{\frac{\partial p'}{\partial x_k}\theta'}^{(B)} = g_\alpha\vartheta\delta_{\alpha k}\overline{\theta'^2} = \frac{1}{3}g_k\vartheta\overline{\theta'^2} \tag{6.125}$$

The simplest assumption for the turbulent-turbulent contribution term is

(Wyngaard, 1982)

$$\frac{1}{\rho_0}\overline{\frac{\partial p'\theta'}{\partial x_k}}^{(TT)} = \frac{C_4}{\tau}\overline{u_k'\theta'} \tag{6.126}$$

where C_4 is a constant and τ is a time scale. As before, it can be assumed that $\tau \sim q^2/\epsilon$. Finally, from (6.122), (6.125), and (6.126) we obtain

$$\begin{aligned}\frac{1}{\rho_0}\overline{\frac{\partial p'}{\partial x_k}\theta'} &= \frac{1}{\rho_0}\overline{\frac{\partial p'}{\partial x_k}\theta'}^{(S)} + \frac{1}{\rho_0}\overline{\frac{\partial p'}{\partial x_k}\theta'}^{(B)} + \frac{1}{\rho_0}\overline{\frac{\partial p'}{\partial x_k}\theta'}^{(TT)} \\ &= \frac{4}{5}\frac{\partial U_\alpha}{\partial x_k}\overline{u_\alpha'\theta'} - \frac{1}{5}\frac{\partial U_k}{\partial x_\alpha}\overline{u_\alpha'\theta'} + \frac{1}{3}\vartheta g_k\overline{\theta'^2} + \frac{C_4\epsilon}{q^2}\overline{u_k'\theta'}\end{aligned} \tag{6.127}$$

A similar expression was accepted by André et al. (1978). Therry and Lacarrére (1983) used a simplified form obtained by omitting the two first terms in (6.127).

6.5.3 The temperature variance equation

The temperature variance equation (3.46) requires that the following additional terms be defined:

$$\boxed{\begin{array}{c} R_i = \overline{u_i'\theta'^2} \\ \hline v_\theta\overline{\dfrac{\partial\theta'}{\partial x_\alpha}\dfrac{\partial\theta'}{\partial x_\alpha}} \end{array}} \tag{6.128}$$

6.5.3.1 The term R_i can be defined on the basis of a differential equation of the form

$$\begin{aligned}\frac{\partial R_i}{\partial t} + U_\alpha\frac{\partial R_i}{\partial x_\alpha} = &-\frac{\partial}{\partial x_\alpha}\overline{u_i'u_\alpha'\theta'^2} + 2\overline{u_i'\theta'}\frac{\partial}{\partial x_\alpha}\overline{u_\alpha'\theta'} + g_{i3}\vartheta\overline{\theta'^3}\delta_{i3} \\ &+ \overline{\theta'^2}\frac{\partial\overline{u_i'u_\alpha'}}{\partial x_\alpha} - 2\overline{u_i'u_\alpha'\theta'}\frac{\partial\Theta}{\partial x_\alpha} - \frac{1}{\rho_0}\overline{\theta'^2\frac{\partial p'}{\partial x_i}}\end{aligned} \tag{6.129}$$

Molecular and Coriolis terms were omitted because they are small. Equation (6.129) is not closed and an additional hypothesis is needed. Analogous to (6.105), it can be accepted that

$$\overline{u_i'u_j'\theta'^2} = \overline{u_i'u_j'}\,\overline{\theta'^2} + 2\overline{u_i'\theta'}\,\overline{u_j'\theta'} \tag{6.130}$$

In analogy to (6.126), it can be written

$$\frac{1}{\rho_0}\overline{\theta'^2\frac{\partial p'}{\partial x_i}} = C_7\frac{\epsilon}{q^2}\overline{u_i'\theta'^2} \tag{6.131}$$

The term before the third moment on the right-hand side of (6.131) is a reciprocal of the time scale defined by (6.102).

Assuming horizontal, horizontally homogeneous, and stationary mean flow, and applying (6.130)–(6.131) and (6.115), equation (6.129) can be written in the form

$$R_i = \overline{u_i'\theta'^2} = -\frac{q^2}{C_7\epsilon}\left[\overline{u_i'u_\alpha'}\frac{\partial\overline{\theta'^2}}{\partial x_\alpha} + 2\overline{u_\alpha'\theta'}\frac{\partial\overline{u_i'\theta'}}{\partial x_\alpha} + \vartheta\overline{\theta'^3}\delta_{i3}\right.$$
$$\left. + \frac{q^2}{3C_3\epsilon}\left(\frac{\partial\overline{u_\alpha'\theta'}}{\partial x_i} + \frac{\partial\overline{u_i'\theta'}}{\partial x_\alpha}\right)\frac{\partial\Theta}{\partial x_\alpha}\right] \tag{6.132}$$

Mellor (1973) used a simpler, flux-gradient parametrization for R_i:

$$\overline{u_i'\theta'^2} = -\frac{q^4}{3C_7\epsilon}\frac{\partial\overline{\theta'^2}}{\partial x_i} \tag{6.133}$$

6.5.3.2 For the last term in (6.128), on the basis of the dimensional analysis, it is commonly accepted that

$$\nu_\theta\overline{\frac{\partial\theta'}{\partial x_\alpha}\frac{\partial\theta'}{\partial x_\alpha}} = C_5\frac{\epsilon}{q^2}\overline{\theta'^2} \tag{6.134}$$

The term before the temperature variance on the right-hand side of (6.134) is a reciprocal of the time scale defined by (6.102). Evaluation of (6.134) is shown in Figure 6.30.

6.5.4 The dissipation rate equation

The balance equation for the dissipation rate ϵ has the form (Davydov, 1961)

$$\frac{\partial\epsilon}{\partial t} + U_\alpha\frac{\partial\epsilon}{\partial x_\alpha} = -2\nu\frac{\partial U_\gamma}{\partial x_\alpha}\left(\overline{\frac{\partial u_\gamma'}{\partial x_\beta}\frac{\partial u_\alpha'}{\partial x_\beta}} + \overline{\frac{\partial u_\beta'}{\partial x_\alpha}\frac{\partial u_\beta'}{\partial x_\gamma}}\right) - 2\nu\frac{\partial^2 U_\gamma}{\partial x_\alpha\partial x_\beta}\overline{u_\alpha'\frac{\partial u_\gamma'}{\partial x_\beta}}$$
$$-\nu\frac{\partial}{\partial x_\alpha}\overline{u_\alpha'\left(\frac{\partial u_\gamma'}{\partial x_\beta}\right)^2} - 2\nu\overline{\frac{\partial u_\alpha}{\partial x_\beta}\frac{\partial u_\alpha'}{\partial x_\gamma}\frac{\partial u_\gamma'}{\partial x_\beta}}$$
$$-2\nu\frac{1}{\rho_0}\overline{\frac{\partial^2 p'}{\partial x_\alpha\partial x_\beta}\frac{\partial u_\alpha'}{\partial x_\beta}} - \nu\vartheta g_\gamma\overline{\frac{\partial u_\gamma'}{\partial x_\alpha}\frac{\partial\theta'}{\partial x_\alpha}} - 2\nu^2\overline{\left(\frac{\partial^2 u_\gamma'}{\partial x_\alpha\partial x_\beta}\right)^2} + \nu\frac{\partial^2\epsilon}{\partial x_\alpha^2} \tag{6.135}$$

Equation (6.135) can be derived by differentiating (3.24) with respect to x_α for $\alpha = 1$, 2, and 3, then multiplying by $2\nu\,\partial u_i/\partial x_\alpha$, and finally, adding the results. The terms on the right of this equation represent production by vortex stretching (first two terms), turbulent transport (third term), generation of turbulence by vortex self-deformation (fourth term), pressure diffusion (fifth term), buoyancy effects (sixth term), molecular destruction (seventh term), and molecular diffusion (the last term). The largest terms are production and molecular destruction.

Equation (6.135) should be closed with respect to the higher-order moments listed below.

$$
\begin{array}{|c|}
\hline
\nu\left(\overline{\dfrac{\partial u'_\gamma}{\partial x_\beta}\dfrac{\partial u'_\alpha}{\partial x_\beta}} + \overline{\dfrac{\partial u'_\beta}{\partial x_\alpha}\dfrac{\partial u'_\beta}{\partial x_\gamma}}\right) \\
\hline
\overline{u'_\alpha \dfrac{\partial u'_\gamma}{\partial x_\beta}} \\
\hline
J_\alpha = \overline{\nu u'_\alpha \left(\dfrac{\partial u'_\gamma}{\partial x_\beta}\right)^2} \\
\hline
\nu \overline{\dfrac{\partial u'_\alpha}{\partial x_\beta}\dfrac{\partial u'_\alpha}{\partial x_\gamma}\dfrac{\partial u'_\gamma}{\partial x_\beta}} \\
\hline
\overline{\dfrac{\partial^2 p'}{\partial x_\alpha \partial x_\beta}\dfrac{\partial u'_\alpha}{\partial x_\beta}} \\
\hline
\nu \vartheta g_\gamma \overline{\dfrac{\partial u'_\gamma}{\partial x_\alpha}\dfrac{\partial \theta'}{\partial x_\alpha}} \\
\hline
\left(\overline{\dfrac{\partial^2 u'_\gamma}{\partial x_\alpha \partial x_\beta}}\right)^2 \\
\hline
\end{array}
\tag{6.136}
$$

The terms listed can be parametrized in the following way (Jones and Launder, 1972).

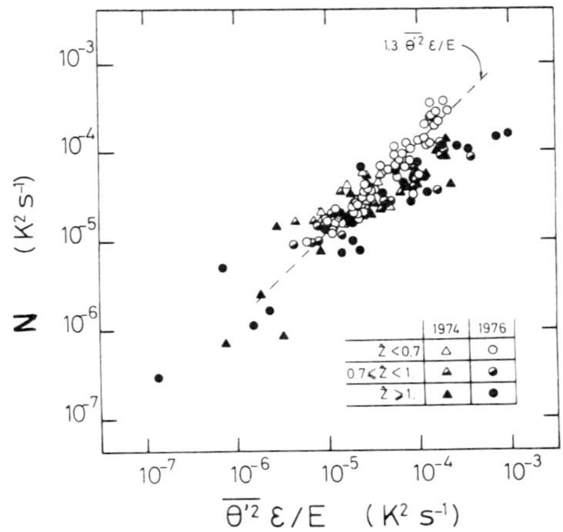

Figure 6.30 Comparison between the destruction rate of temperature variance N and the quantity $\overline{\theta'^2}\epsilon/E$. $\hat{Z}$ is the dimensionless height.

6.5.4.1 The first term of the list (6.136) in the case of anisotropy is parametrized by the following hypothesis:

$$2\nu\left(\overline{\frac{\partial u'_\gamma}{\partial x_\beta}\frac{\partial u'_\alpha}{\partial x_\beta}} + \overline{\frac{\partial u'_\beta}{\partial x_\gamma}\frac{\partial u'_\beta}{\partial x_\alpha}}\right) = C_8\frac{\epsilon}{q^2}\left(\overline{u'_\alpha u'_\gamma} - \frac{1}{3}\delta_{\alpha\gamma}q^2\right) \qquad (6.137)$$

6.5.4.2 The second, fourth, and fifth terms in (6.136) can be omitted because they vanish when turbulence is isotropic.

6.5.4.3 The sixth and seventh terms in (6.136) can be expressed in the form

$$\nu\vartheta g_\gamma\overline{\frac{\partial u'_\gamma}{\partial x_\alpha}\frac{\partial\theta'}{\partial x_\alpha}} = C_{11}g_\gamma\vartheta\epsilon\frac{\overline{u'_\gamma\theta'}}{q^2} \qquad (6.138)$$

$$2\nu^2\overline{\left(\frac{\partial^2 u'_\gamma}{\partial x_\alpha\,\partial x_\beta}\right)^2} = C_9\frac{\epsilon^2}{q^2} \qquad (6.139)$$

6.5.4.4 To obtain an expression for the third term in (6.136), the equation for J_i can be derived. For stationary, horizontally homogeneous and horizontal mean motion, this equation simplifies to the form

$$J_i = -\frac{q^2}{C_{12}\epsilon}\left(C_{14}\overline{u'_i u'_\alpha}\frac{\partial\epsilon}{\partial x_\alpha} + \frac{2}{9}\epsilon\frac{\partial\overline{u'_\alpha u'_i}}{\partial x_\alpha}\right) \qquad (6.140)$$

The characteristic feature of the dissipation rate equation is that every term on the right side of equation (6.135) must be parametrized. By contrast, many of the important terms in wind velocity variance equations can be calculated exactly. Thus one can expect more difficulties in the closure

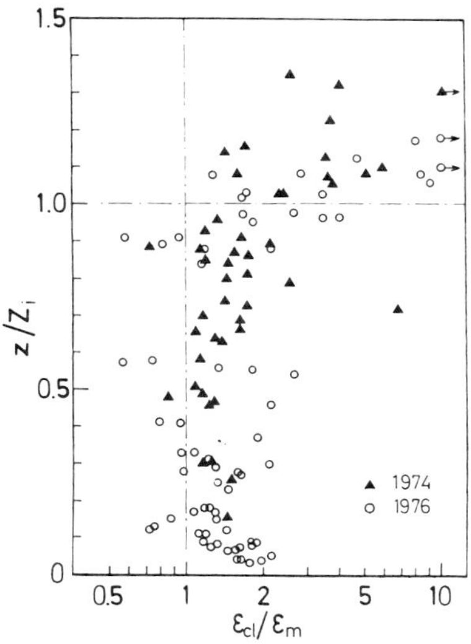

Figure 6.31 Height variation of the ratio between the calculated from (6.18) ϵ_{cl} and measured ϵ_m dissipation rates. Mixing length l_∞ calculated for Mellor and Yamada expression in (6.19). (After Isaka and Guillement, 1983.)

parametrization in the ϵ equation than in the $\overline{u_i'u_j'}$ equation. Because of this, many researchers choose simpler methods to calculate ϵ. The simplest approach is to avoid the ϵ equation entirely and employ the dimensionally derived formula $\epsilon = c_\epsilon q^3/l$, where l is the mixing length (Figure 6.31). The length-scale prescriptions cause, of course, fewer problems in modeling, but often oversimplify the physics by constraining the modeled turbulence to a predetermined or anticipated state.

All of the parametrization formulas discussed in this section contain constants. These constants can be determined from observations in the atmospheric surface layer. For example, to define the constant c_ϵ in the dissipation rate definition $\epsilon = c_\epsilon E^{3/2}/l$, turbulent kinetic energy (3.36) can be considered. In the neutral surface layer, the production term is $u_*^3/\kappa z$, the turbulent transport term is small, the buoyancy term is zero, and, assuming that $E = au_*^2$ (where a is an empirical constant, $a \approx 12.5$), $l = \kappa z$, we obtain $c_\epsilon = a^{-3/2} = 0.02$.

6.6 LARGE EDDY SIMULATION MODELS

Large eddy simulation (LES) modeling was initiated by Deardorff, who in two 1974 papers presented results of a study of the Wangara experiment day 33. LES modeling involves direct numerical integration of the governing equations. In any numerical integration eddies of scales smaller than two grid intervals can never be resolved. Since these small eddies have great importance in atmospheric flows, the governing equations have to be filtered before numerical integration. As mentioned earlier, this can be accomplished by

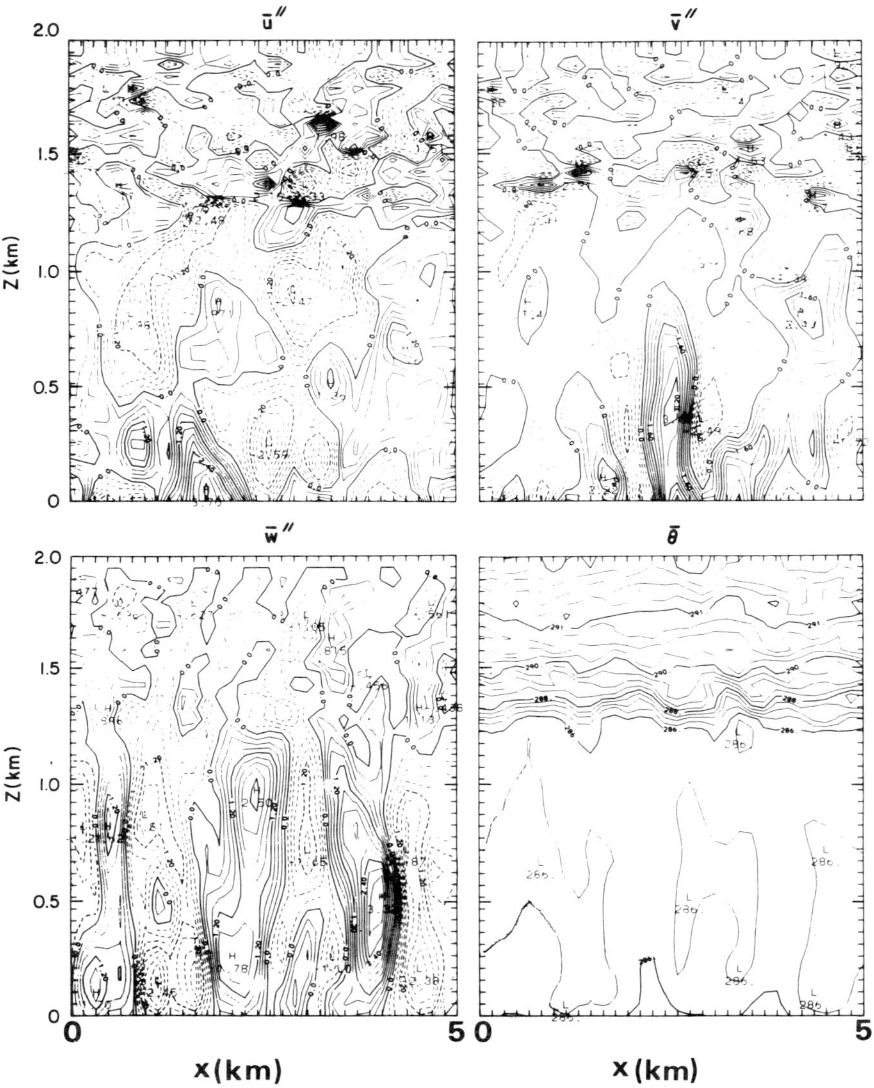

Figure 6.32 The contour plots of $\overline{u}''$, $\overline{v}''$, $\overline{w}''$, and $\overline{\theta}$ in a vertical cross section on an *x-z* plane at hour 14. (After Moeng, 1984. *Journal of Atmospheric Sciences* published by the American Meteorological Society.)

applying the volume-averaging operator (Reynolds averaging). The Reynolds time averaging could also be considered along with the volume averaging. However, this is not necessary because the time-step increment is usually very small.

In LES, the large structures in the flow are computed explicitly from averaged equations and the small ones (the subgrid scale) are modeled. This method has an advantage over methods in which all the eddies are modeled (i.e., ensemble averages models) by having a lower sensitivity to closure inaccuracy. The principal disadvantage of large eddy simulation is that the computations are necessarily three-dimensional and time dependent. This means that the cost is much higher. For example, a 2-hour simulation with 1-hour startup time, a resolution of $\Delta x = \Delta y = 150$ m, $\Delta z = 50$ m, and a domain of $28 \times 59 \times 59$ points will require about 10 hours of computer time (CRAY). However, to obtain statistically significant ensemble-averaged data, several realizations are required. This may increase the CPU time needed by a factor of 10 (Wyngaard, 1984).

The pressure field in an LES model must be evaluated nonhydrostatically by either of two approaches. Neither technique exhibits a clear superiority in terms of computation speed. The method used by Deardorff (1972b) assumes that the flow is incompressible (this eliminates sound waves). Pressure is then evaluated by taking the finite-difference divergence of the equations of motion in the finite-difference form and forming an elliptic or Poisson-type diagnostic, finite-difference equation for pressure. The time steps of explicit integration schemes are limited by slower-moving internal gravity waves and advection time scales. Deardorff (1970) used a staggered grid, which is probably the best choice for solving the incompressible equations, although other grid systems could also be used.

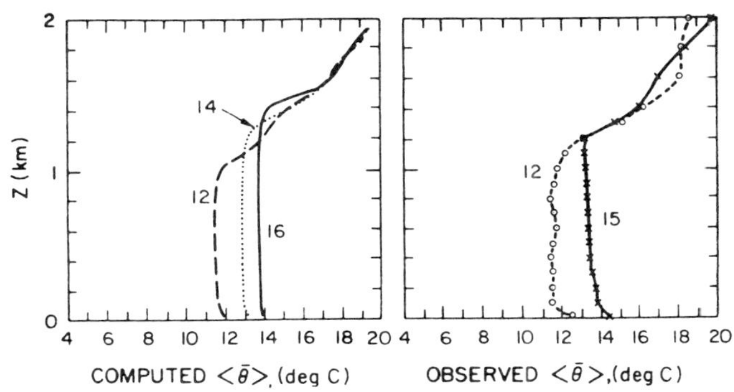

Figure 6.33 Vertical profiles of the horizontal mean potential temperature (a) computed at hours 12, 14, and 16, and (b) observed at hours 12 and 15, during day 33 of the Wangara experiment. (After Moeng, 1984. *Journal of Atmospheric Sciences* published by the American Meteorological Society.)

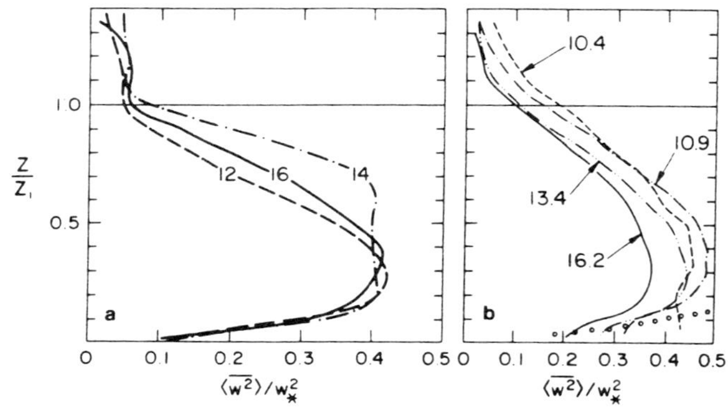

Figure 6.34 Vertical profiles of dimensionless vertical velocity variances; (a) from the mixed pseudospectral finite-difference model of Moeng (1984) (32 × 32 × 40 grid) at hours 12, 14, and 16; (b) from Deardorff's (1974a, b) finite-difference model on a 40 × 40 × 40 grid at hours 10.4, 10.9, 13.4, and 16.2. (After Moeng, 1984. *Journal of Atmospheric Sciences* published by the American Meteorological Society.)

The second approach, used by Klemp and Wilhelmson (1978) and Cotton and Tripoli (1978), is to retain compressibility and to evaluate pressure by a so-called time-splitting procedure. In this procedure, the terms in the equations of motion contributing to sound waves are separated from the long time-scale terms. Sound-wave generating terms and a simplified equation of continuity or pressure tendency are integrated on a short time scale that

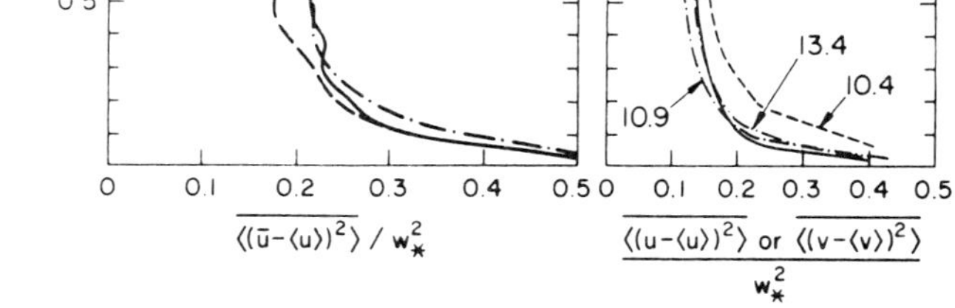

Figure 6.35 Vertical profiles of dimensionless horizontal velocity variances; (a) from the mixed pseudospectral finite-difference model of Moeng (1984) (32 × 32 × 40 grid) at hours 12, 14, and 16; (b) from Deardorff's (1974a, b) finite-difference model on a 40 × 40 × 40 grid at hours 10.4, 10.9, 13.4, and 16.2. (After Moeng, 1984. *Journal of Atmospheric Sciences* published by the American Meteorological Society.)

resolves sound waves, while the terms and equations governed by internal waves and advection are integrated on a time step some 10 to 30 times longer.

LES models allow the use of a simple subgrid-scale closure model. Most of the subgrid schemes presently in use apply second-moment equations for subgrid-scale turbulence. There is concern with the accuracy of simulations near the surface, where some important quantities may reside entirely in subgrid scales. This may cause the second and higher moments to be lower than measured in the atmosphere. For example, a negative third moment of the vertical velocity can be obtained, contrary to the results of measurements in the surface layer (Moeng, 1984).

Lateral boundary conditions are commonly chosen to be periodic. The advantage of this technique is that eddies which propagate through an outflow boundary reenter the model domain at the inflow boundary. Periodic boundary conditions can, however, distort larger-scale eddies such as boundary layer roll vortices, which may be only partially captured in the model domain (Sommeria and LeMone, 1978).

Large eddy simulation has been used to simulate neutral and convective boundary layers. Deardorff (1973) found that stable stratification eliminated the largest eddies and forced the turbulence to be a subgrid, and, therefore, no results could be realized in the stable ABL.

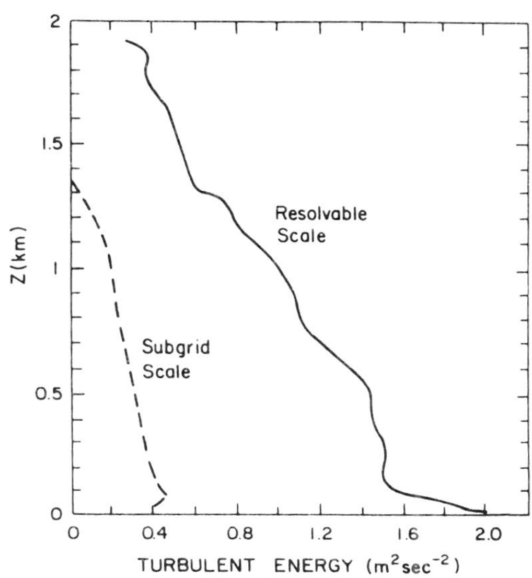

Figure 6.36 Vertical profiles of the horizontal mean turbulent energy of the resolvable-scale field (solid curve) and the subgrid-scale field (dashed curve), respectively. (From Moeng, 1984. *Journal of Atmospheric Sciences* published by the American Meteorological Society.)

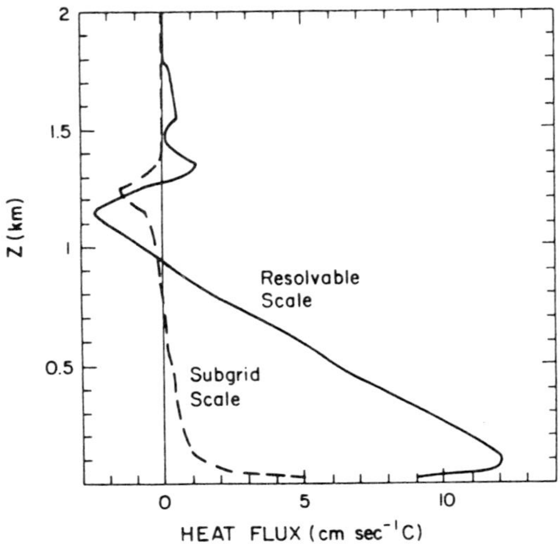

Figure 6.37 Vertical profiles of the turbulent heat flux for resolvable-scale and subgrid-scale fields. (From Moeng, 1984. *Journal of Atmospheric Sciences* published by the American Meteorological Society.)

Results of the LES model (Moeng, 1984) are shown in Figures 6.32 to 6.37. The model outputs were averaged over each horizontal plane. The horizontally averaged quantities (denoted by $\langle\ \rangle$; a bar over a quantity denotes time averaging) were considered to be ensemble-mean properties.

Figure 6.32 shows the x–z plane structure of the large eddies. Quantities depicted in the figure are defined as

$$\overline{U}'' = \overline{U} - \langle \overline{U} \rangle \tag{6.141}$$

The fluctuations visible in the stable layer above the mixed layer are internal gravity waves. Vertical profiles of potential temperature obtained from the model are compared with observational data in Figure 6.33. The dimensionless vertical velocity variance is shown in Figure 6.34. The maximum of this quantity is about $0.42w_*$ and occurs for $z/z_i = 0.3$ to 0.6. Horizontal velocity variance is shown in Figure 6.35. The comparison of Moeng's and Deardorff's model shows differences of horizontal velocity variance profiles close to the top of the mixed layer. Figures 6.36 and 6.37 compare the resolved and subgrid scale amounts of turbulent kinetic energy and heat flux. They were obtained as shown for the momentum and heat fluxes on the right-hand side of (2.54) and (2.55). On the surface, the resolvable term is about five times larger than the subgrid one for the turbulent kinetic energy and two times larger for the heat flux. The subgrid scale terms decrease faster with height than do the resolved terms.

PROBLEMS

6.1. Find the height of the mixed layer z_i and the mixed layer temperature Θ_m

(a) using (6.1), (6.2), and (6.6) and assuming that $\Delta\Theta = \text{const}$

(b) using (6.1), (6.2), and (6.6) and assuming that $z_i\,\Delta\Theta = \text{const}$

(c) using (6.1), (6.2), and (6.6) and assuming that $\overline{w'\theta'}|_i = R\overline{w'\theta'}|_0$ (R is a constant)

(d) using (6.6) and (3.33) written for $z = z_i$, with the mechanical production and dissipation terms omitted and the local term parametrized as $\partial E/\partial t(z_i) = c_T V_s^2/z_i\, dz_i/dt$, and the transport term parametrized as V_s^3/z_i; where V_s is the velocity scale, $V_s = w_* + c_t u_*$, and c_T and c_t are constants (Zeman and Tennekes, 1977).

6.2. Compare results obtained in Problem 6.1(a), (b) plotting the time evolution of the mixed layer for $\gamma = 0.005$ K/m, $Q_0 = 0.15 \sin(2\pi t/24)$, $\theta_m(0) = 273$ K, $\Delta\Theta_0 = 0.2$ K, $z_i = 100_m$.

6.3. Assuming that the potential temperature is independent of time and that the advection term is independent of height, prove that in the mixed layer the turbulent heat flux is a linear function of height. Use (2.56) and assume that molecular and sink/source terms can be omitted.

6.4. Find the profile of wind velocity in the surface layer, assuming that:

(a) $k_m(z) = \kappa u_* z$

(b) $k_m(z) = k_\infty[1 - \exp(-mz)]$; k_∞ and m are constants

6.5. Find the profile of wind velocity in the surface layer assuming in (6.13) that $u_t = u_*$ and

(a) $l = \kappa z \exp(-mz)$; m is a constant

(b) $l = \kappa z/[1 + \kappa z/l_\infty]$; l_∞ is a constant

6.6. Find surface layer profiles of wind velocity and temperature in extremely stable, neutral, and free convection regimes using the results of Section 6.3.1. Make a plot and compare it with the plot in Problem 4.6.

6.7. Solve (6.74) for $\Lambda = 0$. Assume that $U_g + iV_g = (\nu + f)/fA_g \exp(i\nu t)$, where ν and A_g are the frequency and amplitude of a simple periodic wave [this expression can be obtained from (2.69) assuming zero stress and periodic wind in the free atmosphere: $U + iV = A_g \exp(i\nu t)$]. Plot a time hodograph for $z = 200$ m, $k_m = 1\ \text{m}^2/\text{s}$ and $\nu = 0.1\ \text{h}^{-1}$. Find the height of the ABL. Discuss the results for the geographical latitude for which $f = -\nu$. [*Hint:* Assume a solution in the form $U + iV = A(z)\exp(i\nu t)$ with $A(0) = 0$ and $A(\infty) = A_g$. Employ (6.33) to find a solution for A.]

6.8. Using (6.47), show that the thermal wind is perpendicular to the temperature gradient.

6.9. Assuming constant k_m, solve (6.32) with the lower boundary condition in the form:

(a) For $z = 0$: $U = V = 0$ and in addition that for $z = u_*/f$ flow is geostrophic.

(b) For $z = h_s$: $U = U_h$, $V = V_h$ (i.e., at the top of the surface layer h_s the wind velocity is given).

(c) For $z = h_s$: the surface stress is parallel to the surface wind and given is the cross-isobar angle α_0.

6.10. Assuming that constant k_m and $U_g = V_g = 0$:

(a) Solve (6.32) with boundary conditions in the form (Ekman's problem for wind-induced ocean currents):

$$\text{For } z = 0: \qquad k_m \frac{dU}{dz} = u_*^2 \qquad k_m \frac{dV}{dz} = 0$$

$$\text{For } z = -\infty: \qquad U = V = 0$$

(b) Show that at $z = 0$, $U_0^2 + V_0^2 = u_*^4/2fk_m$.

(c) Show that the largest velocity is at $z = 0$ and 45° to the right of the wind stress (Batchelor, 1967).

6.11. Find the order of magnitude of:

(a) The cross-isobar angle α_0

(b) The geostrophic drag coefficient c_g

(c) The drainage acceleration, when the slope angle $\psi = 0.001$

6.12. Employing (6.32), derive the formula

$$k_m(z) = \frac{f}{U(dV/dz) - V(dU/dz)} \int_{z_0}^{z} \left(U^2 + V^2 - UU_g - VV_g\right) dz$$

6.13. Employing (6.32), derive the formula

$$u_*^2 = \int_{z_0}^{z_m} f\left(V - V_g\right) dz$$

where z_m is the height for which $dU/dz = 0$. Give a geometrical interpretation of the result.

6.14. Employing (6.32), derive the formula

$$u_*^2 = \int_{z_0}^{h} f\left(V - V_g\right) dz$$

where h is the height of the ABL. Give a geometrical interpretation of the result.

6.15. Employing (6.32), derive the formula

$$-\overline{u'w'} = u_*^2 - \int_{z_0}^{z} f\left(V - V_g\right) dz$$

Give a geometrical interpretation of the result.

6.16. Employing (6.32), derive the formula

$$-\overline{v'w'} = \int_{z_0}^{z} f\left(U - U_g\right) dz$$

Give a geometrical interpretation of the result.

6.17. Prove (6.109).

6.18. Write the full form of:

(a) ϵ_{ij} in (6.90)

(b) $\Pi_{\alpha\alpha}^{(2)}$ in (6.94)

(c) P_{ij} in (6.98)

6.19. Define the difference between resolvable and subgrid heat fluxes in Figure 6.37.

6.20. From the Ekman solution, determine:

(a) The value of the drag coefficient c_g.

(b) The value of the production of turbulent kinetic energy.

6.21. Very close to the surface, the equation of motion has the form

$$\nu \frac{\partial U}{\partial z} = u_*^2$$

where ν = const. Find the velocity profile in this layer.

6.22. Show that the angle ψ between the tangent to the Ekman hodograph and the vector $\mathbf{V} - \mathbf{G}$ equals 45° for all z.

6.23. The direction and velocity of the wind change slightly as air flows over the lake.

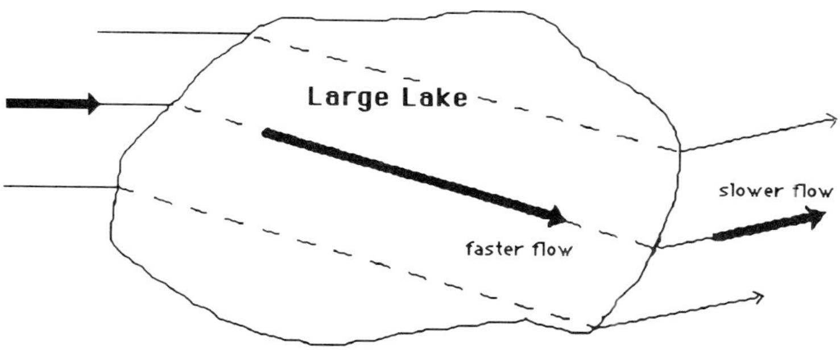

Explain this phenomenon using Figure 6.9. Show regions where subsidence and uplift occur.

6.24. Assuming constant k_m, solve (6.32) numerically (see the Appendix). Compare analytical and numerical solutions.

6.25. Flow within the layer of obstacles (plants, urban buildings) causes a portion of kinetic energy of the flow near the surface to be lost on work against the friction force generated by the obstacles (see also Problem 4.8). This effect can be modeled by including on the right-hand side of (6.41) an additional term $[C_r|\mathbf{V}|\mathbf{V}I(z)]$, where C_r is a coefficient (for urban area $C_r \sim 0.01\text{–}0.001\ \text{m}^{-1}$), $\mathbf{V}$ is a wind velocity vector, $|\mathbf{V}|$ is a modulus of the wind velocity, and $I(z)$ is equal to 1 in the obstacle layer and equal to 0 above it (Sorbjan, 1979). Assuming constant k_m, solve the modified (6.41) numerically (see the Appendix).

6.26. Consider a flow over a sloping terrain described by the following equations (Prandtl's problem):

$$k_m \frac{d^2 U''}{dz^2} + \beta \sin \psi \Theta'' = 0$$

$$k_m \frac{d^2 \Theta''}{dz^2} - \Gamma \sin \psi U = 0$$

where Γ is the temperature lapse rate and U'' and Θ'' are perturbations imposed on the reference state. Discuss the physical sense of the above equations. Solve the equations introducing the complex function $F = U'' + i(\beta/\Gamma)^{1/2}\Theta''$. Plot the solutions for $\sin \psi = 0.01$, $\Gamma = 1°\text{C}/100$ m, and $\Theta''(z = 0) = 0.5°\text{C}$.

CHAPTER SEVEN

DIFFUSION

7.1 GENERAL CONSIDERATIONS

Research on atmospheric turbulence has a very important practical application in the control and management of air quality. Its significance was realized when the growth of large metropolitan areas and the increasingly heavy use of coal for home heating and industrial power led to episodes of extreme sulfur pollution. The most famous case occurred in December 1956 in London, where approximately 800 people died as a direct result of pollution. Since that time, the problem has grown. High air pollution concentrations are no longer restricted to urban areas with high source density, but can be transported for long distances by large-scale weather patterns.

Dispersion of pollutants in the atmosphere depends strongly on meteorological conditions. The vertical appearance of a real plume offers considerable information as to how the thermal and dynamic state of the lower atmosphere influences the transport of atmospheric pollutants. Some idealized patterns of vertical smoke spreading are displayed in Figure 7.1.

So-called *fanning* occurs in the stable ABL. Since vertical mixing is very small, the plume expands very little in depth over long distances. Lateral diffusion may proceed quite differently. In inversion close to the surface, the instantaneous plume may be quite narrow, but slow meandering of the wind results in a similar meandering of the plume. Therefore, the time-averaged plume may appear quite broad. The fanning behavior of plumes is not considered an unfavorable condition for tall stacks. This situation might be unfavorable, however, where the stack is short.

The second case is called *fumigation.* Shortly after the sun rises on a

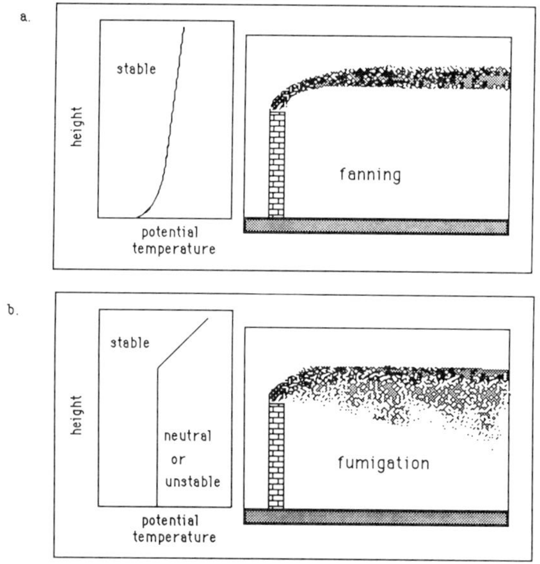

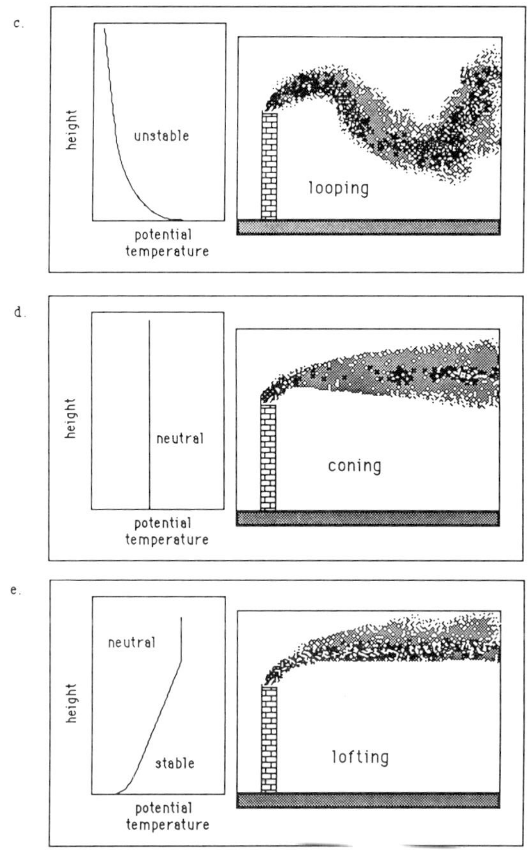

Figure 7.1 Idealized types of smoke-plume patterns in the atmosphere: (a) fanning; (b) fumigation; (c) looping; (d) coning; (e) lofting.

clear morning, the existing radiational inversion begins dissipating and is slowly replaced by a neutral or unstable layer. At some time after sunrise, inversion is present just above the top of the stack and acts as a lid while the newly developed convective eddies mix the pollutants within the shallow unstable layer next to the ground. This condition causes a sudden rise in ground-level concentrations.

Looping is frequently evident by midday, when the temperature structure is very unstable, especially in the lower layers. This generates large, vertical, thermally induced eddies that bring the plume to the ground. The concentration that reaches the ground decreases very rapidly with distance from the stack.

Coning takes place when a vertical temperature gradient is adiabatic. The plume is shaped like a cone with a horizontal axis. The distance from the stack at which the smoke first comes to the ground is greater than it would be in looping conditions because of a lower level of thermally induced turbulence. Coning usually accompanies cloudy, windy conditions and can occur either during the day or night.

Lofting occurs during the transition from unstable to inversion conditions and is most often observed near sunset. Depending on the stack height and the deepening rate of the inversion layer, the lofting condition may be very transitory or could persist for several hours. When the source is above the surface inversion, lofting may be considered the most favorable condition.

7.2 THE STATISTICAL THEORY OF DIFFUSION

The Brownian motion of a dust particle in a fluid provides the classic example of a random process. This motion is a result of the bombardment of this particle by molecules of the fluid in thermal motion. Molecules are not visible and exist in great number. The resulting process is called *molecular diffusion.*

It seems intuitively obvious that random movements of certain numbers of particles in the laminar or turbulent medium should lead to their dispersion. This random process can be described in terms of a probability function $P(\mathbf{x}, \mathbf{x}', t)$, where $\mathbf{x}$ is the position of the particle at the moment t and $\mathbf{x}'$ is its initial position. The function P can be obtained empirically by counting the number of particles found in a volume around a given point $\mathbf{x}$ at a certain moment of time. The concentration of particles at point $\mathbf{x}$ is defined as

$$C(\mathbf{x}, t) = QP(\mathbf{x}, \mathbf{x}', t) \tag{7.1}$$

where Q is the total mass of particles. The variance is defined as

$$\sigma_x^2 = \iiint_v x^2 P(\mathbf{x}, t)\, dx\, dy\, dz = \frac{1}{Q} \iiint_v x^2 C(\mathbf{x}, t)\, dx\, dy\, dz \tag{7.2}$$

where $\mathbf{x}'$ was assumed to be zero.

It is convenient to illustrate the foregoing idea by considering the simple model of a parcel that randomly moves along the x-axis (Chandrasekhar, 1943). Each parcel movement has the same finite length Δx and is performed in a time interval Δt ("drunkard's walk"). In this way, the parcel jumps between points, which can be marked by integer numbers: $-N$, $-N+1, \ldots, -1, 0, 1, \ldots, N-1, N$, with the release point at 0 (Figure 7.2). The probability of a step to the left and to the right is assumed to be the same and equal to $\frac{1}{2}$.

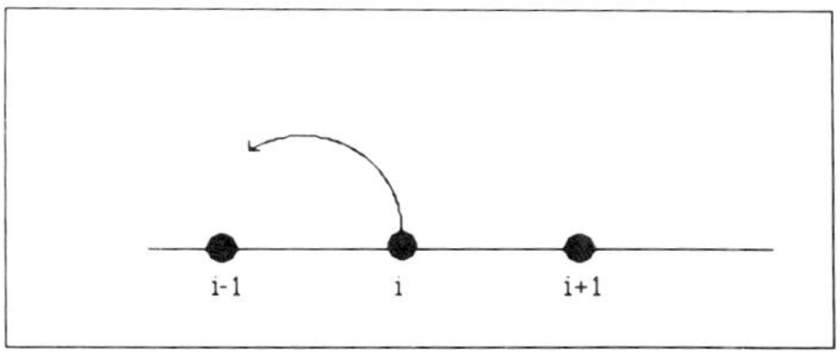

Figure 7.2 Random walk on an axis.

Now one might ask about the position of the parcel after N jumps. Since the movement of the parcel is random, we can only look for the probability $P(M, N)$ that the parcel will be at the certain point M after N jumps. The finite position of the parcel must be in one of the following points:

$$\begin{array}{lll} -2, 0, 2 & \text{for } N = 2 & \\ -3, -1, 1, 3 & \text{for } N = 3 & \\ -4, -2, 0, 2, 4 & \text{for } N = 4 & \text{etc.} \end{array}$$

If p is the number of jumps forward and b is the number of jumps backward, then

$$\begin{aligned} p + b &= N \\ p - b &= M \end{aligned} \tag{7.3}$$

The probability of one particular trajectory is $(\frac{1}{2})^p(\frac{1}{2})^b = (\frac{1}{2})^N$. The number of trajectories leading to the same point M and consisting of p jumps forward and b jumps backward is equal to a number of combinations $C_N^p = \binom{N}{p}$. Thus the probability $P(M, N)$ is given by

$$P(M, N) = \binom{N}{p}\left(\frac{1}{2}\right)^N = \binom{N}{\frac{N+M}{2}}\left(\frac{1}{2}\right)^N$$

$$= \frac{N!}{\left(\frac{N-M}{2}\right)!\left(\frac{N+M}{2}\right)!}\left(\frac{1}{2}\right)^N \tag{7.4}$$

Applying Stirling's formula $n! = \sqrt{2\pi n}\, n^n e^{-n}$, we obtain

$$\lim_{N\to\infty} P(M, N) = \lim \sqrt{\frac{2\pi N}{2\pi\left(\frac{N-M}{2}\right)2\pi\left(\frac{N+M}{2}\right)}}$$

$$\times \frac{\left(\frac{1}{2}\right)^N N^N \exp(-N)}{\left(\frac{N-M}{2}\right)^{(N-M)/2}\left(\frac{N+M}{2}\right)^{(N+M)/2} \exp\left(-\frac{N-M}{2}\right)\exp\left(-\frac{N+M}{2}\right)}$$

$$= \frac{1}{\sqrt{2\pi N/4}} \exp\left(-\frac{M^2}{2N}\right) \tag{7.5}$$

The above result was obtained using the following identities valid for $-1 < t < 1$,

$$\ln(1+t) = t - \frac{t^2}{2} + \frac{t^3}{3} - \frac{t^4}{4} \cdots \tag{7.6}$$

and

$$\ln(1-t) = -t - \frac{t^2}{2} - \frac{t^3}{3} - \frac{t^4}{4} \cdots \tag{7.7}$$

which yields

$$\frac{N+M}{2}\ln\left(1+\frac{M}{N}\right) + \frac{N-M}{2}\ln\left(1-\frac{M}{N}\right)$$

$$= \frac{N+M}{2}\left[\frac{M}{N} - \frac{1}{2}\left(\frac{M}{N}\right)^2 + \frac{1}{3}\left(\frac{M}{N}\right)^3 - \cdots\right]$$

$$- \frac{N-M}{2}\left[\frac{M}{N} \cdot \frac{1}{2}\left(\frac{M}{N}\right)^2 + \frac{1}{3}\left(\frac{M}{N}\right)^3 + \cdots\right]$$

$$= \frac{M^2}{N} - \frac{M^2}{2N} + \frac{M^4}{3N} \approx \frac{M^2}{2N} \tag{7.8}$$

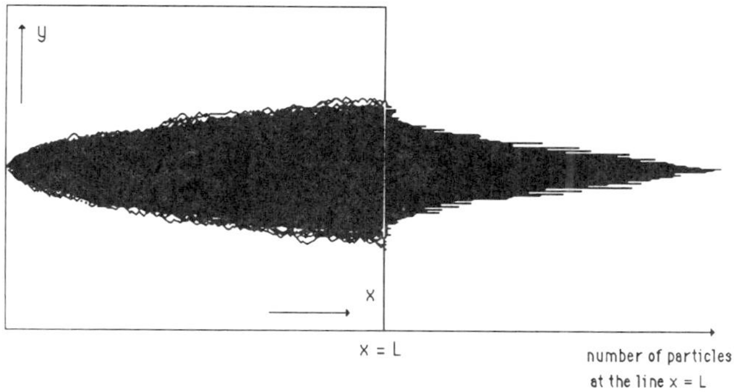

Figure 7.3 Result of a computer simulation of a two-dimensional random walk.

The resulting probability of the parcel's position in (7.5) is described by Bernoulli's distribution.

Now we will find how this result changes as the distance between points decreases and the number of points increases.

Assuming that $M = x/\Delta x$, $N = t/\Delta t$, and the increments Δx and $\Delta t \to 0$ so that $\Delta x^2/\Delta t \to 2k$ = constant, we obtain from (7.5)

$$f(X) = \frac{P(x, \infty)}{2\Delta x} = \frac{1}{2\sqrt{\pi k t}} \exp\left(-\frac{x^2}{4Kt}\right) \tag{7.9}$$

The result is the Gaussian (or normal) distribution. This is not surprising since it is known that for large N, the Bernoulli distribution approaches the Gaussian distribution. The resulting normal distribution is characterized by a zero mean value and a variance equal to $2kt$.

The result of a computer-simulated two-dimensional random walk is shown in Figure 7.3. Each of the 1000 trajectories in the figure was generated by moving a particle one step along the x-axis (nonrandom motion) and m steps along the y-axis, where m is equal to 1 or 0 or -1 with an equal probability equal to $\frac{1}{3}$. On the right-hand side of the plot, a frequency $P(y)$ of particles crossing line $x = L$ at the point y is also shown. The frequency obtained is described approximately by the Gaussian distribution.

The expression for the variance of a fluid particle that moves along an x-axis with an average velocity U and whose movement along a y-axis is governed only by turbulent velocity fluctuations v' was first derived in 1921 by Taylor. To derive this expression, the turbulent movement along the y-axis can be considered:

$$\frac{dy}{dt} = v'(t) \tag{7.10}$$

During a time interval t, the parcel will be carried a distance

$$y = \int_0^t v'(\tau)\, d\tau \tag{7.11}$$

The product of (7.10) and (7.11) results in

$$y\frac{dy}{dt} = v'(t)\int_0^t v'(\tau)\, d\tau \tag{7.12}$$

and

$$y^2 = 2\int_0^t v'(\tau_1)\int_0^{\tau_1} v'(\tau_2)\, d\tau_2\, d\tau_1 \tag{7.13}$$

From (7.13), we obtain

$$\sigma_y^2(t) = \overline{y^2} = 2\int_0^t \int_0^{\tau_1} \overline{v'(\tau_1)v'(\tau_2)}\, d\tau_1\, d\tau_2 \tag{7.14}$$

where σ_y^2 is a horizontal variance.

Now, it is convenient to introduce the Lagrangian correlation coefficient, defined in the following way:

$$R_L(\tau) = \frac{\overline{v'(t)v'(t+\tau)}}{\overline{v'^2(t)}} \tag{7.15}$$

Using this definition and assuming homogeneity allows (7.14) to be rewritten in the form

$$\sigma_y^2(t) = 2\overline{v'^2}\int_0^t \int_0^{\tau_1} R_L(\tau)\, d\tau\, d\tau_1 \tag{7.16}$$

Integrating (7.16) by parts gives

$$\int_0^t \int_0^{\tau_1} R_L(\tau)\, d\tau\, d\tau_1 = \tau_1 \int_0^{\tau_1} R_L(\tau)\, d\tau\Big|_0^t - \int_0^t \tau_1 R_L(\tau_1)\, d\tau_1$$

$$= \int_0^t (t-\tau) R_L(\tau)\, d\tau \tag{7.17}$$

Substituting this result in (7.16) yields

$$\sigma_y^2(t) = 2\overline{v'^2}\int_0^t (t-\tau) R_L(\tau)\, d\tau \tag{7.18}$$

This equation was first obtained by Kampé de Fériet in 1939. The function R_L can be expressed in the form of a Taylor series. Taking only the first three terms and considering that $R_L(0) = 1$ and that $\partial R_L/\partial\tau|_0 = 0$ (symmetry) results in

$$R_L(\tau) = R(0) + \tau\frac{dR_L}{d\tau}(0) + \frac{\tau^2}{2}\frac{dR_L}{d\tau^2}(0) + \cdots \approx 1 - \frac{\tau^2}{\lambda^2} \tag{7.19}$$

where λ is the *Lagrangian time microscale,* defined as

$$-\frac{2}{\lambda^2} = \frac{d^2R_L}{d\tau^2}(0) \tag{7.20}$$

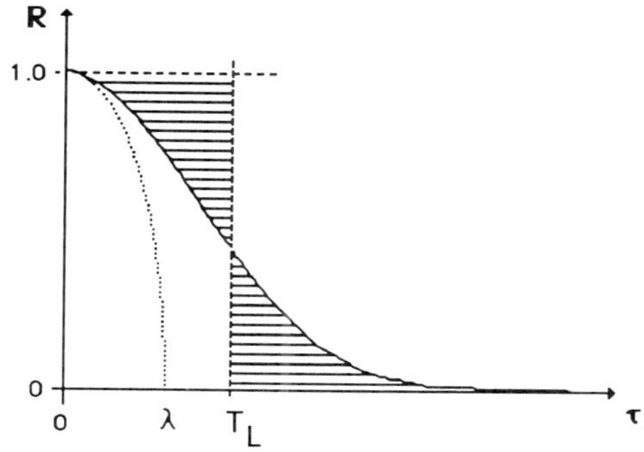

Figure 7.4 Lagrangian autocorrelation function R_L, Lagrangian time microscale λ, and Lagrangian time scale T_L defined by equating the shaded areas.

From (7.19) it follows that the Lagrangian time microscale is the intercept of the parabola that matches R_L at the origin (Figure 7.4).

Substituting (7.19) into (7.18) yields

$$\sigma_y^2(t) = \overline{v'^2}t^2\left(1 - \frac{t^2}{6\lambda^2}\right) \tag{7.21}$$

For short time intervals, (7.21) simplifies to

$$\sigma_y^2(t) = \overline{v'^2}t^2 \tag{7.22}$$

It is also convenient to introduce another time scale, the *Lagrangian integral time scale*:

$$T_L = \int_0^\infty R_L(\tau)\,d\tau \tag{7.23}$$

This scale defines a time interval in which there is a substantial correlation between the velocity of different parcels of air in their turbulent movement. For $\tau > T_L$, $R_L(\tau) \to 0$.

Equation (7.23) is equivalent to

$$\int_0^{T_L}[1 - R_L(\tau)]\,d\tau = \int_{T_L}^\infty R_L(\tau)\,d\tau \tag{7.24}$$

This equality allows T_L in Figure 7.4 to be the point that equates the shaded areas above and below the R_L curve. Integrating (7.18) for $\tau \to \infty$ gives

$$\sigma_y^2(t) = 2\overline{v'^2}T_L t - 2\overline{v'^2}\int_0^\infty \tau R_L(\tau)\,d\tau \approx 2\overline{v'^2}T_L t + \text{const} \approx 2\overline{v'^2}T_L t \tag{7.25}$$

because for $\tau \to \infty$, $R_L \to 0$, the integral in (7.25) is finite.

Both asymptotics (7.22) and (7.25) can be united in the following formula:

$$\sigma_y^2(t) = \frac{\overline{v'^2}t^2}{1 + t/(2T_L)}$$

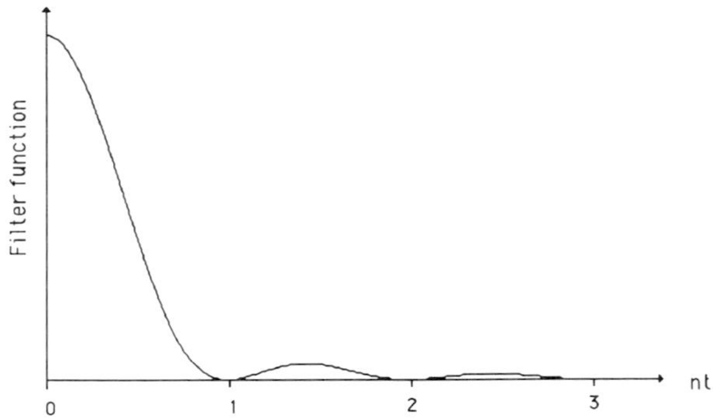

Figure 7.5 Filter function affecting averaging over a time T.

or (7.26)

$$\sigma_y^2(x) = \frac{\overline{v'^2}(x/U)^2}{1 + x/(2T_L U)}$$

where time was equated to $t = x/U$.

Substituting R_L in (7.18) by its Fourier transform (5.6) leads, after some calculations, to a useful result:

$$\sigma_y^2(t) = \overline{v'^2}t^2 \int_0^\infty S_L(n) \frac{\sin^2 \pi nt}{(\pi nt)^2}\, dn \tag{7.27}$$

where $S_L(n) = 2\pi\Phi(\omega)$ is the spectral density of the Lagrangian velocity component perpendicular to the mean wind, with $\omega = 2\pi n$ [see (5.8)]. Notice that because of a difference in definitions (5.1) and (7.15), S_L must be normalized by $\overline{v'^2}$ and, by this fact, differs from S in (5.8). The quantity $(\sin \pi nT/\pi nt)^2$ in (7.27) can be understood as a low-pass filter, so that the integrand represents the spectrum of averages of S_L over period t. When diffusion time is small, $\sigma_y(t)$ depends on the entire energy spectrum S_L [since $(\sin x/x)^2 \to 1$ for $x \to 0$]. It can be seen from Figure 7.5 that as diffusion time t increases, high frequencies (small-scale eddies) progressively decrease in their influence on $\sigma_y(t)$. This means that as the plume reaches greater and greater distances and grows in size, larger and larger turbulent eddy sizes become effective in diffusing the cloud, while smaller eddies become increasingly ineffective.

If t is very large, the filter is very narrow, because its first zero occurs at $nt = \pi$ (i.e., when $n = \pi/t \to 0$). In this case, $S_L(n) \approx S_L(0) = 2/\pi T_L$ and (7.27) is equivalent to (7.25) (since the integral of the filter function is equal π/t). If t is small, the filter function and the integral tend to 1 and (7.27) yields (7.22).

The spectrum and the autocorrelation in the formulas above are Lagrangian. In practice, only Eulerian parameters (fixed-point instruments) are

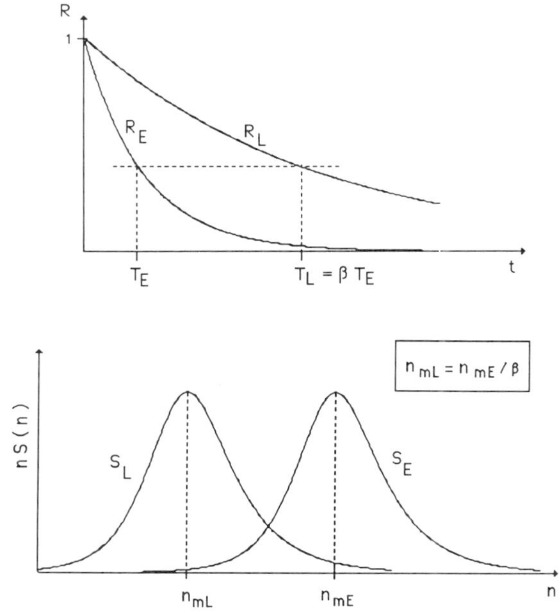

Figure 7.6 Eulerian and Lagrangian (a) autocorrelograms and (b) spectra.

measured, and the relation between Eulerian and Lagrangian turbulence needs to be developed.

Intuitively, velocities of particles following the airflow (Lagrangian) are more slowly varying than those measured by a fixed anemometer (Eulerian). Gifford (1955) and Hay and Pasquill (1959) suggested that Lagrangian and Eulerian spectra and autocorrelations are similar in shape but are displaced from each other by a factor β (Figure 7.6):

$$\begin{aligned} R_L(\beta t) &= R_E(t) \\ nS_L(k) &= \beta n S_E(\beta n) \end{aligned} \tag{7.28}$$

The last formula is obtained from the definition (5.3) for $S_E(n) = 2\pi\Phi_E(\omega)$ [see (5.8)], where $\omega = 2\pi n$. Parameter β can be formally defined from (7.23) as the ratio of the Lagrangian and Eulerian time scale $\beta = T_L/T_E$. On the average in the planetary boundary layer, β equals 4 for neutral conditions, about 10 for stable conditions, and about 1 for convective conditions (Pasquill, 1974).

The value of σ_y^2 depends strongly on the averaging procedure. The consequences are shown in Figure 7.7. The left side of Figure 7.7 depicts the boundaries of a smoke plume at any instant (such as in a snapshot) or as they would appear having been averaged over 15 minutes or 1 hour (as in a time exposure). The right side of the figure represents the distributions of the

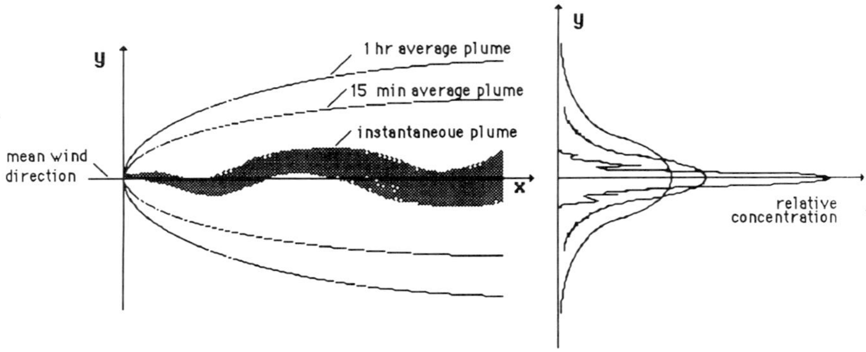

Figure 7.7 The diagram on the left represents the approximate outlines of a smoke plume observed instantaneously and of plumes averaged over 15 minutes and 1 hour. The diagram on the right shows the corresponding cross-plume distribution patterns.

concentration for each of the three plume types. The snapshot shows an irregular, varying plume, which gradually expands with distance from the source. The corresponding concentration distribution is irregular and quite peaked. The 15-minute average shows a smooth and regular boundary that can be considered an envelope containing all the short-period fluctuations of the plume. The 15-minute distribution resembles the Gaussian curve, as shown on the right side of the figure. The 1-hour cross-plume average concentration distribution is broader and smoother than in the instantaneous case but has a lower peak value.

In the random walk model, not only one, but many randomly moving particles can be considered. If N_i^k is the number of particles at point i at the moment of time k and the probability of the jump to the right is equal to the probability of jumping to the left and equal to $\frac{1}{2}$, then in the next moment of time, denoted by $k + 1$, the particle amount at the point i is

$$N_i^{k+1} = \frac{1}{2}N_{i-1}^k + \frac{1}{2}N_{i+1}^k \tag{7.29}$$

This formulation is equivalent to a Markov chain (Moore and Yackel, 1974). Equation (7.29) can be transformed to

$$\frac{C_i^{k+1} - C_i^k}{\Delta t} = \frac{\Delta x^2}{2\Delta t} \frac{C_{i+1}^k - 2C_i^k + C_{i-1}^k}{\Delta x^2} \tag{7.30}$$

where concentration C_i^k is defined as $C_i^k = N_i^k/\Delta x$. Assuming that the increments Δx and $\Delta t \to 0$ so that $\Delta x^2/\Delta t \to 2k = \text{const}$, it can be said that

$$\frac{\partial C}{\partial t} = k\frac{\partial^2 C}{\partial x^2} \tag{7.31}$$

This result shows that the formulation in terms of the random walk model is mathematically equivalent to the problem described by the diffusion equation. When the probability of the jump to the left is q and to the right is p, then

$$N_i^{k+1} = qN_{i-1}^k + pN_{i+1}^k \tag{7.32}$$

which leads to

$$\frac{\partial C}{\partial t} + U\frac{\partial C}{\partial x} = k\frac{\partial^2 C}{\partial x^2} \tag{7.33}$$

where

$$\begin{aligned} U &= (p - q)\frac{\Delta x}{\Delta t} \\ k &= q\frac{\Delta x^2}{2\Delta t} \end{aligned} \tag{7.34}$$

The quantity U can be interpreted as the flow velocity and k is the diffusivity. If $p = q$, then $U = 0$ and only diffusion is effective. If $p > q$, the particles move to the right along the x-axis.

7.3 K-THEORY MODELS

The turbulent diffusion equation was derived in Chapter 3 from the principle of conservation of mass. The turbulent diffusion equation is "unclosed" with respect to Reynolds terms. According to the K-theory, these terms can be defined as

$$\overline{u_i'c'} = -k_i\frac{\partial C}{\partial x_i} \tag{7.35}$$

To investigate the properties of the diffusion equation, a simple, one-dimensional diffusion equation based on K-parametrization with constant k will be considered:

$$\frac{\partial C}{\partial t} = k\frac{\partial^2 C}{\partial z^2} \tag{7.36}$$

Initial and boundary conditions are of the form

$$\begin{aligned} C(0, z) &= f(z) \qquad \text{for } t = 0 \\ C(t, z) &= 0 \qquad \text{for } z \to \pm\infty \end{aligned} \tag{7.37}$$

It will also be assumed that all space derivatives of C are zero on the edges of the domain considered.

In general, a well-defined boundary condition for a diffusion equation can be written in the form

$$a\frac{\partial C}{\partial z} + bC = c \tag{7.38}$$

where a, b, and c are coefficients or given functions. If $a = 0$ as in (7.37), function C is given (a boundary condition of the *first type*). This can be used to describe a *continuous point source* in three-dimensional case:

$$C = \frac{Q}{U}\delta(z - H_e)\delta(y - y_0) \tag{7.39}$$

where δ is Dirac's delta, which has the following properties:

$$\int_{-\infty}^{\infty} \delta(x)\, dx = 1 \tag{7.40}$$

$$\delta(z) = \begin{cases} 0 & \text{for } z \neq 0 \\ \infty & \text{for } z = 0 \end{cases} \tag{7.41}$$

Equation (7.39) can be obtained from the mass balance:

$$\int_{-\infty}^{\infty}\int_{-\infty}^{\infty} CU\, dy\, dz = Q \tag{7.42}$$

where Q is the continuous source strength in units of mass by units of time and U is the wind velocity at the height of the source. A continuous plume can be considered to be made up of an infinite number of puffs released sequentially with a vanishingly small interval of time between puffs (instantaneous sources).

The instantaneous point source is the conventional approximation to the type of release associated with a very short venting of material to the atmosphere (e.g., an explosion). The term "instantaneous point" is, of course, a mathematical simplification since even a small rapid explosion will have finite time and space dimensions. For an *instantaneous point source*, we find that

$$C = Q\delta(x - x_0)\delta(y - y_0)\delta(z - H_e) \tag{7.43}$$

The mean speed and direction of the puff can be expected to change as the wind pattern in which it is embedded changes with time. As the puff moves, it will expand about its center due to the action of turbulent fluctuations. First, the puff will grow slightly as its edges are mixed with the pure air by the small eddies, proportionally decreasing the concentration of the smoke. Finally, when a puff is embedded in eddies of approximately the same size, diffusion will be quite rapid and the concentration within the puff will decrease rapidly.

If in (7.38) the coefficient $a \neq 0$, the first derivative term describes the flux through the boundary as a result of the presence of a sink or source of mass.

Let us consider some properties of solutions of the diffusion equation (7.36). The first property results from the linearity: If $C_1, \ldots, C_n$ are solutions of this equation, their sum is also a solution of (7.36). The second property is expressed by the equation

$$\int_{-\infty}^{\infty} C\,dz = M_0 = \text{const} \tag{7.44}$$

which indicates that the total mass in the considered domain is constant. To prove this, integrate (7.44) with respect to time:

$$\frac{\partial}{\partial t}\int_{-\infty}^{\infty} C\,dz = \int_{-\infty}^{\infty} \frac{\partial}{\partial t} C\,dz = \int_{-\infty}^{\infty} k\frac{\partial^2}{\partial z^2} C\,dz = k\frac{\partial C}{\partial z}\bigg|_{\infty} - k\frac{\partial C}{\partial z}\bigg|_{-\infty} = 0 \tag{7.45}$$

According to the third property, the first moment of the solution C is constant:

$$\int_{-\infty}^{\infty} zC\,dz = \text{const} \tag{7.46}$$

This property can be derived by differentiation with respect to time.

From equations (7.44) and (7.46), it can be concluded that the gravity center of the puff is constant:

$$z_c = \frac{\int_{-\infty}^{\infty} zC\,dz}{\int_{-\infty}^{\infty} C\,dz} = \text{const} \tag{7.47}$$

The second moment of the solution has the following property:

$$\frac{\partial}{\partial t}\int_{-\infty}^{\infty} z^2 C\,dz = k\int_{-\infty}^{\infty} z^2\frac{\partial^2}{\partial z^2} C\,dz = -k\int_{-\infty}^{\infty} z\frac{\partial C}{\partial z}\,dz = 2k\int_{-\infty}^{\infty} C\,dz = 2kM_0 \tag{7.48}$$

which yields

$$\int_{-\infty}^{\infty} z^2 C\,dz = 2kM_0(t - t_0) \tag{7.49}$$

The result obtained simply means that the moment of inertia for the puff increases in time due to the spatial dispersion of pollutants.

A Fourier transform can be used to solve (7.36). Applying the transform operator to both sides of (7.36) yields

$$\frac{\partial \mathbb{F}}{\partial t} = k(i\omega)^2 \mathbb{F} \tag{7.50}$$

where $\mathbb{F}(t, \omega) = \mathscr{F}[C(z, t)]$ is the Fourier transform of the function C. The solution of the ordinary differential equation above has the form

$$\mathbb{F} = Ae^{-k\omega^2 t} \tag{7.51}$$

Notice that for $t = 0$, $\mathbb{F} = A$, which, together with the initial condition (7.37), yields

$$A = \mathscr{F}[C(z,0)] = \mathscr{F}[f(z)] \tag{7.52}$$

Applying inverse Fourier transform and the *convolution theorem* (according to the convolution theorem $\mathscr{F}[f^*g] = \mathscr{F}[f]\mathscr{F}[g]$) yields

$$C = f(z)^*\left[\frac{1}{\sqrt{4\pi kt}}\exp\left(-\frac{z^2}{4kt}\right)\right] = \frac{1}{\sqrt{4\pi kt}}\int_{-\infty}^{+\infty} f(s)\exp\left[-\frac{(z-s)^2}{4kt}\right]ds \tag{7.53}$$

where "*" is the convolution operator and the second term in the convolution is equal to $\mathscr{F}^{-1}[\exp(-k\omega^2 t)]$.

If the function f is defined as $f(z) = Q\delta(z - H_e)$, where Q is a mass of pollutants emitted from an instantaneous point source, then (7.53) simplifies to

$$C = \frac{Q}{\sqrt{4\pi kt}}\exp\left[-\frac{(z-H_e)^2}{4kt}\right] \tag{7.54}$$

Formula (7.54) describes the normal distribution with a mean value equal to H_e and a variance $\sigma^2 = 4kt$. In the case of three-dimensional diffusion equations, the solution takes the form

$$C = \frac{Q}{\sqrt{(4\pi kt)^3}}\exp\left(-\frac{r^2}{4kt}\right) \tag{7.55}$$

where $r^2 = x^2 + y^2 + (z - H_e)^2$. Integrating (7.55) with respect to time produces the formula for the concentration emitted from a continuous point source:

$$C = \frac{Q}{\sqrt{(4\pi k)^3}}\int_0^t \exp\left[-\frac{r^2}{4k(t-\tau)}\right]\frac{d\tau}{(t-\tau)^{3/2}} = \frac{Q}{4\pi kr}\operatorname{erfc}\left(-\frac{r}{\sqrt{4kt}}\right) \tag{7.56}$$

where $\operatorname{erfc}(p) = 1 - \operatorname{erf}(p)$ and

$$\operatorname{erf}(p) = \frac{2}{\sqrt{\pi}}\int_0^p \exp(-q^2)\,dq \tag{7.57}$$

When a reflecting surface is located on the level $z = \eta$, the flux through this surface is zero and (7.54) has to be modified to

$$C = \frac{Q}{\sqrt{4\pi kt}}\left\{\exp\left[-\frac{(z-H_e)^2}{4kt}\right] + \exp\left[-\frac{(z+H_e-2\eta)^2}{4kt}\right]\right\} \tag{7.58}$$

Formula (7.58) was obtained by placing an additional, but fictitious, source at the point $z = -H_e + 2\eta$, which yields the same effect as the reflecting surface.

The Gaussian function is the fundamental solution of the diffusion equation with the initial condition given by Dirac delta at the origin of the coordinate system. This means that a "velocity of diffusion" or speed with which a solution propagates in space is infinite. The velocity of air particles in Brownian motion is typically two orders of magnitude higher than in turbulent motion. As a result, the diffusion equation better describes molecular diffusion than turbulent diffusion.

A more complex problem of transport along the x-axis with a mean velocity U, and diffusion along the y and z axes perpendicular to the wind direction can be described by the equation

$$U\frac{\partial C}{\partial x} = k_y\frac{\partial^2 C}{\partial y^2} + \frac{\partial}{\partial z}k_z\frac{\partial C}{\partial z} \pm I \tag{7.59}$$

where k_x and k_z are eddy-diffusivity coefficients and I is a source/sink function. Coefficient k_y is a constant, while k_z is a function of height. The sink term is very often modeled as

$$I = -\kappa C \tag{7.60}$$

where coefficient κ is related to the *half-life time*, $\tau_{1/2}$, in the following way:

$$\kappa = \frac{\ln 2}{\tau_{1/2}} \tag{7.61}$$

To solve the problem above, it is convenient to assume that the solution can be split into two components in the following way:

$$C(x, y, z) = C_1(x, z)C_2(x, y) \tag{7.62}$$

where functions C_1 and C_2 are defined by the following equations:

$$U\frac{\partial C_1}{\partial x} = \frac{\partial}{\partial z}k_z\frac{\partial C_1}{\partial z} \pm I \tag{7.63}$$

$$U\frac{\partial C_2}{\partial x} = k_y\frac{\partial^2 C_2}{\partial y^2} \tag{7.64}$$

with boundary conditions $C_2(0, y) = \delta(y)$ and $C_2(x, \pm\infty) = 0$. It was shown earlier that the solution to (7.64) has the form

$$C_2 = \frac{Q}{\sqrt{2\pi\sigma_y^2}}\exp\left(-\frac{y^2}{2\sigma_y^2}\right) \tag{7.65}$$

where $\sigma_y^2 = 2k_y x/U$. This simply means that the solution in the y direction is described by a normal distribution. Linear dimensions of the plume perpendicular to its axis are often given in terms of the standard deviation of the concentration distribution, which is in good agreement with the empirical data. The function C_1 can be obtained by a numerical solution of (7.63) (see the Appendix).

For ground releases, the K-theory can successfully predict the vertical spread of a plume at short range. For example, Nieuwstadt and van Ulden (1978) numerically solved the diffusion equation with wind and diffusivity profiles specified by the Monin-Obukhov expressions and found agreement between their predictions and field measurements.

7.4 GAUSSIAN MODELS

It was shown that the Gaussian distribution provides a solution for the random walk problem and is also a fundamental solution to the diffusion equation. Many experimental studies indicate that the Gaussian formula has a practical applicability for constructing atmospheric diffusion models (Pasquill, 1974). Models based on the assumption that the concentration can be described by normal distribution are called *Gaussian models.* The most commonly used model in engineering practice was developed by Pasquill (1961). It is based on the equation

$$C(x, y, z) = \frac{Q}{2\pi\sigma_y(x)\sigma_z(x)U} \exp\left\{-\frac{1}{2}\left[\frac{y}{\sigma_y(x)}\right]^2\right\} \times\left(\exp\left\{-\frac{1}{2}\left[\frac{z-H_e}{\sigma_z(x)}\right]^2\right\} + \exp\left\{-\frac{1}{2}\left[\frac{z+H_e}{\sigma_z(x)}\right]^2\right\}\right) \tag{7.66}$$

where H_e is an effective height of the source, calculated as the sum of the geometric height of the source h_e and the plume rise Δh resulting from the buoyancy effect in the plume (e.g., Briggs, 1973, 1985; Manins, 1979; Berkowicz et al., 1985). The coefficient σ_y was derived empirically and σ_z theoretically, as functions of distance and stability. Equation (7.66) includes the reflection of pollutants from the ground surface.

The surface-concentration pattern from an elevated source described by (7.66) is typically like the one shown in Figure 7.8. The concentration is essentially zero near the base of the chimney, rises rapidly to a peak value at some downstream distance, and decreases regularly thereafter. The concentration is always highest along the projection of the plume axis on the ground.

Since the time-averaged plume from a continuous point source expands both laterally and vertically with downwind distance from the source, the centerline concentration within the plume will decrease continuously with distance. The rate at which the centerline concentration decreases varies with the atmospheric stability.

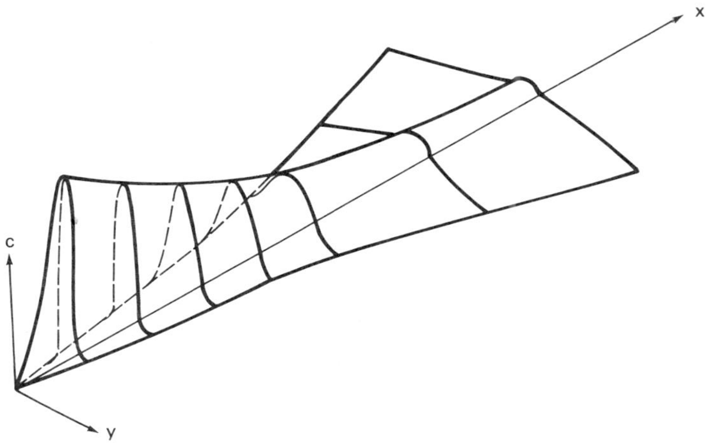

Figure 7.8 The surface-concentration pattern downwind from an elevated source. C, concentration; Y, cross-wind distance; X, downwind distance.

On the basis of empirical data, Pasquill (1961) and Gifford (1961) developed a practical scheme for the estimation of thermal stability. Pasquill established the following stability classes:

A extremely unstable

B moderately unstable

C slightly unstable

D neutral

E slightly stable

F moderately stable

TABLE 7.1 Definition of Pasquill's Stability Classes[a]

	Daytime Insolation			Nighttime Conditions	
Surface Wind Speed (m/s)	Strong	Moderate	Slight	Thin Overcast or 4/8 Low Clouds	3/8 Cloudiness
< 2	A	A–B	B		
2, 3	A–B	B	C	E	F
4, 5	B	B–C	C	D[a]	E
6	C	C–D	D	D	D
> 6	C	D	D	D	D

[a] Class D is applicable to heavy overcast day or night, or strong winds. The cloudiness is defined as that fraction of the sky that is covered by clouds.

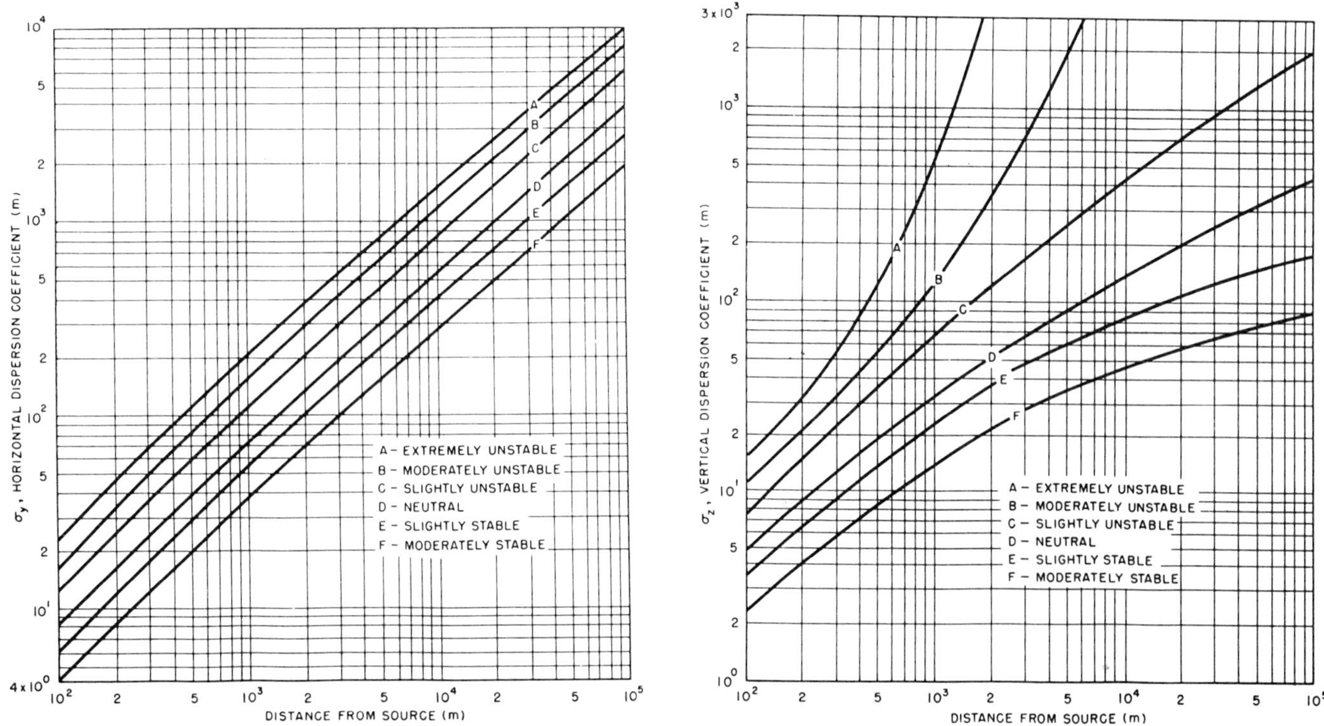

Figure 7.9 Pasquill–Gifford dispersion coefficients σ_y and σ_z versus downwind distance and stability. (After Gifford, 1961, "Nuclear Safety.")

TABLE 7.2 Briggs' Formulas for σ_y and σ_z

Pasquill Class	σ_y (m)	σ_z (m)
	Rural Conditions	
A	$0.22x(1 + 0.0001x)^{-0.5}$	$0.20x$
B	$0.16x(1 + 0.0001x)^{-0.5}$	$0.12x$
C	$0.11x(1 + 0.0001x)^{-0.5}$	$0.08x(1 + 0.0002x)^{-0.5}$
D	$0.08x(1 + 0.0001x)^{-0.5}$	$0.06x(1 + 0.0015x)^{-0.5}$
E	$0.06x(1 + 0.0001x)^{-0.5}$	$0.03x(1 + 0.0003x)^{-1}$
F	$0.04x(1 + 0.0001x)^{-0.5}$	$0.016x(1 + 0.0003x)^{-1}$
	Urban Conditions	
A–B	$0.32x(1 + 0.0004x)^{-0.5}$	$0.24x(1 + 0.0001x)^{-0.5}$
C	$0.26x(1 + 0.0004x)^{-0.5}$	$0.20x$
D	$0.16x(1 + 0.0004x)^{-0.5}$	$0.14x(1 + 0.0003x)^{-0.5}$
E–F	$0.11x(1 + 0.0004x)^{-0.5}$	$0.08x(1 + 0.00015x)^{-0.5}$

The stability classes are determined by easily measured variables such as cloud cover, wind speed, and insolation (Table 7.1). This approach, however, ignores the vertical structure of the boundary layer and is strongly biased toward neutral stability when convective conditions actually exist.

For each class, Pasquill (1961) and Gifford (1961) provided graphs of σ_y and σ_z as functions of distance from the source. These curves represent diffusion from ground sources over flat and smooth terrain (Figure 7.9). The Pasquill and Gifford curves were revised by Briggs (1973) and modified by differentiating between rural and urban conditions (Table 7.2).

Many other similar models are reported in literature. However, at present, there is a dearth of accurate measurements (especially in the vertical) and graphs of σ_y and σ_z, and their dependence on a state of the ABL is not very reliable. As a result, the operational Gaussian diffusion models estimate measured concentrations within a factor of 2 or 3 at best.

7.5 LABORATORY AND LARGE EDDY SIMULATION EXPERIMENTS

The understanding of diffusion in the convective boundary layer was advanced significantly by the laboratory water tank experiments of Willis and Deardorff (1976, 1978, 1981) (Figure 7.10). Their investigations illustrated that for elevated sources ($h_e > 0.1z_i$), the average plume centerline, defined as the mean maximum concentration, descended within a short distance from the source until it reached the ground. In contrast, the average centerline from near surface releases ($h_s < 0.1z_i$) ascended after a short downwind distance (Figure 7.11a). This result strongly differed from the conclusions of the

Figure 7.10 Tank used by Deardorff (pictured above) and Willis in their laboratory experiments. (Deardorff and Willis.)

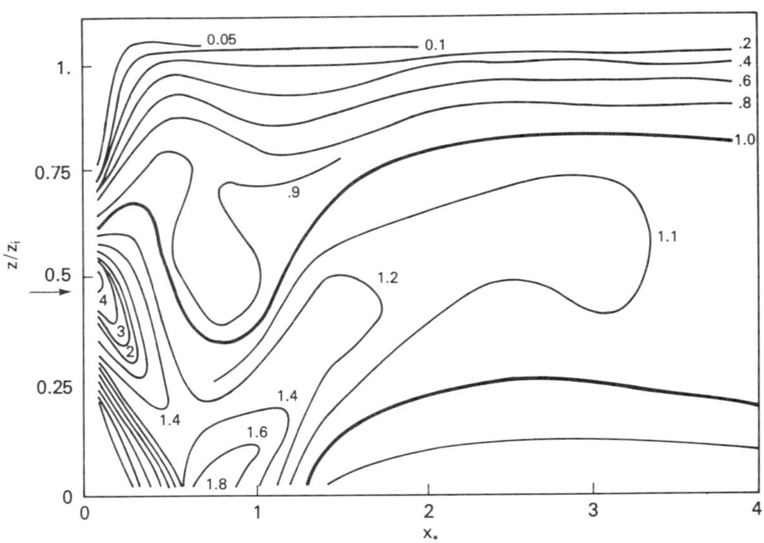

(a)

Figure 7.11 Dimensionless cross-wind integrated concentrations C_y^* simulated in the laboratory experiments for point sources: (a) $z_s/z_i = 0.49$; (b) $z_s/z_i = 0.24$; (c) $z_s/z_i = 0.067$. (Reprinted with permission from *Atmospheric Environment*, vol. 12, 1978, and vol. 15, 1981, Willis and Deardorff, Pergamon Journals, Ltd.)

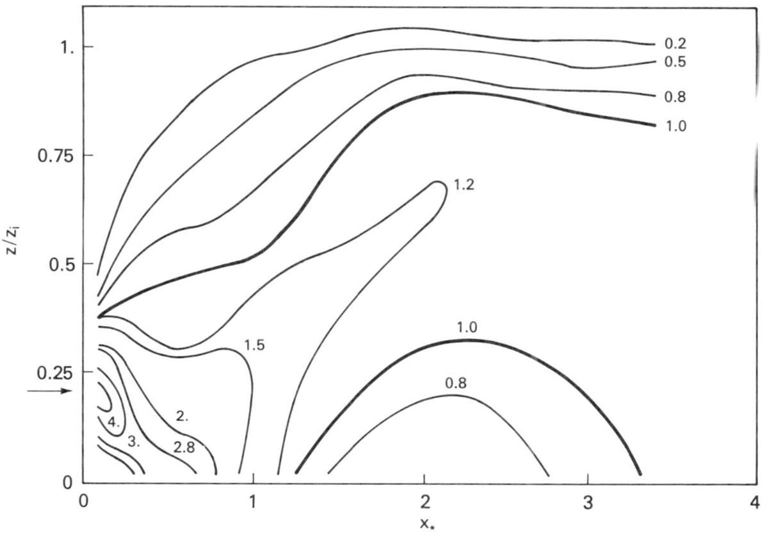

(b)

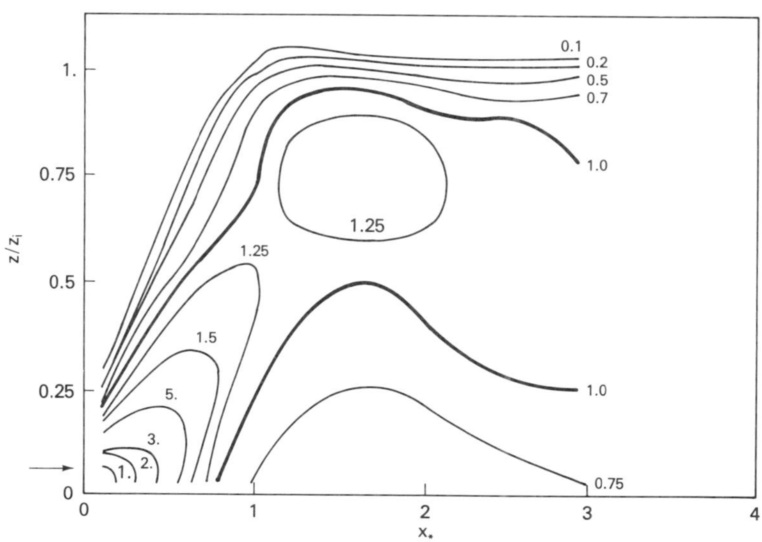

(c)

Figure 7.11 *continued*

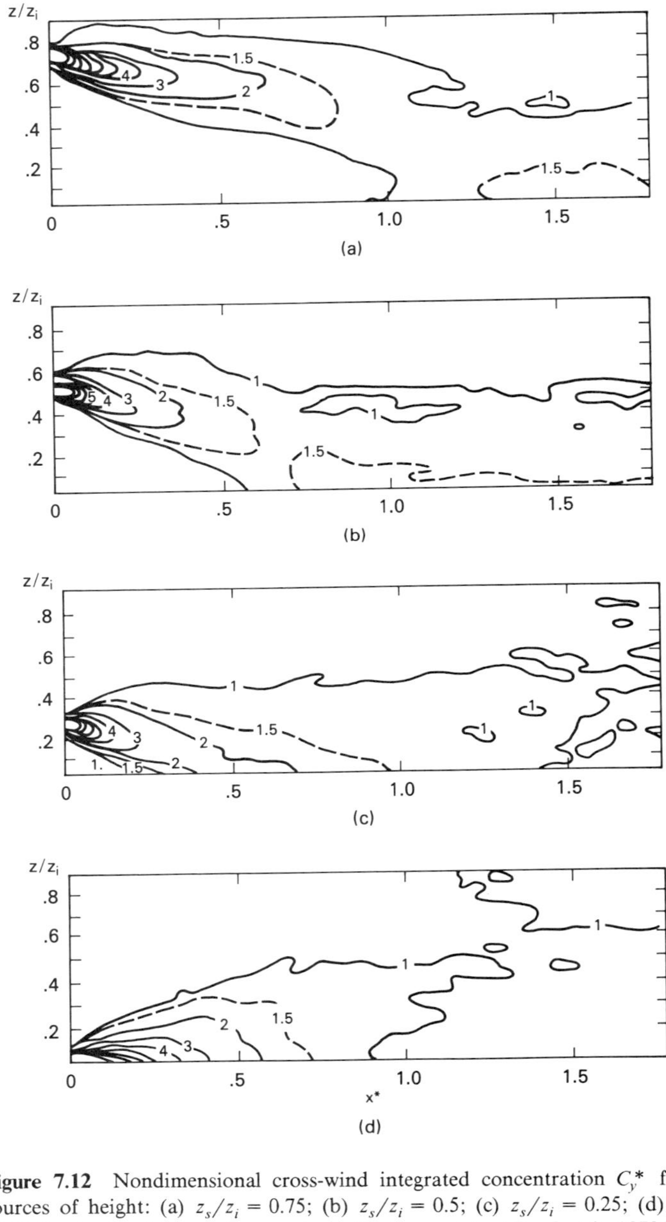

Figure 7.12 Nondimensional cross-wind integrated concentration C_y^* for point sources of height: (a) $z_s/z_i = 0.75$; (b) $z_s/z_i = 0.5$; (c) $z_s/z_i = 0.25$; (d) $z_s/z_i = 0.025Z_i$ calculated from a Lagrangian diffusion model. (After Lamb, 1979.)

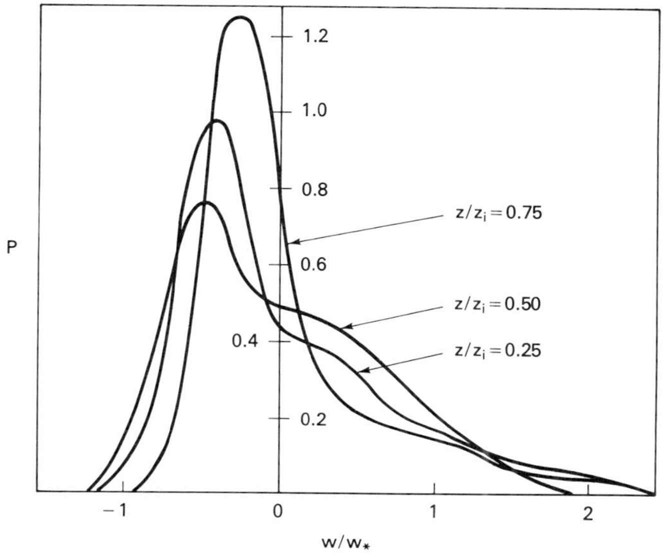

Figure 7.13 Probability density distribution of vertical velocity at three levels of a convective mixed layer derived by Lamb (1982) using the numerical model of Deardorff (1974a,b).

Gaussian plume model. The laboratory experiments also demonstrated that turbulent flux in certain regions of the mixed layer is opposed to the local concentration gradient.

The first support for Willis and Deardorff's laboratory observations was obtained from the numerical experiments of Lamb (1979, 1982). Lamb used the results from the large-eddy simulation model of Deardorff (1972b) to trace the motions of thousands of particles "released" into a numerical flow field. Source heights h_e were chosen to be $h_e/z_i = 0.067$, 0.24, and 0.49 (Figure 7.12).

Position of maximum concentrations can be explained by the probability density distributions of the vertical velocity shown in Figure 7.13 (compare with Figures 3.12 and 3.13). Because downdrafts cover more than half the area of the horizontal plane over the bulk of the mixed layer depth, a majority of material released by an elevated source starts descending and continues to descend for a significant time. On the other hand, material released at the surface can only ascend or move horizontally. Particles released into the base of an updraft begin to rise immediately, while those emitted into a downdraft move approximately horizontally until they encounter updrafts and are transferred upward. After a sufficiently long travel time (about z_i/w_*, which in a typical daytime CBL corresponds to about 15 minutes), when a majority of particles have entered updrafts, the location of the maximum particle concentration lifts off the ground and rises toward the top of the mixed layer.

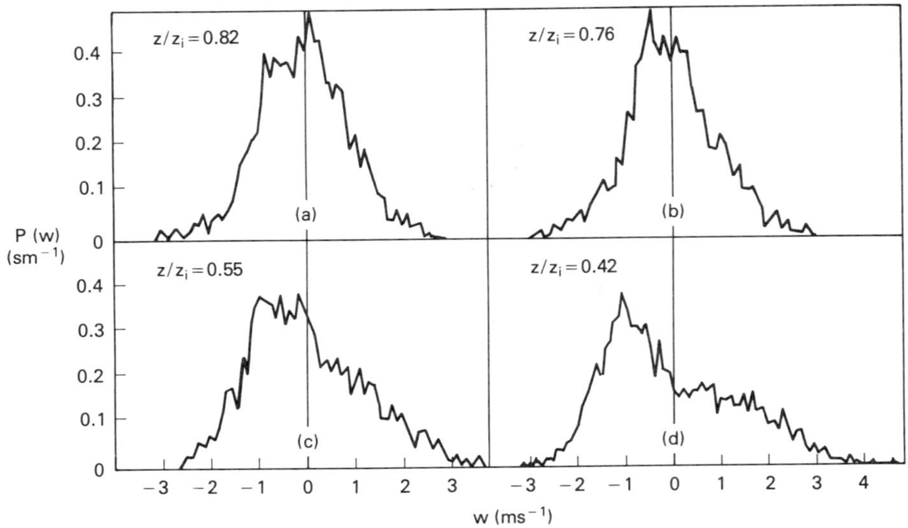

Figure 7.14 Distribution as in Figure 7.13 obtained experimentally by Caughey et al. (1983).

These conclusions were confirmed by the ABL measurements shown in Figure 7.14. The distributions of the vertical velocity above the surface layer (i.e., $z/L > 1$, $z/z_i > 0.1$) are positively skewed with a negative mode, except for the run at $z/z_i = 0.82$, which has a mode close to zero. The fraction of total run length covered by downdrafts decreases with height and is 0.58 at $z/z_i = 0.42$, 0.56 at $z/z_i = 0.55$, 0.52 at $z/z_i = 0.76$, and 0.50 at $z/z_i = 0.82$. The skewness is marked mostly in the middle of the mixed layer, where turbulence is generated primarily by buoyancy effects or by the upward transport of buoyancy-produced turbulence from the surface layer. At $z/z_i = 0.42$, the most probable velocity is a downdraft of 1 m s^{-1}. Near the upper and lower boundaries of the CBL, the probability density distributions have a mode near zero and show a much more symmetrical form. This is due to the generation of turbulent energy at the boundaries (i.e., mechanical production in the surface layer and breaking wave at the upper boundary).

The dispersion parameters found in laboratory and numerical simulations were in agreement with one another when suitably nondimensionalized (e.g., Lamb, 1982). For elevated releases, the dispersion parameter σ_y/z_i is dependent on the dimensionless distance $x_* = x/z_i\ w_*/U$:

$$\frac{\sigma_y}{z_i} = \frac{0.6x_*}{(1 + \alpha x_*)^{1/2}} \tag{7.67}$$

where α depends on the proportionality coefficient between T_L and z_i/w_*. Field observations (Briggs, 1985; Weil, 1983) suggest α ranges from 0 to ~ 2, with a typical value of ~ 0.7. Deardorff and Willis's laboratory data yield $\alpha = 0.91$.

For σ_z, Deardorff and Willis (1975) suggested an interpolation formula based on their laboratory data. The formula matches the neutral and free convection limits. It is valid for surface sources ($h_e < 0.1z_i$), for short ranges ($x_* < 0.5$) and its form is

$$\frac{\sigma_z}{z_i} = \left[\left(\beta \frac{u_*}{w_*} x_*\right)^2 + 0.25x_*^3 + 1.8h_e^{*2/3}x_*^2\right]^{1/2} \tag{7.68}$$

where

$$\beta = \frac{0.4 + 0.8w_*h_e^*/(u_*x_*)}{1 + w_*h_e^*/(u_*x_*)} \tag{7.69}$$

and $h_e^* = h_e/z_i$.

The results obtained can be used to improve the Gaussian models. When applied to the Gaussian models, the formulas above approximate ground-level concentrations for $X < 0.5$, but beyond that range overestimate them somewhat (Weil, 1985).

7.6 HIGHER-ORDER CLOSURE MODELS

The Gaussian and K-closure models fail to describe diffusion under convective conditions. Convective diffusion is much better modeled by employing higher-order closures. This type of modeling can be examined using a model proposed by Sun and Chang (1986b). The diffusion equation in this model has the form

$$\frac{\partial C}{\partial t} + U\frac{\partial C}{\partial x} + V\frac{\partial C}{\partial y} = -\frac{\partial \overline{v'c'}}{\partial y} - \frac{\partial \overline{w'c'}}{\partial z} \tag{7.70}$$

This form was obtained from (2.56) by omitting the molecular diffusion terms and assuming zero vertical velocity. In addition, the magnitude of the diffusive term in the x direction was assumed to be much less than that of the advection term. Similar models were developed by Enger (1983) and Sykes et al. (1986).

To close (7.70), second-moment equations were used [compare with (3.39) and (3.49)]:

$$\begin{aligned}\frac{\partial \overline{w'c'}}{\partial t} + U\frac{\partial \overline{w'c'}}{\partial x} + V\frac{\partial \overline{w'c'}}{\partial y} &= -\overline{u'w'}\frac{\partial C}{\partial x} - \overline{v'w'}\frac{\partial C}{\partial y} - \overline{w'^2}\frac{\partial C}{\partial z} \\ &+ \frac{g}{T_o}(1 - a_2)\overline{c'\theta'} - \frac{\partial \overline{v'w'c'}}{\partial y} + A\frac{\partial}{\partial z}\left(ql\frac{\partial \overline{w'c'}}{\partial z}\right) - a_1\frac{q}{l}\overline{w'c'}\end{aligned} \tag{7.71}$$

and

$$\frac{\partial \overline{\theta' c'}}{\partial t} + U\frac{\partial \overline{\theta' c'}}{\partial x} + V\frac{\partial \overline{\theta' c'}}{\partial y} = -\overline{u'\theta'}\frac{\partial C}{\partial x} - \overline{v'\theta'}\frac{\partial C}{\partial y} - \overline{w'\theta'}\frac{\partial C}{\partial z}$$
$$-\overline{c'w'}\frac{\partial \Theta}{\partial z} - \frac{\partial \overline{v'\theta' c'}}{\partial y} + A\frac{\partial}{\partial z}\left(ql\frac{\partial \overline{\theta' c'}}{\partial z}\right) - a_3\frac{q}{l}\overline{\theta' c'} \tag{7.72}$$

Moreover, higher-order terms were parametrized in the manner described in Chapter 6. Terms multiplied by the constants a_1 and a_2 were obtained as a result of parametrization of the pressure term. The flux-gradient formula was accepted as a parametrization of the transport term. Molecular terms were neglected. The term multiplied by the constant a_3 parametrizes the molecular destruction rate. It can be noticed that the closure scheme employed is based on the mixing length l and doubled turbulent kinetic energy $q^2 = 2E$, instead of the dissipation rate ϵ.

To simplify the problem, a cross-wind integration was performed. Since the mean and turbulence fields of meteorological variables are horizontally homogeneous, (7.70)–(7.72) can be integrated along the y direction, yielding

$$\frac{\partial C_y}{\partial t} + U\frac{\partial C_y}{\partial x} = -\frac{\partial \overline{w'c'_y}}{\partial z}$$

$$\frac{\partial \overline{w'c'_y}}{\partial t} + U\frac{\partial \overline{w'c'_y}}{\partial x} = -\overline{u'w'}\frac{\partial C_y}{\partial x} - \overline{w'^2}\frac{\partial C_y}{\partial z} + \frac{g}{T_o}(1 - a_2)\overline{\theta' c'_y}$$
$$+A\frac{\partial}{\partial z}\left(ql\frac{\partial \overline{w'c'_y}}{\partial z}\right) - a_1\frac{q}{l}\overline{w'c'_y} \tag{7.73}$$

$$\frac{\partial \overline{\theta' c'_y}}{\partial t} + U\frac{\partial \overline{\theta' c'_y}}{\partial x} = -\overline{w'\theta'}\frac{\partial C}{\partial z} - \overline{c'_y w'}\frac{\partial \Theta}{\partial z} + A\frac{\partial}{\partial z}\left(ql\frac{\partial \overline{\theta' c'_y}}{\partial z}\right) - a_3\frac{q}{l}\overline{\theta' c'_y}$$

where C_y and c_y are the cross-wind averaged concentration and its fluctuation.

The model was tested for wind, potential temperature, and turbulence fields generated by the ABL model discussed in Section 6.3.6. The system of governing equations was solved numerically for five cases, which corresponded to pollution sources at the heights 0.75, 0.50, 0.25, 0.083, and $0.021z_i$, where z_i was the height of the inversion.

The cross-wind integrated quantities of concentration C_y, concentration flux $\overline{w'c'_y}$, and covariance of temperature and concentration $\overline{\theta' c'_y}$ were transformed to the dimensionless form

$$C_y^* = \frac{C_y U_m z_i}{Q} \tag{7.74}$$

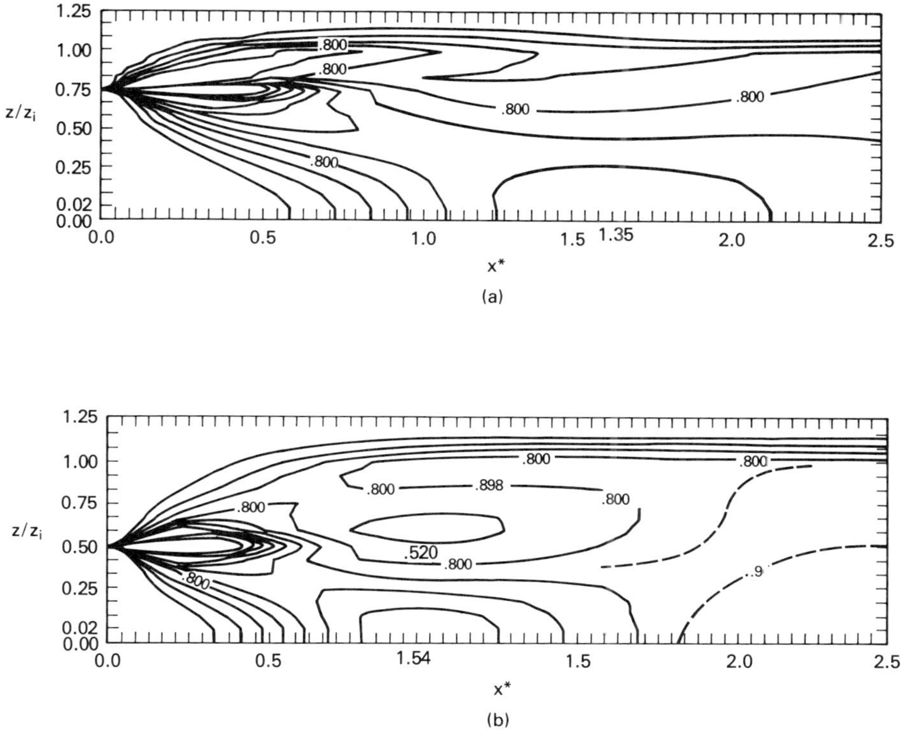

Figure 7.15 Calculated dimensionless cross-wind integrated concentrations C_y^* (with a contour interval of 0.2) for different dimensionless source heights: (a) $z_s/z_i = 0.75$; (b) $z_s/z_i = 0.50$; (c) $z_s/z_i = 0.25$; (d) $z_s/z_i = 0.083$. (Sun and Chang, 1986. *Journal of Climate and Applied Meteorology* published by the American Meteorological Society.)

and

$$\left[\overline{w'c'_y}\right]_* = \frac{\overline{w'c'_y}U_m z_i}{Qw_*}$$
$$\left[\overline{\theta'c'_y}\right]_* = \frac{\overline{\theta'c'_y}U_m z_i}{Q\Theta_*} \qquad (7.75)$$

where Q is the strength of the continuous source, U_m is a mean wind velocity, and w_* and θ_* are the convective wind velocity and temperature scales. The results of this model are shown in Figure 7.15. The dimensionless downwind distance x_* in this figure is defined as $x_* = x/z_i w_*/U_m$.

For an elevated point source, the cross-wind integrated concentration centerline was found to descend until it intercepted the ground. Material

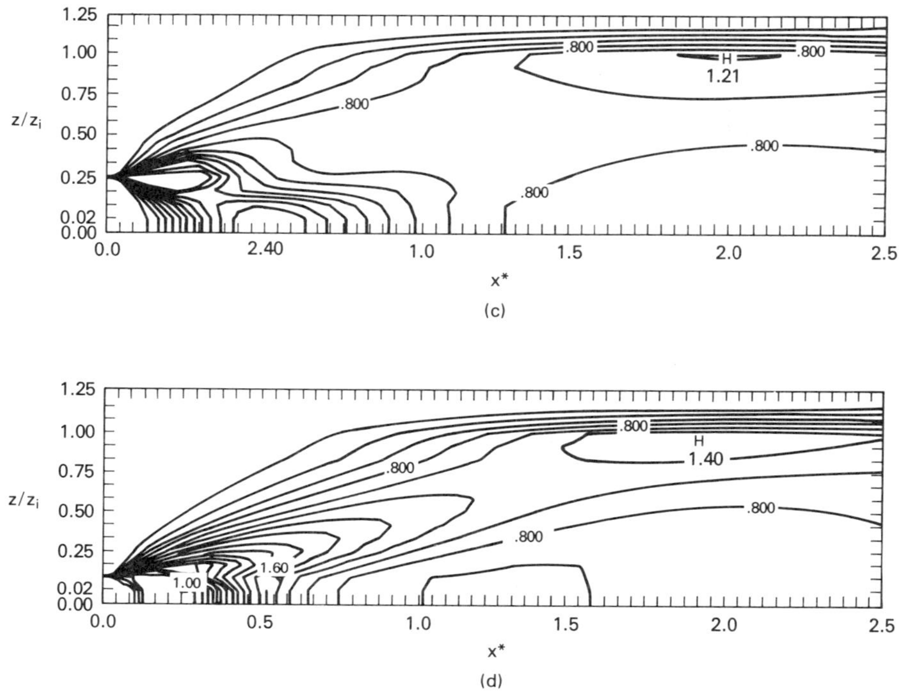

Figure 7.15 *continued*

released by a surface source gradually moved upward into the mixed layer, carried aloft by the concentrated but small convective updrafts. The major deficiency is that the tongue of high concentration extends to a higher altitude in the numerical simulations than in the laboratory (Figure 7.11), which may be caused by insufficient turbulent mixing in the upper mixed layer.

7.7 MONTE CARLO METHODS

The Monte Carlo method was initiated in Obukhov's (1959) brief paper, where it was proposed that the evolution of a diffusing air particle's motion in the atmosphere formed a Markov process. The method is based on Langevin's equation, which can be derived from the simple hypothesis that the velocity W for the diffusing air particle at a time τ after the present time t is given by the sum of a correlated part $R(\tau)W(t)$ and a purely random, uncorrelated part $\eta(t)$ (Smith, 1968):

$$W(t + \tau) = R(\tau)W(t) + d\eta(t) \tag{7.76}$$

For a stationary, homogeneous turbulent flow, $R(\tau)$ can be shown to be the Lagrangian velocity autocorrelation function. The random increment to the wind speed $d\eta(t)$ satisfies the conditions that $\overline{d\eta} = 0$ and $\overline{d\eta W} = 0$, where the overbar indicates ensemble averaging.

The turbulent wind speed $W(t + \tau)$ can be expanded into a Taylor's series:

$$W(t+\tau) = W(t) + dt\frac{dW}{dt} + r \tag{7.77}$$

where r are higher-order terms. For a small enough time interval τ_s, the higher-order terms can be neglected in (7.77) which, together with (7.76), forms Langevin's equation:

$$dW(t) + \{[1 - R(\tau)]\}W(\tau) = d\eta(\tau) \tag{7.78}$$

Langevin's equation is usually presented in the form

$$dW + \beta W(t) = d\eta(t) \tag{7.79}$$

from which, with (7.78), it follows that $\beta = (1 - R)$. Assuming that $R(t) = \exp(-t/T_L) \approx 1 - dt/T_L$ gives $\beta = dt/T_L$, where T_L is the Lagrangian integral time scale. Equation (7.79) can be interpreted as describing the particle motion, influenced by a small, random force $d\eta$ which arises from pressure terms and by a smaller force $-\beta W$ associated with viscous terms (Lin and Reid, 1962). Statistical properties of $d\eta$ differ for inhomogenous turbulence.

The Monte Carlo models have provided a number of interesting and potentially quite useful results. As an example of such a model, consider the model developed by De Baas et al. (1986). The model is based on Langevin's equation with $\beta = dt/T_L = \text{const}$. The $d\eta$ distribution is skewed in the mixed layer and can be constructed from two Gaussian distributions P_1 and P_2 such that the resulting distribution has the given moments dependent on $\overline{W'^2}$, $\overline{W'^3}$, and T_L:

$$\overline{d\eta}, \quad \overline{d\eta^2}, \quad \overline{d\eta^3} \tag{7.80}$$

There are four requirements for moments of P_1 and P_2. The first is

$$a_1\int P_1(x) + a_2\int P_2(x)\,dx = 1 \tag{7.81}$$

where a_1 and a_2 are coefficients and each of the two distributions P_i have unknown yet mean m_i and variance σ_i^2. This leads to

$$a_1 + a_2 = 1 \tag{7.82}$$

The three other requirements are

$$\begin{aligned} a_1 m_1 + a_2 m_2 &= \overline{d\eta} \\ a_1\left(m_1^2 + \sigma_1^2\right) + a_2\left(m_2^2 + \sigma_2^2\right) &= \overline{d\eta^2} \\ a_1\left(m_1^3 + 3m_1\sigma_1^2\right) + a_2\left(m_2^3 + 3m_2\sigma_2^2\right) &= \overline{d\eta^3} \end{aligned} \tag{7.83}$$

The equations above still have two degrees of freedom and can be closed assuming that $m_i^2 = \sigma_i^2$, where $i = 1, 2$. For the given moments of $d\eta$, all six unknowns a_1, a_2, m_1, m_2, σ_1^2, and σ_2^2 can be calculated. The probability that a first distribution will be chosen is a_1. Since the probability of downdrafts is greater, the larger of a_1 and a_2 defines the direction of random motion along the z-axis.

The Monte Carlo computer program first generates a random number to determine the direction of the random motion. Next, the computer generates a second number. This number $d\eta$ defines a random increment along the z-axis in (7.79). The motion in the model is obtained by the use of the following scheme:

$$\begin{aligned} x(t + \Delta t) &= x(t) + U(z)\,\Delta t \\ z(t + \Delta t) &= z(t) + \frac{\Delta t\left[W(t + \Delta t) + W(t)\right]}{2} \\ W(t + \Delta t) &= \frac{W(t)(1 - 0.5\,\Delta t/T_L)}{1 + 0.5\,\Delta t/T_L} + \frac{d\eta}{1 + 0.5\,\Delta t/T_L} \end{aligned} \tag{7.84}$$

where the last equation was obtained from (7.79). The number of particles released was set at 2×10^4 and the time step at $\Delta t/T_L = 0.05$. At the top and bottom, reflective boundary conditions were imposed on the particle motion.

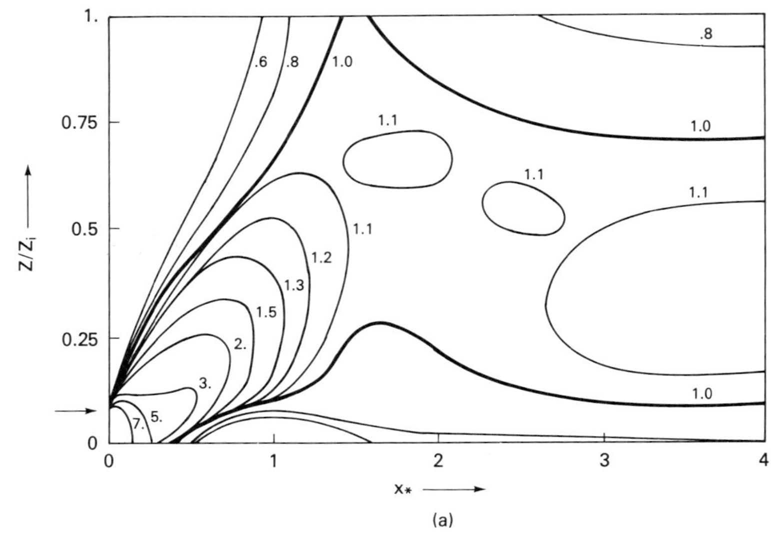

Figure 7.16 Contours in the vertical x, z plane of the dimensionless concentration presenting the results of the Langevin model for different source heights: (a) $z_s/z_i = 0.067$; (b) $z/z_i = 0.24$; (c) $z/z_i = 0.49$. Source height is indicated by arrow on ordinate. (After De Baas et al., 1986.)

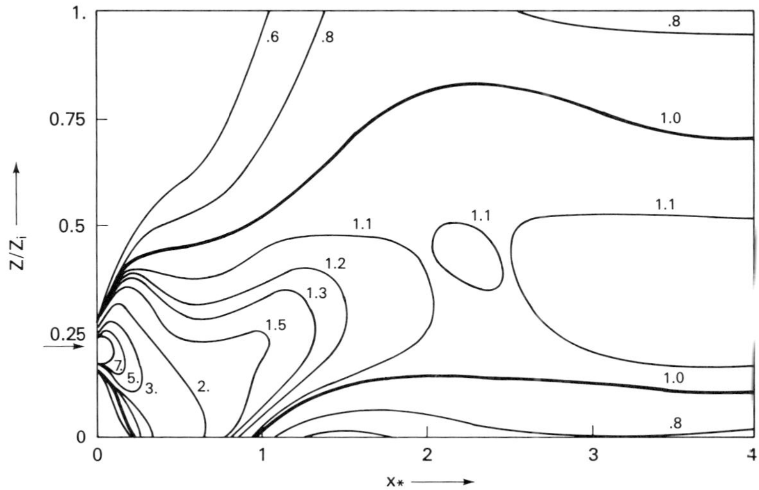

(b)

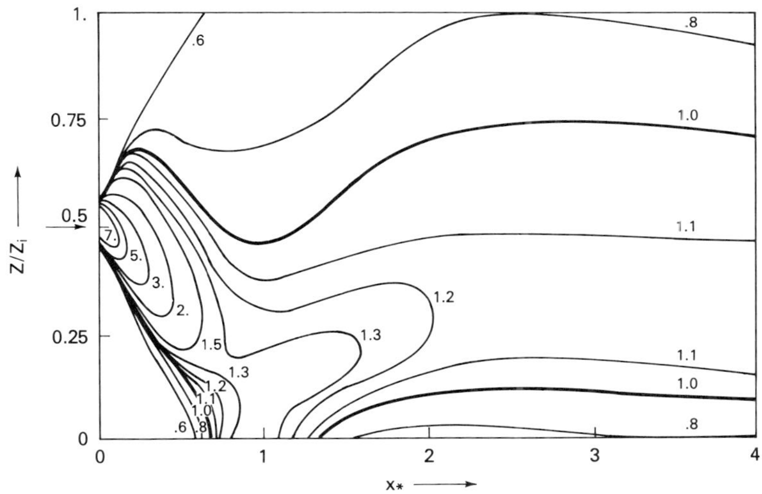

(c)

Figure 7.16 *continued*

The resulting concentration profiles were nondimensionalized with Q/z_iU and given as functions of a nondimensional distance $x_* = (w_*/U)(x/z_i)$, where x is the distance over which particles are advected by the uniform mean wind U in time t. Contour plots were made of the concentration as a function of height and time or downwind distance (Figure 7.16) until the distribution reached a steady state. The simulations described the significant features found in the water-tank experiments, shown in Figure 7.11. Particles released at ground level remained at the ground at first, but then rose above midlevel before reaching the equilibrium state. Particles released at one-fourth of the boundary layer height and at midlevel were first swept downward, remained a short time at the surface, and then rose to midlevel. The flaw of the model is that concentration profiles obtained for large x_* values are not uniform. This can be corrected by applying a slightly different form of governing equations.

PROBLEMS

7.1. Write a program simulating the random walk problem shown in Figure 7.3.

7.2. Find the Lagrangian time scale assuming that $R_L(t) = \exp(-at^2)$.

7.3. Adopting the solution of the diffusion equation for an instantaneous release in a uniform flow,

$$C = \frac{Q}{(2\pi)^{3/2}\sigma_x\sigma_y\sigma_z}\exp\left[-\frac{(x-Ut)^2}{2\sigma_x^2} - \frac{y^2}{2\sigma_y^2} - \frac{z^2}{2\sigma_z^2}\right]$$

where σ_x, σ_y, and σ_z are functions of time:

(a) Find

$$\frac{\partial C}{\partial t} + U\frac{\partial C}{\partial x}$$

(b) Find the second derivatives of C with respect to x, y, and z.

(c) Use the results above to show that

$$k_x = \frac{1}{2}\frac{\partial}{\partial t}\sigma_x^2 \qquad k_y = \frac{1}{2}\frac{\partial}{\partial t}\sigma_y^2 \qquad k_z = \frac{1}{2}\frac{\partial}{\partial t}\sigma_z^2$$

where k_x, k_y, and k_z are diffusion coefficients.

(d) Show that if $t \ll T_L$, then

$$k_y = \overline{v'^2}t$$

(e) Show that if $t \gg T_L$, then

$$k_y = \overline{v'^2}T_L$$

7.4. Assuming that $T_L = a\overline{v'^2}/\epsilon$ and $\epsilon = E^2/k_m$, from (7.26b), find $\sigma_y^2(x)$ for the neutral surface layer. Assume that $z = 10$ m, $\overline{v'^2} = 1.2$ m^2/s^2, $E = 2$ m^2/s^2, $U = 10$ m/s and $a = 1$. Plot the result.

7.5. Solve (7.36)–(7.37) numerically (see the Appendix). Compare the numerical and analytical results.

7.6. Find x for which $C(x, 0, 0)$ in (7.66) reaches maximum. Assume stability class A and rural conditions. Use Table 7.2. Assume $H_e = 100$ m.

7.7. Derive (7.27).

CHAPTER EIGHT

MEASUREMENTS

8.1 INTRODUCTION

Measurements of temperature, wind, humidity, and other meteorological parameters are obtained by positioning the appropriate sensor instrumentation at a chosen sight. Sensors are usually located in instrument shelters, on masts or towers, or carried aloft in aircrafts, on balloons, or as radiosondes. Measurements made in this manner are called *in situ measurements.* Data obtained by in situ measurements, however, are insufficient to meet all needs in the field of boundary layer meteorology. Therefore, new technologies such as *ground-based remote sensing systems* have recently been developed. Ground-based remote sensing systems include sodars, lidars, and radars. All these devices measure the parameters of the atmosphere by emitting electromagnetic or acoustic waves and then analyzing the returned signal.

This chapter presents a short review of the methods used in detecting atmospheric turbulence. It is intended to be a supplement to the material described in previous chapters. A more detailed description of sensing instrumentation and measurement techniques can be found in Dobson et al. (1980), Brock and Nicolaidis (1984), and Lenschow (1986).

8.2 SENSORS

8.2.1 Wind sensors

Measurement of wind speed and direction requires sensors that operate reliably at wide ranges of speed. *Cup anemometers* or *propeller anemometers* and *wind vanes* are traditionally used by meteorologists to measure wind in the atmosphere (Figures 8.1 and 8.2). Vanes are used to direct cup, propeller, or hot-wire anemometers into the wind.

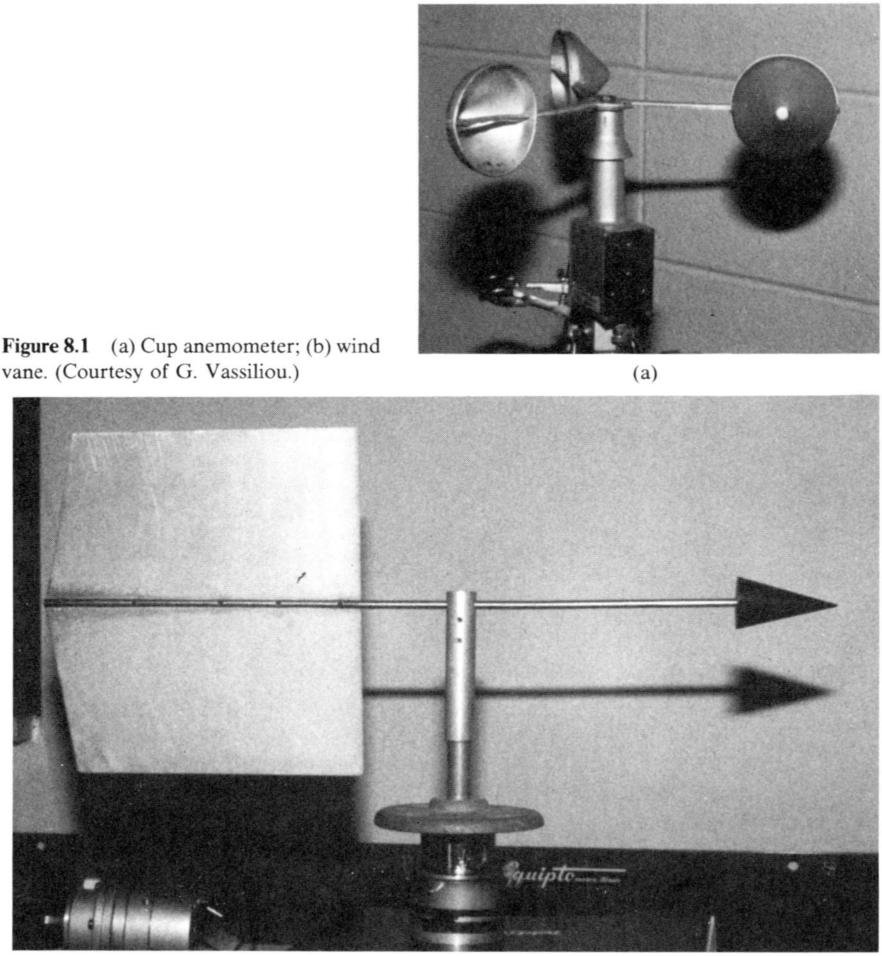

Figure 8.1 (a) Cup anemometer; (b) wind vane. (Courtesy of G. Vassiliou.)

(a)

(b)

Cup anemometers consist of three or four hemispherical cups mounted around a vertical shaft. Air striking the concave sides of the cups exerts more force than air hitting the convex sides and causes them to rotate. The number of rotations per unit of time is a measure of the wind speed. A standard cup anemometer measures the horizontal wind speed and fluctuations up to about 0.5 Hz.

Most propeller anemometers have helicoid rotors with two or more blades. They are used in a propeller–vane combination or in two-component or three-component orthogonal configurations. A propeller's speed of revolution is linearly related to the magnitude of the wind component directed along

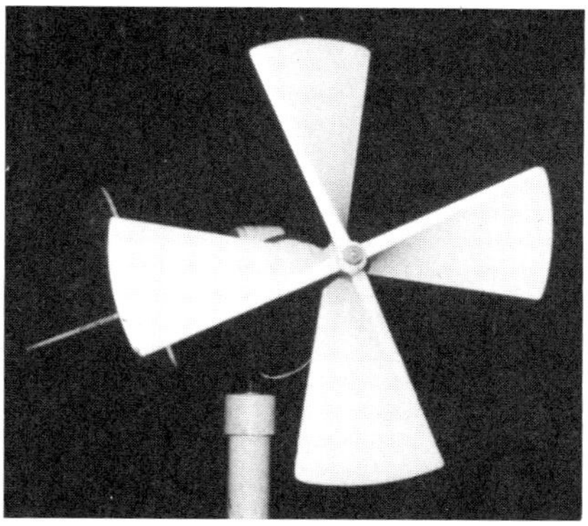

(a)

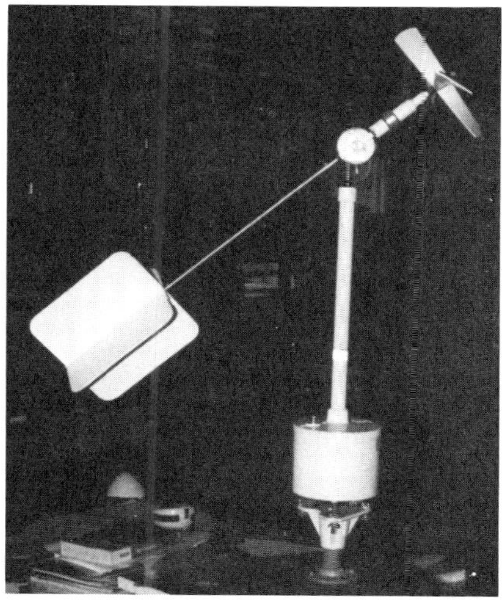

(b)

Figure 8.2 (a) Helicoidal four-bladed Gill propeller; (b) a bivane. (Courtesy of G. Vassiliou.)

the propeller's axis. Propeller anemometers, employed in the orthogonal *U-V* or *U-V-W* configuration, have a special source of error due to the lack of response when there is an angle between the wind direction and the axis of the instrument. When the wind vector is off the propeller axis by an angle α, the response is less than the expected $\cos\alpha$ factor. The *Gill unit* is perhaps the most common propeller anemometer used in research (Figure 8.2). Propeller anemometers combined with sensors for temperature or humidity allow routine measurements of turbulent fluxes.

Most wind vanes are single flat fins. However, three-dimensional wind direction can be assessed by the use of bivanes. Bivanes consist of a bidirectional vane mounted on an arm and balanced by a counterweight.

A number of techniques exist for evaluating turbulent velocity fluctuations (e.g., Wyngaard, 1981). *Hot-wire* and *hot-film anemometers* have been used extensively for the past four decades to measure turbulent velocity fluctuations. Figure 8.3a shows a hot-wire anemometer probe. A wire sensor is typically 0.004 to 0.005 mm in diameter and 1 to 2 mm long. The film type of sensor is shown in Figure 8.3b. In addition to the cylindrical shape shown in the figure, hot films have been made on cones, wedges, parabolas, and flat surfaces.

A hot wire or hot film placed in a three-dimensional turbulent flow measures the magnitude of the velocity vector normal to the sensor axis, independent of the azimuthal angle about the axis. An array of several wires inclined at different angles can resolve all three components of the wind vector. The system requires the calibration of each sensor for accurate work.

There are two types of hot-wire anemometry. In the first, the hot wire is maintained at constant temperature by increasing or decreasing the current to

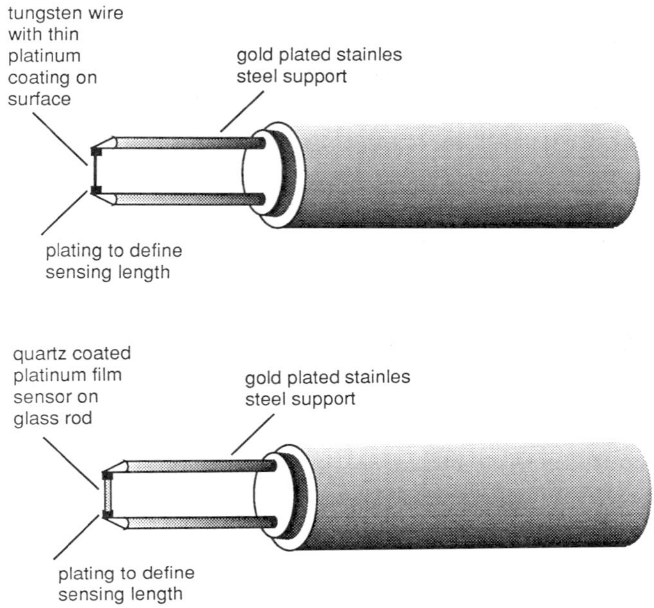

(b)

Figure 8.3 Hot-wire and hot-film anemometers: (a) tungsten hot-wire sensor and support;
(b) cylindrical hot-film sensor and support.

keep a *Kelvin bridge* in balance. The second type allows the temperature of the hot wire to vary and the output voltage is read on a *Wheatstone* or *Kelvin bridge*.

Sonic anemometers, developed in the 1950s, employ the propagation of a sound wave to measure the velocity and temperature of a gas flow. The basic concept of the sonic anemometer is demonstrated in Figure 8.4.

In the figure, the wind velocity has two components: along the acoustic path V_d and normal V_n. Sound is generated by two transmitters, T_1 and T_2, in opposite directions. Receivers R_1 and R_2 detect these sound impulses after two different intervals of time t_1 and t_2, due to velocity differences:

$$t_1 = \frac{d}{c \cos \gamma + V_d}$$
$$t_2 = \frac{d}{c \cos \gamma - V_d} \tag{8.1}$$

where c is the sound speed, which for a given temperature T can be obtained from the formula $c^2 = kRT$ ($k = c_p/c_v$ is the ratio of specific heats, R is the

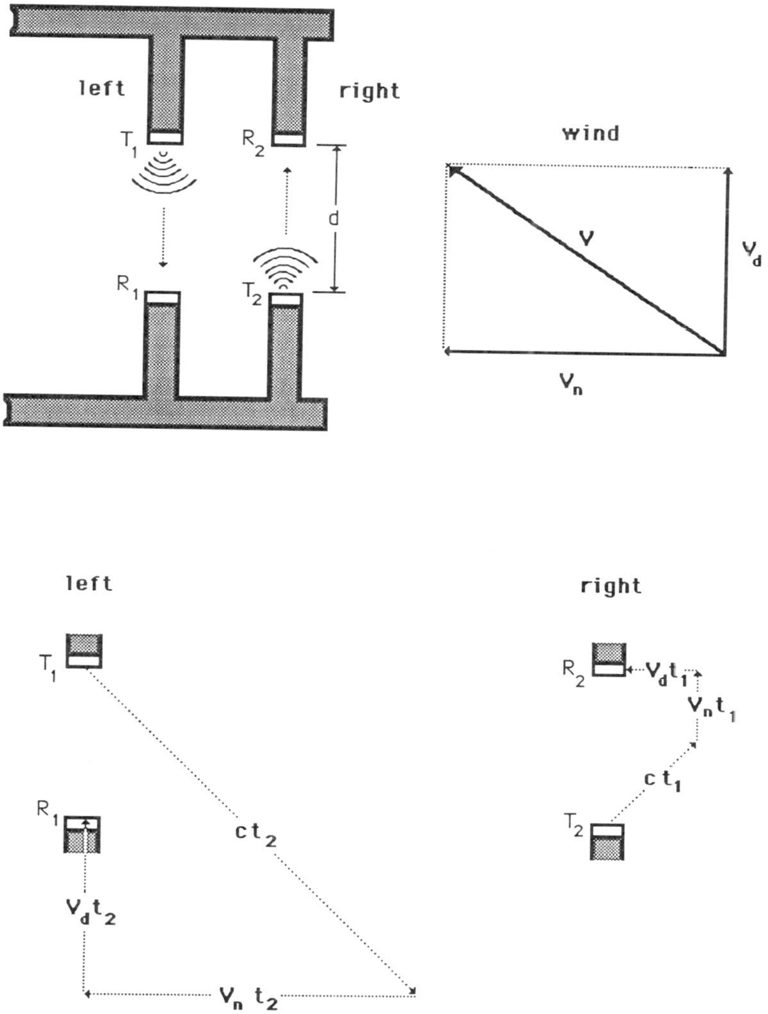

Figure 8.4 Principle of operation for a single-axis anemometer.

gas constant), d is the path length, and $\sin \gamma = V_n/c$. The derivation of (8.1) is explained on the left and right vector diagrams in Figure 8.4. Assuming that $c^2 \gg V^2$, it is possible to evaluate the transit-time difference as

$$\Delta t = \frac{2d}{c^2} V_d \tag{8.2}$$

The influences of temperature and humidity fluctuations on sound speed and calculated wind velocity are negligible. A three-dimensional sonic array makes

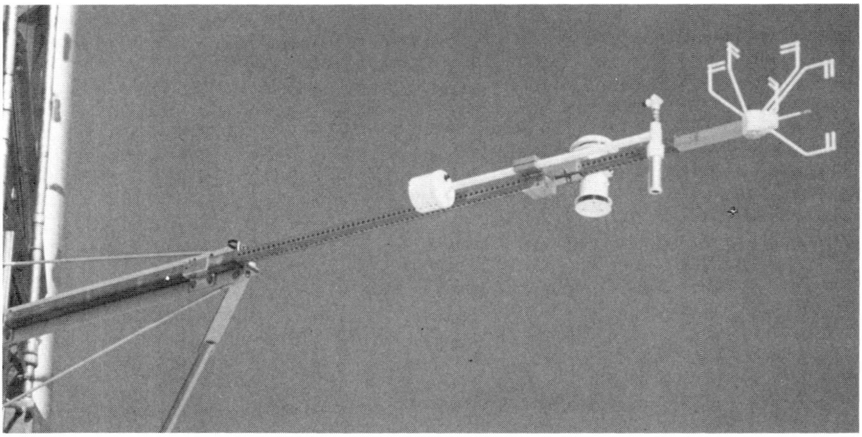

Figure 8.5 Sonic anemometer. (Courtesy of Dr. J. C. Kaimal.)

it possible to measure all components of the wind velocity with a frequency up to 10 Hz (Figure 8.5).

8.2.2 Temperature sensors

Temperature sensors include thermocouples, resistance elements, thermistors, electronic diodes, and transistors. *Thermocouples* utilize an effect based on the fact that two dissimilar metals formed into two junctions at different temperatures produce a voltage linearly dependent on the temperature difference between the two junctions. Since thermocouples naturally measure temperature difference, one of the junctions must be a reference junction at a known temperature (Figure 8.6). A thermocouple junction is formed by butt-welding very fine wires together. This type of sensor has a very low thermal mass and therefore a quick response.

Electrical resistance thermometers consist of a transducer whose resistance is a function of temperature. Electrical resistance thermometers include both conductive and semiconductive types. The first group includes *platinum resistance thermometers*, which are available in a wide variety of forms and resistance values. The diameter of a platinum resistance wire is very small, on the order of 1 μm. Platinum resistance thermometers are usually operated in a *Wheatstone bridge*.

Semiconductive sensors are called *thermistors* (thermally sensitive resistors). There are two general types: positive temperature coefficient thermistors (PTC) and negative temperature coefficient thermistors (NTC). When the temperature increases, the electrical resistance of PTC thermistors increases and the electrical resistance of NTC thermistors decreases.

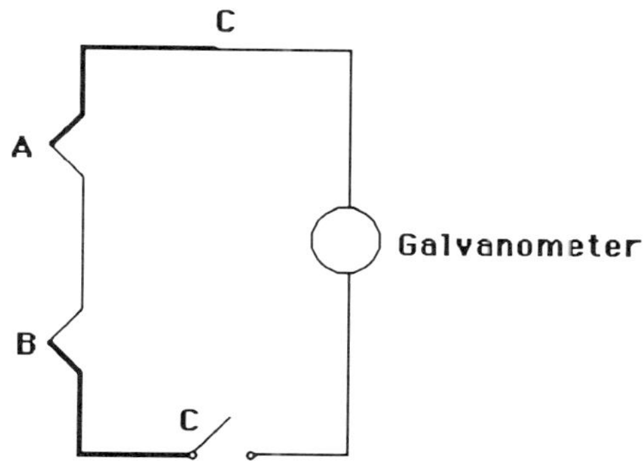

Figure 8.6 Thermocouple. The junctions A and B are at different temperatures, one of which is usually a known reference temperature. Junction C should be at the same temperature.

Electronic diodes and *transistors* can also be used for temperature measurements since their electrical characteristics can be dependent on temperature.

8.2.3 Measurements of humidity fluctuations

Various instruments are available for measuring water vapor concentration. Among them are the dew point hygrometer, the Lyman-alpha hygrometer, the infrared hygrometer, and the microwave refractometer. Atmospheric humidity can also be measured by a system of two thermocouples, one of which must be the "wet bulb."

Dew-frost point hygrometers consist of a mirror that is cooled until a thin water or frost layer forms (Wexler, 1965). This changes the reflectivity of the mirror surface and is detected optically (Figure 8.7). The mirror is then heated until the water or frost starts to disappear. If the system controlling the mirror temperature is adjusted properly, the temperature reaches an equilibrium which is at the dew-frost point temperature.

The *Lyman-alpha hygrometer* has an extremely short response time and can be used for measurements of turbulence. This instrument is based on the fact that H_2O strongly absorbs short ultraviolet radiation (the Lyman-alpha line is an electromagnetic emission line found in the far ultraviolet at 121.56 nm). This absorption is described by *Beer's law*:

$$I = I_0 \exp(-k\rho r) \tag{8.3}$$

where I and I_0 are the detected and the initial intensity of radiation, k the absorption coefficient, ρ the absorber concentration, and r the distance. Rearranging this equation allows finding the absolute humidity ρ (grams per cubic meter). The influence of two other species, O_2 and O_3, absorbing radiation in the short-wave range of the spectrum is negligible. Detection of Lyman-alpha radiation is accomplished in an ion chamber filled with nitric oxide (Figure 8.8).

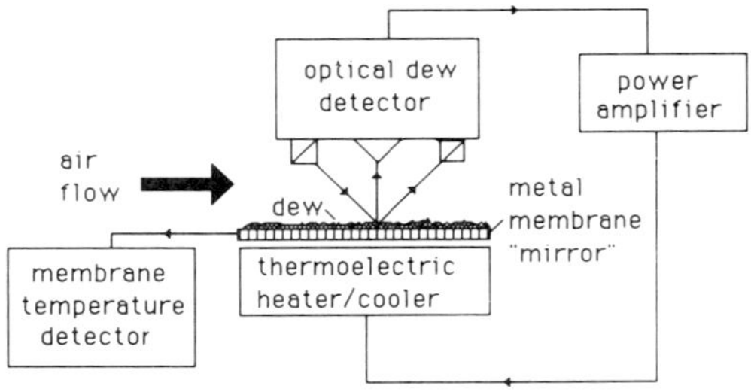

Figure 8.7 Schematic diagram of the dew-frost point hygrometer. (After Dobson, et al., 1980)

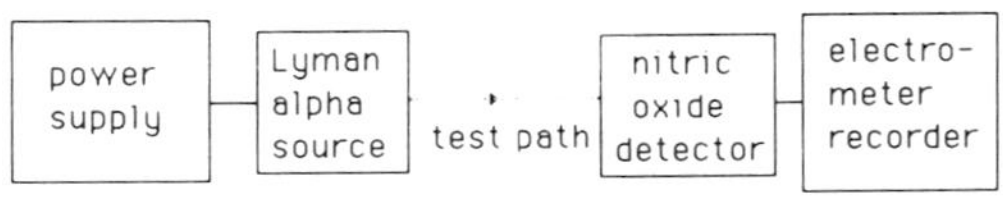

Figure 8.8 Basic components of the Lyman-alpha hygrometer. (After Dobson, et al., 1980)

Infrared hygrometers involve the absorbtion of infrared radiation by water vapor, while the *microwave refractometer* is based on the dependence of microwave refractivity on humidity.

8.2.4 Measurements of concentration

There are two reasons for measuring air pollutants: qualitative, to determine the chemical nature of pollutants, and quantitative, to determine the concentration of pollutants. The two most common units of concentration are parts per million (ppm $= 10^6 V_p/(V_p + V_a)$, where V_p is the volume of gaseous pollutant and V_a is the volume of air) and micrograms per cubic meter (μg m^{-3}).

Air pollutant measurements can be classified into: the direct methods, which require physical collection or handling samples, and the remote-sensing methods. Remote sensing techniques are discussed later.

Direct measurement methods can be divided into two main categories. The first consists of methods that measure physical properties of the pollutants in situ. Normally, the sample volume or sampling rate is unimportant for such measurements. Many in situ measurements employing spectroscopic techniques fall into this category. The second category includes methods that require extraction before physical or chemical measurements are made. This is done by drawing atmospheric air through a suitable gas absorbent which is later analyzed in the laboratory. It is important to know the sample volume or sampling rate when using such methods. This approach gives the mean concentrations of pollutants. Measurements of concentration fluctuations have been attempted in only a limited number of cases (Fackrell and Robins, 1982; Wesely et al., 1985).

8.2.5 Static and dynamic characteristics of sensors

Characterization of how well the sensor measures the primary inputs and how it responds to spurious inputs requires examination of its *static* and *dynamic performance.* Static performance characteristics are obtained when the input and the sensor are not changing in time. Dynamic characteristics describe how the sensor responds to temporal changes in the input.

Analysis of static characteristics of a sensor allows us to determine bias, imprecision, inaccuracy, and static sensitivity and resolution. *Bias* is systematic error that can be removed by calibration. *Imprecision* is a random error that cannot be removed, but can be defined during the *calibration* process. Calibration is a comparison of the instrument against a standard of previously defined accuracy. *Inaccuracy* is a combination of bias and imprecision. *Static sensitivity* is the slope of the curve obtained by relating different values of input and responses obtained after equilibrium is achieved. The *resolution* of a sensor is the smallest change in input that produces a defined change in output.

Generally, the dynamic response of a sensor is modeled by a differential equation. The dynamic response of cup or propeller anemometers, as well as thermal sensors, can be represented by a first-order ordinary differential equation of the form

$$\tau \frac{dX}{dt} + X = X_e \tag{8.4}$$

where τ is the time constant, X is a measured quantity (temperature or wind velocity), and X_e is the forcing function. The wind vane response is modeled with the second-order differential equation.

The *time constant* is a function of many parameters, depending on the type of sensor. For cup and propeller anemometers, another parameter, called the *distance constant*, is used. The distance constant is defined as the product of the time constant and the wind velocity.

Equation (8.4) can be solved for a given initial condition. If the instrument is a thermal element at an initial temperature T_0 which is placed in an environment of constant temperature T_e, the response obtained by a solution of (8.40) is described by

$$T - T_e = (T_0 - T_e) \exp\left(-\frac{t}{\tau}\right) \tag{8.5}$$

At time $t/\tau = 1$, the relative change $(T - T_e)/(T_0 - T_e) = e^{-1} = 35.8\%$. For large values of t/τ the difference $(T - T_e)$ tends to zero (Figure 8.9a).

When the temperature of the air changes linearly with time as $T_e = T_{e0} + at$, the response is

$$T - T_{e0} = -a\tau + a\tau \exp\left(-\frac{t}{\tau}\right) \tag{8.6}$$

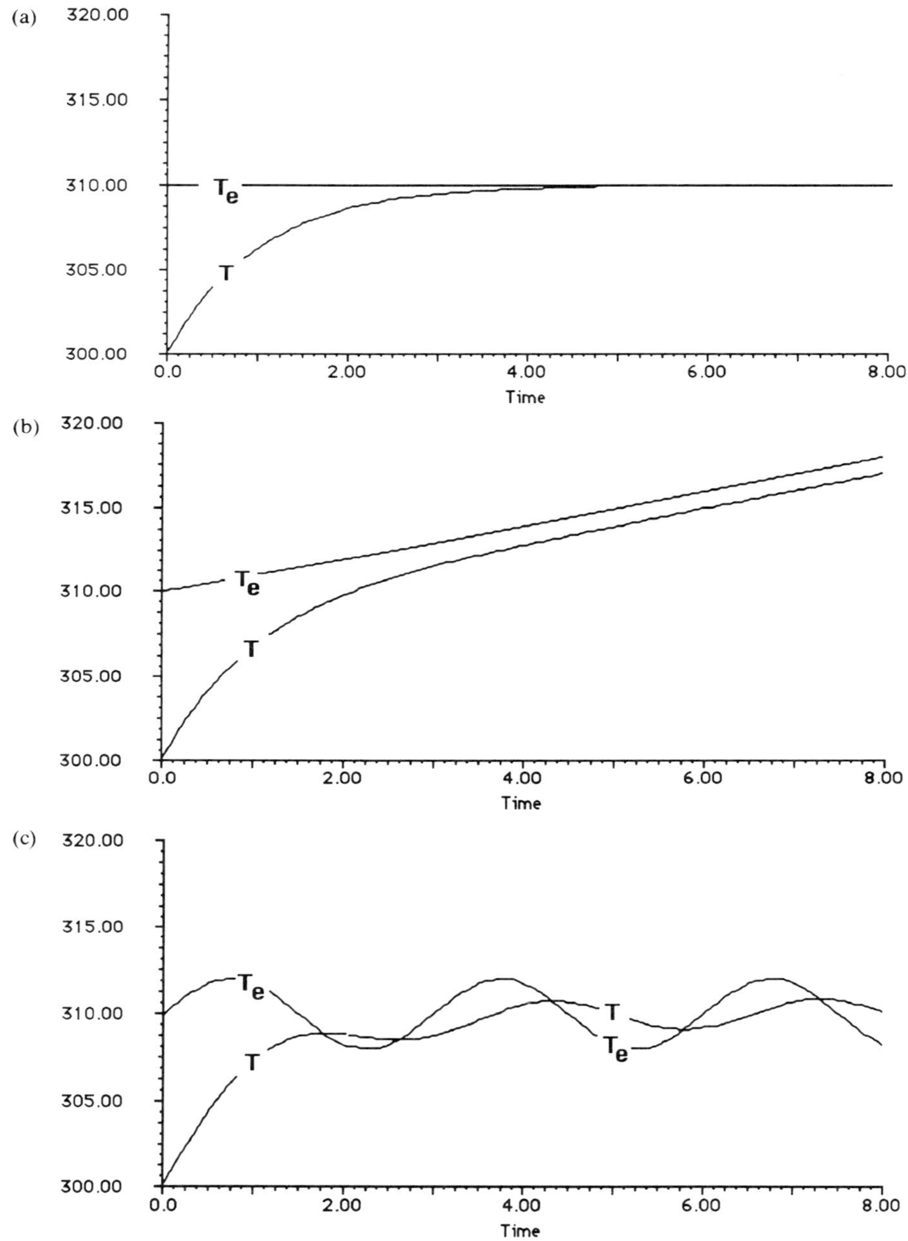

Figure 8.9 Dynamic response of the instrument with the time constant $\tau = 1$ second to the forcing function: (a) constant forcing function, $T_e = 310$; (b) linear forcing function, $T_e = 310 + t$; (c) simulated forcing function, $T_e = 310 + 2\sin(2\pi t/3)$.

It can be seen that for large t/τ, the thermometer will lag the actual temperature by a difference equal to $-a\tau$ (Figure 8.9b).

When the temperature of the air changes as a sinusoidal function of time as $T_e = T_{e0} + A\sin(2\pi t/P)$, then

$$T - T_{e0} = c\exp\left(-\frac{t}{\tau}\right) + \frac{A}{\left[1 + (2\pi\tau/P)^2\right]^{1/2}}\sin\left(\frac{2\pi}{P}t - \beta\right) \quad (8.7)$$

where A is the amplitude, β the phase angle, $\beta = \arctan(2\pi\tau/P)$, and c is a constant. The response and forcing functions are shifted and the amplitude of the response function is reduced (Figure 8.9c).

Figure 8.9c indicates that the averages of the response function and the forcing function are the same. However, in measurements of wind velocity, this is not strictly correct because the anemometer speeds up faster than it slows down and, therefore, gives a mean that is higher than the mean wind. This also indicates that the analysis described is only approximate.

For measurements of turbulent fluctuations, the time constant has to be very small (1 second or less). When temperature and wind velocity are measured simultaneously, the wind and temperature sensor response times must be matched.

8.3 PLATFORMS

Platforms are used for mounting meteorological sensors whenever elevated measurements of meteorological parameters are required. Platforms include masts, towers, aircrafts, and balloons. Towers and masts must be rigidly designed so that they will cause only a small distortion of the wind field (Figure 8.10). The highest towers are 200 to 300 m in height (e.g., Boulder, Colorado; Cobauw, Holland; and Voeikovo, USSR). Measurements from movable platforms (planes and balloons) are more complex and require additional comments.

8.3.1 Aircraft measurements

Planes have been used for performing meteorological measurements since the 1940s. Their major advantage is mobility. An aircraft can be used for either vertical profiling or horizontal traverses. It can measure along a path at any arbitrary angle with respect to the wind. An aircraft can also obtain statistically significant measurements much faster than can devices using direct fixed-point methods.

Unfortunately, an aircraft's mobility also creates some disadvantages. Aircraft motion needs to be measured accurately. Corrections must be applied

Figure 8.10 Boulder Aerological Observatory (BAO) tower in Boulder, Colorado. (Courtesy of J. C. Kaimal.)

to the data for compressibility, adiabatic heating, and flow distortion induced by the aircraft. This complicates the analysis of aircraft data. Moreover, aircraft are not suitable for long-time-series measurements. Thus fixed-point measurements are, in many cases, complementary to aircraft measurements.

One of the most basic and difficult measurements to obtain from an airplane is that of the three components of air velocity. The velocity of the air with respect to the earth is obtained by adding the velocity of the aircraft with respect to the earth V_e and the velocity of the air with respect to the aircraft, V_a:

$$V = V_e + V_a \tag{8.8}$$

The components of V_a are measured in the airplane's coordinate system by sensors mounted on the aircraft, usually on a boom in front to reduce aircraft-induced airflow distortion.

The external sensors used to measure the velocity of the air relative to the aircraft are shown in Figure 8.11. Longitudinal air velocity, or true airspeed, is calculated from the pitot-static pressure difference, static pressure, and temperature. The two lateral air velocity components are calculated from the attack and sideslip angle vanes and the true airspeed. These values are constrained from rotating. The force exerted on the vane by the airstream is

Figure 8.11 Air-sensing probe and nose boom on an NCAR aircraft (Beechcraft King Air). 1, Pitot-static tube; 2, side slip and attack vanes; 3, Rosemount (resistance) thermometer; 4, Lyman-alpha hygrometer; 5, reverse-flow thermometer. (National Center for Atmospheric Research/National Science Foundation photo archive.)

measured by means of strain gauges. To relate the airplane's coordinate system with the local earth coordinate system, the airplane's attitude angles (pitch, roll, and yaw) and the angular velocity of the airplane (i.e., the pitch, roll, and yaw rates) are measured by resolvers mounted on platform gimbals.

Aircraft velocity can be measured up to sufficiently high frequencies by integrating the output of accelerometers from an inertial navigation system (INS). The INS contains three accelerometers which are orthogonally mounted on a gyroscopically stabilized platform. The INS measures the three components of the airplane velocity and the three attitude angles (Figure 8.12). The INS is aligned on the ground. Once in the air, the system maintains its

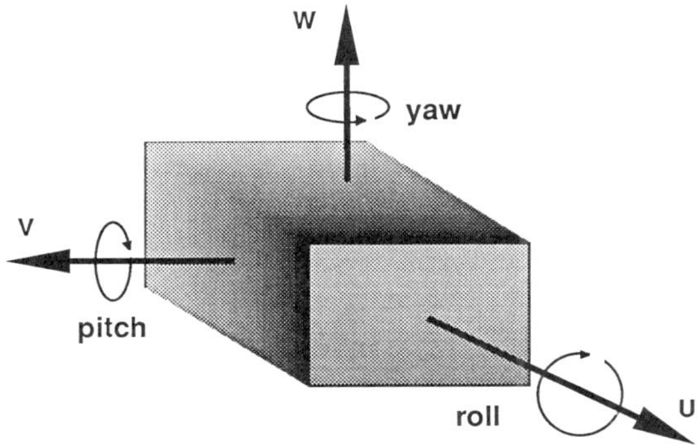

Figure 8.12 Diagram of the inertial airplane guidance system.

alignment by twice integrating the accelerometer to determine its position continuously. A more detailed description of air motion sensing instrumentation and measurement techniques can be found in Lilly and Lenschow (1971) and Lenschow (1970, 1972, 1986).

8.3.2 Balloons

Balloon systems play an important role in studies concerning the structure of the planetary boundary layer. Four primary types of balloon systems are used in the ABL: constant-density level balloons, pilot balloons, radiosonde balloons, and tethered balloons.

Constant-density level balloons are made of an inelastic film with a lighter-than-air gas mixture. They are designed to float at a constant-density level. They are produced in a variety of shapes, but the tetrahedron shape, called tetroon, has been used most often. Radars can be used to track the motion of these balloons.

Pilot balloons are used together with one or two theodolites to measure wind velocity profiles in the ABL. Accuracy of the measurement is poor when only one theodolite is used.

Radiosonde balloons carry a device measuring pressure, temperature, and relative humidity and transmit the data by radio to a receiving station. In addition, a radiosonde can be tracked by radar to provide wind data.

Tethered balloons provide a convenient method for obtaining vertical profiles of mean wind, temperature, and humidity as well as the characteristics

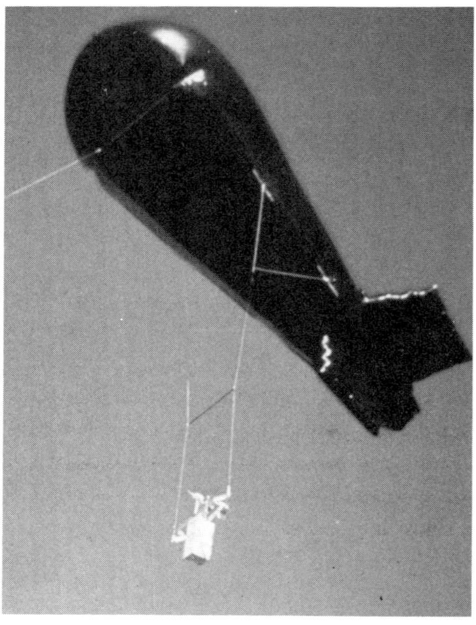

Figure 8.13 Tethered balloon. (National Center for Atmospheric Research/National Science Foundation photo archive.)

of turbulence up to a few hundred meters above the surface. Tethered balloons are preferably filled with helium (there are risks involved in the use of hydrogen) and are attached to a tethering line by a harness rigged to produce a significant kiting angle (Figure 8.13). The kiting is necessary to obtain a dynamic lift. In strong winds the angle of the tether line of the balloon increases to about 20° from the vertical. The lower angle produces a more stable flying behavior but reduced dynamic lift.

The characteristic motion of a tethered balloon is a slow periodic migration crosswind with additional superimposed longitudinal and vertical movements caused by changes of wind, and vertical gusts. The tether line may also oscillate at relatively high frequencies. These vibrations are capable of causing significant errors in the measured variances of horizontal and vertical wind speed and the vertical flux of momentum.

During profiling, the balloon is usually winched up or down at 1 to 2 m/s with a transit time of about 10 minutes from the surface to 1000 m. A single profile does not adequately represent the mean vertical structure. Horizontal winds are exaggerated during descent and reduced in ascent by alongwind displacement of the sensors produced by profiling.

8.4 REMOTE GROUND-BASED OBSERVING SYSTEMS

Remote sensing interprets the interaction of electromagnetic or acoustic waves with the atmosphere. In the majority of these systems, a signal source generates a short pulse of energy. The pulse is radiated in a narrow beam that can be directed to the region of the atmosphere under study. As the pulse of energy propagates in the atmosphere, some of its energy is lost through absorption and some through various scattering processes. A small portion of this scattered energy is captured by the receiving antenna of the remote sensor. This energy is converted by a detector to an electrical signal that is processed to obtain information describing the atmosphere. In many systems, the transmitting antenna is also used for reception.

Three major remote-sensing techniques are used in boundary layer research: radars, lidars, and sodars. The word *radar* is formed from the initial letters for the compound term "radio detecting and ranging." Radars use a microwave signal source. Similarly, *lidar* refers to a system that utilizes a laser as the signal source and is formed from "light detecting and ranging." *Sodar* is an acoustical system based on "sound detecting and ranging." Parameters of the three remote-sensing systems are listed in Table 8.1.

Remote-sensing methods have a number of notable advantages over in situ measurements. One of them is their ability to obtain volume averages, which are more representative than point values. The characteristics and performance capabilities of each sensor class are described next.

TABLE 8.1 Parameters of Remote-Sensing Systems

System	Maximum Range (km)	Range Resolution (m)	Beam Width (mrad)	Wavelength	Cost (thousands of dollars)
Radar	50–300	1–1000	10–30	0.3–30 cm and longer	100–500
Lidar	1–20	10–300	0.1–3	0.3–10 μm	100–500
Sodar	1.5	5–50	150	15–30 cm	50

8.4.1 Radars

Meteorological radars can be grouped into two classes, termed *coherent* and *noncoherent* radars. The primary difference between them is that in noncoherent radar, the phase of the transmitted signal is not monitored by the radar receiver (Doviak and Jobson, 1979). Noncoherent radar measures the range dependence of the average power backscattered to the radar. It is often referred to as conventional radar, because it is routinely used at the present time. Sensitive conventional radars have been used to study convective thermals, Kelvin–Helmholtz waves, and sea-breeze fronts.

In coherent radar, also called *Doppler radar*, the change of phase with time can be interpreted as a Doppler-shifted signal caused by the radial motion of the scattering particles. Coherent radars have significantly greater sensitivity than that of conventional radars. This sensitivity permits the detection of scattering from refractive index inhomogeneities under most atmospheric conditions. Such radars can map boundary layer winds and detect regions of wind shear (Doviak and Jobson, 1979).

Both coherent and noncoherent radars operate in bands covering the microwave portion of the electromagnetic spectrum. In a typical system, the radar transmitter generates a stream of pulses one-half to several microseconds in duration at a repetitive frequency of 800 Hz. These pulses, approximately 100 to 1000 kW in power, are directed by an antenna in a beam of 1° width (Figure 8.14). The antenna can scan 360° in azimuth and 90° in elevation. The receiver has a minimum signal level of about 10^{-13} W, which permits the detection of precipitation over a range of 300 km (Schotland, 1985).

When the antenna is used to scan in azimuth and the elevation angle is fixed, the result is a plane section of the atmosphere with the radar located at the center of the display. The system can also be operated in elevation angle for a given azimuthal direction to provide a vertical slice of the atmosphere. Most modern radars employ advanced digital data processors and display units that improve the interpretation and understanding of the radar signals. Color displays on regular TV monitors are often used. They provide contours of reflectivity, velocity fields, and other meteorological variables.

Figure 8.14 NCAR Doppler radar. (Courtesy of Dr. J. C. Kaimal.)

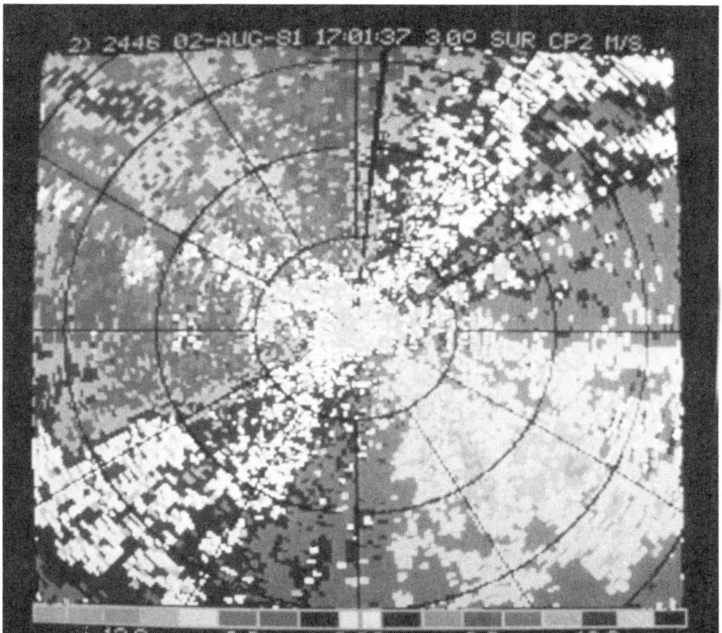

Figure 8.15 Doppler radar measurement of wind. In the original color photograph, blue and green colors, indicate flow to the radar; and yellow and red colors, indicate flow from the radar. (Courtesy of Dr. J. Wilson, N.C.A.R.)

Radars have been reported to measure wind velocity, temperature (with sodars), concentration, stress, eddy dissipation rate, and the ABL height. Figure 8.15 shows a display from the Doppler lidar.

8.4.2 Lidars

Lidar, also referred to as *optical radar* or *laser radar*, is an analog of microwave radar. Both systems are functionally identical. The difference between them is that radars operate in the centimeter range, while lidars operate in the micrometer range. Discovered in the early 1960s, lidars have been used to measure wind velocity and direction in storms near airports. Lidars have also proved useful in defining the temporal and spatial distribution of airborne particles over a wide range of meteorological situations. Studies have been made of the transport of natural atmospheric particulates, as well as of the pollutants from industrial stacks.

The primary influence of the atmosphere on a low-power laser beam is through scattering and absorption. Both processes cause an attenuation of the beam according to *Brouguer's law*:

$$I = I_0 \exp(-\alpha r) \tag{8.9}$$

where I is the intensity of the optical beam after transmission over a distance r, α the atmospheric extinction coefficient, and I_0 the initial intensity of the beam. Coefficient α is the sum of four coefficients:

$$\alpha = \alpha_{\text{Ray}} + \alpha_{\text{Mie}} + \alpha_{\text{Ram}} + \alpha_{\text{abs}} \tag{8.10}$$

where α_{Ray}, α_{Mie}, and α_{Ram} are the *extinction coefficients* related to *Rayleigh*, *Mie*, and *Raman scattering*, respectively, and α_{abs} is the *molecular absorption* coefficient. Rayleigh scattering is due to particles that are smaller than the optical wavelength λ of the laser. Mie scattering is associated with larger particles whose sizes are on the order of λ. Raman scattering involves excitation of a molecule and reradiation of energy at different wavelengths. Absorption of laser beams is a resonant interaction causing excitation of molecules and possibly fluorescence.

Lidars can be used to measure the components of wind velocity, temperature, humidity, concentration, and ABL height. The wind vector can be determined by lidar from observations of the time variation of the spatial distribution of particulate matter. The schematic of the pulsed lidar system, where the laser and detector are located together, is shown in Figure 8.16. In this case the returned laser radiation is due to backscattering from aerosols or dust in the atmosphere.

Two general approaches of lidar measurement are typically employed. In the first, a lidar is elevated by a small angle to scan between closely spaced azimuth angles at ranges as great as 6.5 km. The lidar output consists of a time series of the backscattering coefficient as a function of range and

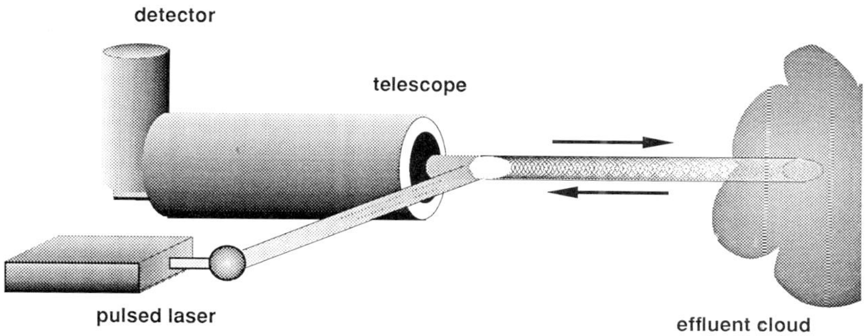

Figure 8.16 Schematic of a lidar.

azimuthal angle. The horizontal wind perpendicular to the lidar beam is determined by the time required for particles to move from one azimuthal direction to the next. In an analogous way, the radial wind component is determined by radial transport. Typical measurements report wind vectors, averaged over 1 to 5 minutes, with 1 m/s and 7° azimuth uncertainties.

The second category of lidar wind sensors analyzes the Doppler shift, which occurs when the laser signal is backscattered from moving particles. The system is able to measure the radial wind vector to a height of 1 km, with an accuracy of 0.25 m/s (Schotland, 1985).

The Raman component of scattered laser energy is a unique function of the species of the scattering molecule and could therefore be used to identify and measure the distribution of individual gases such as N_2 and O_2, to heights of 3 km (Cooney, 1972). Raman observations of water vapor profiles to altitudes of 1.5 km have also been reported.

Figures 8.17 and 8.18 show a sequence of lidar pictures obtained by E. Eloranta during the BLX-83 field experiment, using the ruby lidar of the University of Wisconsin in Madison. The lidar is able to provide nearly instantaneous (60 s), two-dimensional (x–y or x–z) pictures of convective plumes within 7 to 10 km of the instrument.

8.4.3 Sodars

Sodar, also referred to as *echosonde* or *acoustic sounder,* was initially developed to investigate sound propagation in the atmosphere. Later, this application was extended to allow investigation of the spatial and temporal distribution of velocity and thermal parameters characterizing atmospheric turbulence. Sodar can now provide a vertical profile of the wind vector to altitudes in excess of 500 m, with a resolution of 1 m/s. Sodars are also used to measure characteristics of temperature, concentration, ABL height, and turbulent kinetic energy (McAlister et al., 1969).

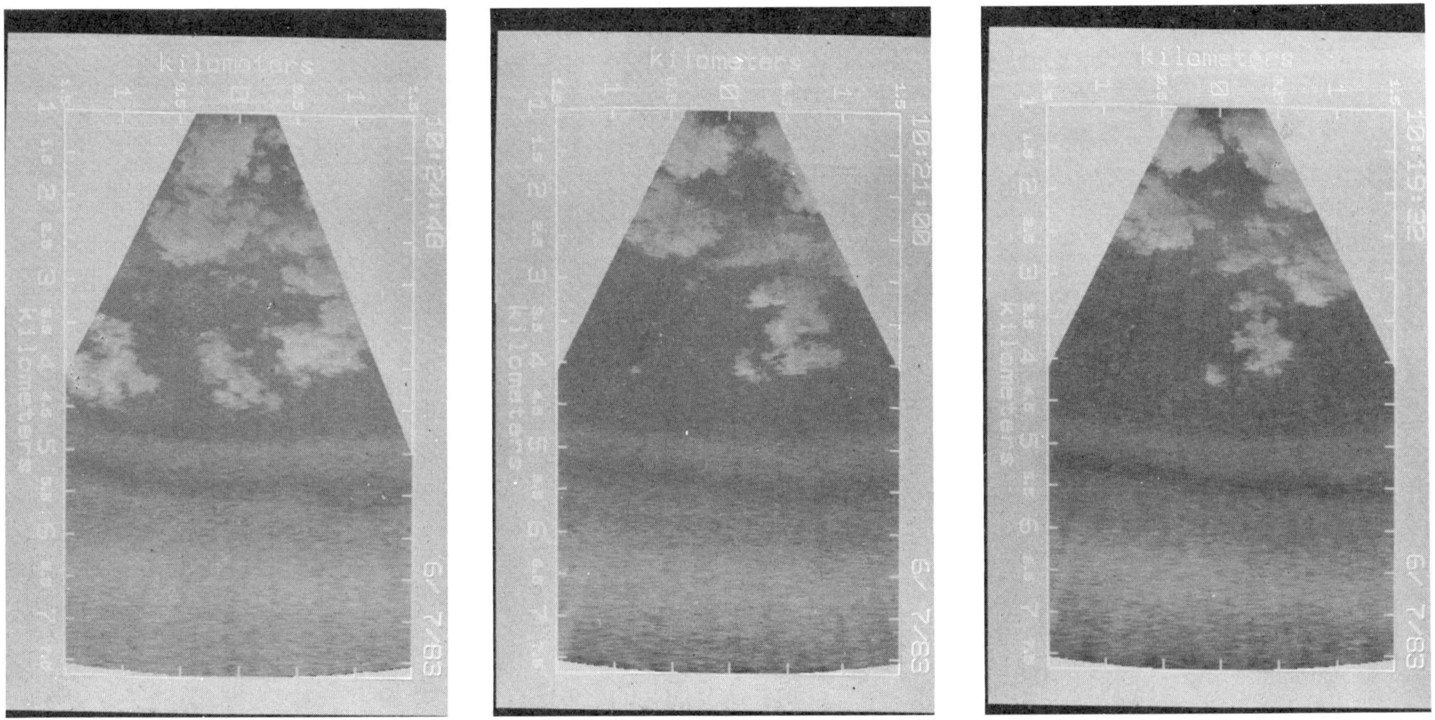

(a) (b) (c)

Figure 8.17 Display of the University of Wisconsin–Madison lidar. Horizontal view during BLX-83 experiment. (Courtesy of Dr. E. Eloranta.)

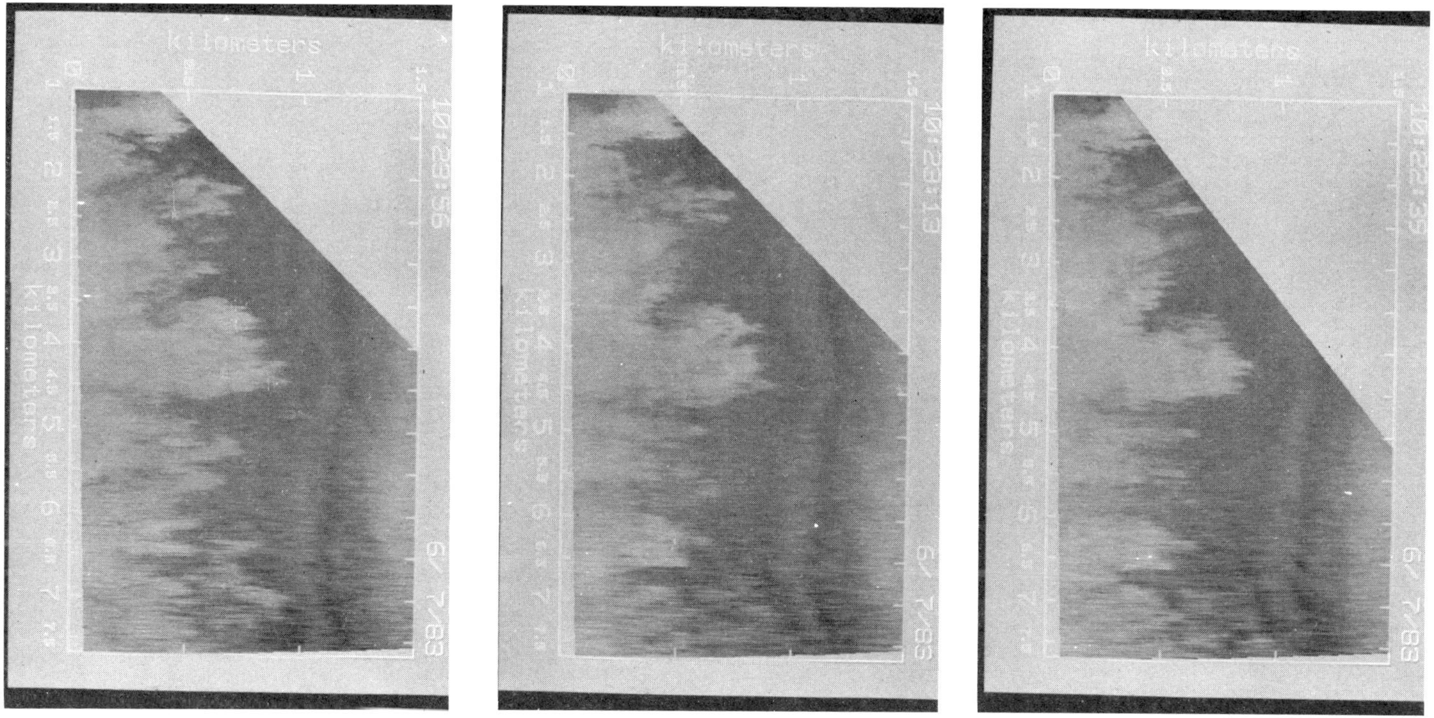

(a) (b) (c)

Figure 8.18 Display of the University of Wisconsin–Madison lidar. Vertical view during BLX-83 experiment. (Courtesy of Dr. E. Eloranta.)

Figure 8.19 Monostatic sodar at Boulder Aerological Observatory in Boulder, CO. (Courtesy of Dr. J. C. Kaimal.)

Monostatic sodar systems transmit and receive sound through a common horn antenna, parabolic dish, or loudspeaker array. The antenna is normally surrounded by a shield in order to suppress ambient noise (Figure 8.19). *Bistatic sodar* systems transmit and receive sound through horn antennas and loudspeakers, which are separated.

In a typical system, the oscillator-amplifier operates at a frequency of 1500 Hz and produces pulses 0.1 s in duration and 100 W in power. These pulses are converted into acoustical waves by electromagnetic transducers. Their sound is directed by a parabolic reflector in the form of a beam approximately 15° in width. The signal received, after amplification and filtering, is presented graphically on a TV screen or paper on which the degree of darkness or color is proportional to the signal power. In some systems, the amplified received signal is processed directly by a microcomputer and displayed as the vertical distribution of the wind vector or other quantities (Schotland, 1985).

Scattered acoustic power is proportional to the temperature structure parameter C_T^2 and a product $C_V^2 \cos(\theta_s/2)$, where C_V is the wind velocity structure parameter and θ_s is a scattering angle from a transmitter beam (e.g., Lenschow, 1986). For the monostatic scattering, $\theta_s = 180°$ (backscattering) and the scattered acoustic power depends solely on the temperature structure function C_T^2. The major limitation in the application of monostatic sodar to wind sensing is that C_T^2 is variable in strength and falls off strongly with height. Bistatic sodars have an advantage in that the scattered signal depends on both C_T^2 and C_V^2. To measure three wind velocity components, three adjacent antennas are used. If N-S and W-E components of the wind vector are to be measured, one antenna tilts south usually 10 to 30° from the vertical

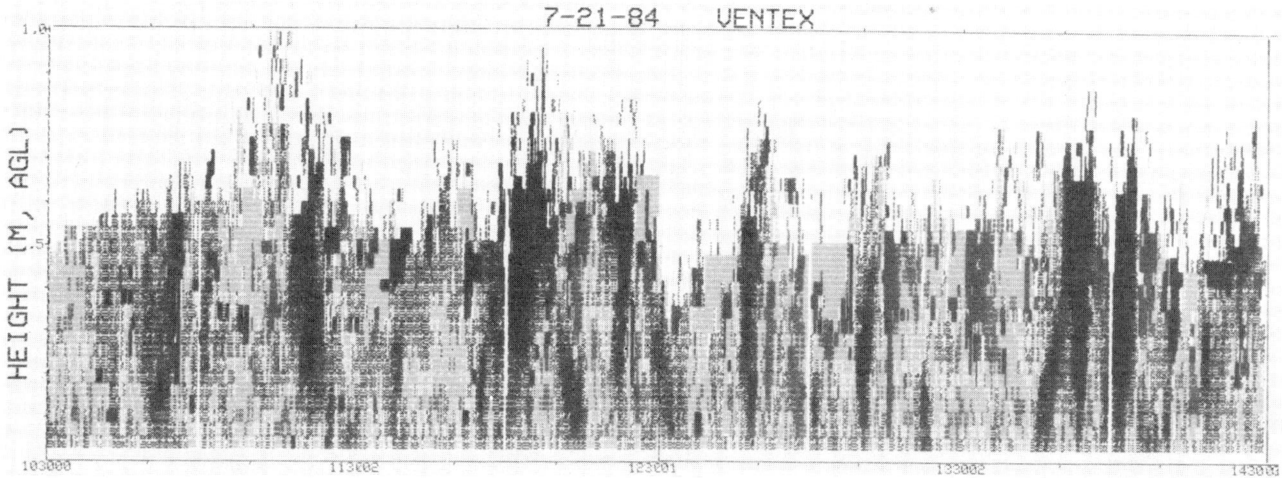

Figure 8.20 A display of the Argonne National Laboratory sodar. In the original color photograph updrafts are red and downdrafts are blue. (Courtesy of Dr. R. L. Coulter, Argonne National Laboratory.)

to be sensitive to the N-S component, the second tilts west with the same angle from the vertical to be sensitive to the W-E component, and the third aims straight up to be sensitive to the vertical component.

Sodars frequently reveal layered surfaces that are identified with regions of wind shear or thermal inhomogeneities. These regions are characterized by substantial values of C_T^2. For example, convective plumes that rise from the surface of the earth during morning hours can be observed very clearly by sodar as vertical shafts whose darkness varies with the magnitudes of C_T^2 (see Figure 1.7).

The operation of a sodar is limited to relatively noise-free locations. Sodars are affected by the noise produced by falling raindrops. With special transmit-receive transducers having sufficient noise immunity, sodars can operate at somewhat reduced performance levels even in urban areas. The operating range of vertically pointing sodars located in a quiet region is typically 1 to 2 km.

Sodars have proven useful in defining those meteorological conditions which are associated with high air pollution potential. Sodars can be used to measure the spatial location and duration of regions of convective instability and stability in the boundary layer. Sodar can also be used to observe the size and location of heated plumes emitted from power plants.

A display obtained from a monostatic sodar is shown in Figure 8.20. The photograph of thermals and rising motion shows the vertical velocity field as a function of time and height above the surface. In the original color photograph red color indicates rising motion and blue indicates descending motion with the most intense red/blue indication corresponding to ± 4m/s velocity and the least intense indicating ± 0.5m/s velocity.

8.5 AVERAGING OF MEASURED QUANTITIES

All measured instantaneous characteristics of turbulence have to be averaged. As mentioned in Chapter 2, the averaging procedure can be defined as time, space, or ensemble averaging. The ensemble average is useful in obtaining equations of motion for mean quantities and is rarely used to calculate a mean value of any measured quantity. In situ measurements are time averaged, while aircraft observations are averaged over the flight path.

Meteorological measurements are based on the ergodic hypothesis, according to which in the stationary case the time and ensemble averages coincide. This assumption, however, creates an essential difficulty as random characteristics of the turbulent flow in the ABL are not strictly stationary and the time averages depend on the interval over which they are averaged. The time averages are relatively "stable" if the intervals of averaging exceed the time scales of eddies containing the main part of the turbulent energy.

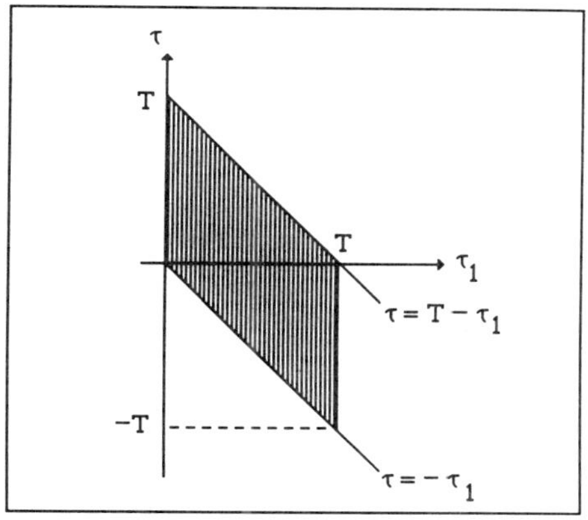

Figure 8.21 Interchange of integral limits in (8.14).

To evaluate the time interval of averaging, consider the ensemble variance of the time average about the ensemble mean value:

$$\sigma^2 = \left\langle \left[\frac{1}{T} \int_0^T f(t+\tau)\, d\tau - \langle f(t) \rangle \right]^2 \right\rangle \tag{8.11}$$

where $\langle\ \rangle$ is an ensemble averaging operator. This can be expanded as

$$\begin{aligned} \sigma^2 &= \left\langle \frac{1}{T^2} \int_0^T \int_0^T f(t+\tau_1) f(t+\tau_2)\, d\tau_1\, d\tau_2 \right. \\ &\qquad \left. - \frac{2}{T} \langle f(t) \rangle \int_0^T f(t+\tau)\, d\tau + \langle f(t) \rangle^2 \right\rangle \\ &= \left\langle \frac{1}{T^2} \int_0^T \int_0^T f(t+\tau_1) f(t+\tau_2)\, d\tau_1\, d\tau_2 - \langle f(t) \rangle^2 \right\rangle \end{aligned} \tag{8.12}$$

Introducing the autocorrelation function of the form

$$C(t) = \frac{\langle f(t+\tau_1) f(t+\tau_2) \rangle - \langle f(t) \rangle^2}{\langle f'^2 \rangle} \tag{8.13}$$

and using the interchangeability of integration (Figure 8.21) and the symmetry properties of autocorrelation functions, we obtain

$$\begin{aligned} \sigma^2 &= \frac{\langle f'^2 \rangle}{T^2} \int_0^T \int_0^T C(\tau_2 - \tau_1)\, d\tau_1\, d\tau_2 = \frac{\langle f'^2 \rangle}{T^2} \int_0^T \int_{-\tau_1}^{T-\tau_1} C(\tau)\, d\tau\, d\tau_1 \\ &= \frac{\langle f'^2 \rangle}{T^2} \left[\int_{-T}^0 (T+\tau) C(\tau)\, d\tau + \int_0^T (T-\tau) C(\tau)\, d\tau \right] \\ &= \frac{2\langle f'^2 \rangle}{T} \int_0^T \left(1 - \frac{\tau}{T}\right) C(\tau)\, d\tau \underset{T\to\infty}{\rightarrow} \frac{2\langle f'^2 \rangle}{T} \int_0^T C(\tau)\, d\tau \end{aligned} \tag{8.14}$$

If the integral of the function C exists, it is called an integral time scale and denoted as $\mathscr{T}$. In this case, (8.14) can be rewritten in the form

$$\sigma^2 = \frac{2\langle f'^2\rangle}{T}\mathscr{T} \tag{8.15}$$

If $T \to \infty$, then $\sigma^2 \to 0$.

A similar result can be obtained by changing the time average into a one-dimensional space average:

$$\sigma^2 = \frac{2\langle f'^2\rangle}{L}\mathscr{L} \tag{8.16}$$

where $\mathscr{L}$ is a space integral scale. For two-dimensional space averaging or the average over an area, the result is

$$\sigma^2 = \frac{2\langle f'^2\rangle}{L^2}\mathscr{L}^2 \tag{8.17}$$

For the volume average, the power of $\mathscr{L}/L$ becomes 3.

Assuming an acceptable error $\mathscr{E} = \sigma/\langle f\rangle$, the required averaging time can be obtained from (8.15):

$$T = 2\frac{\langle f'^2\rangle}{\langle f\rangle^2}\frac{\mathscr{T}}{\mathscr{E}^2} \tag{8.18}$$

As an example, this formula is applied to calculate the averaging time for wind velocity. Assuming that $\mathscr{T} = z/U$ (Wyngaard, 1973) and also that $\langle u'^2\rangle/\langle U\rangle^2 = 0.05$, $\mathscr{E} = 1\%$, $z = 5$ m, and $U = 5$ m/s, the result is $T = 15$ minutes. For $f = uw$,

$$L = 2\frac{\left\langle (uw - \langle uw\rangle)^2\right\rangle}{\langle uw\rangle^2}\frac{\mathscr{L}}{\mathscr{E}^2} \tag{8.19}$$

which is equivalent to

$$\frac{L}{\mathscr{L}} = 2\left(\frac{1 - r^2}{r^2}\right)\frac{1}{\mathscr{E}^2} \tag{8.20}$$

where $r^2 = \langle uw\rangle^2/\langle (uw)^2\rangle$.

If $r = 0.3$ and $\mathscr{E} = 10\%$, then $L/\mathscr{L} = 2000$. If, in addition, $\mathscr{L} = 500$ m, then $L = 1000$ km. This is much longer than a typical flight path. Equation (8.13) indicates that for an area average, $L = 22$ km. This shows that averaging over a 22-km square (e.g., by the use of a radar or lidar) is equivalent to line averaging over a 1000-km distance.

8.6 MAJOR FIELD EXPERIMENTS

On many pages of this book and in many figures the reader has found references to field experiments that have been of great importance in providing us with knowledge and understanding of the atmospheric boundary layer. Below, short overviews of some major experiments are given.

8.6.1 The 1953 Great Plains experiment

One of the first boundary layer experiments was the so-called Great Plains project (Lettau and Davidson, 1957). The Great Plains Turbulence Field Program was performed in the open prairie country near O'Neill, Nebraska, between August 1 and September 10, 1953. The topography of the site was relatively flat. The maximum height difference in the entire mile area was not more than 3 m.

The Great Plains expedition provided a great deal of information. Published materials from the experiment include synoptic information (synoptic charts, geostrophic wind profiles), subsoil measurements (temperature, moisture, thermal conductivity, heat capacity, moisture tension, heat flux), interface measurements (radiation fluxes, shear stress, and standard meteorological measurements), profile data from a 16-m mast (wind velocity, temperature, water vapor pressure, ozone concentration), and free air observations (double theodolite and rawinsonde measurements of wind velocity and direction, and radiosonde measurements of pressure, temperature, and relative humidity).

A broad range of unique instrumentation was used for collecting data. It included thermistors, thermocouples, and thermometers measuring temperature, different types of anemometers (including a sonic anemometer) measuring wind velocity, different types of radiometers measuring radiation, boundary shear stress recorders and drag plates measuring surface stress, micro-barographs, and ozone recorders.

As a result of the Great Plains project, some very important features of the ABL were revealed. One of them was the phenomenon of the nocturnal low-level jet. The data obtained were also used to calculate the values of the universal functions of the Rossby similarity theory. The results of the experiment are still valuable, but are restricted in use because of their fragmentary nature.

8.6.2 The 1967 Wangara experiment

An experiment named "Wangara" was conducted July 15 to August 27, 1967 in Australia (Clarke et al., 1971). The area chosen for this experiment was a large, flat land with a low vegetative cover and a slope of about $2\frac{1}{2}$ parts in 10,000.

Hourly pilot balloon winds were measured up to 2 km and averaged over five stations. Four of these stations were located approximately at the corners of a square having sides of 60 km and the fifth was located near the center. The mean winds were presented as westerly and southerly components at 50-m height intervals up to 1 km, with 100-m intervals at higher levels. Radiosonde temperatures and mixing ratios at 3-hour intervals were measured at the central station.

Micrometeorological measurements on a mast provided information on surface layer winds at 1, 2, 4, 8, and 16 m at two of the five stations: the smooth, central site (station 5) and station 4. At the former site, winds at 0.5 m were also given. Temperature differences were measured between 1 and 2 m and 2 and 4 m, with the use of matched pairs of resistance thermometers.

The data obtained were of great scientific importance. The Wangara measurements provided the basis for determination of the Monin–Obukhov similarity functions. They were also used to calculate the values of the universal functions of the Rossby similarity theory and to determine the most appropriate choice of scaling parameters. Convective data obtained during the famous day 33 of the experiment were used to test many numerical models of the atmospheric boundary layer. Finally, results of the Wangara experiment pointed out the difficulties and limitations of obtaining accurate measurements of thermal winds, vertical velocities, and representative spatially averaged fluxes.

8.6.3 The 1968 Kansas experiment

The next intensive field measurement program, limited to surface layer measurements was the 1968 Kansas experiment (Izumi, 1971; Kaimal et al., 1972). It was designed to study the behavior of the surface layer, especially to verify the validity of the Monin-Obukhov hypothesis.

The experimental site for the field program was extremely flat farmland in the southwestern region of Kansas. The local height variations were less than ± 0.5 m/km, with no prominent obstacles. Analysis of the profile data suggested a surface roughness of ~ 2.4 cm.

The focal point of the micrometeorological site was a 32-m instrument tower in the center of the square-mile section. The instrumentation for the field program consisted of fast-response instruments for direct measurements of heat and momentum fluxes and slow-response instruments for measurements of vertical profiles of wind speed and temperature. The observation period for individual runs was limited to 1 hour and the basic averaging period for collected data used was 15 minutes.

Wind speed was measured with three-cup anemometers, mounted at heights of 2, 4, 5.66, 8, 11.31, 16, 22.63, and 32 m on the tower. To measure wind direction, wind vanes were mounted at 5.66, 11.31, and 22.63 m on the tower. The wind speeds and the wind directions were sampled at the rate of one per second. Three-component sonic anemometers with path lengths of 20 cm were used to measure fluctuating wind components. These were mounted at the levels 5.66, 11.31, and 22.63 m on the tower. Measurements of the steamwise wind fluctuations were also made with constant-temperature linearized hot-wire anemometers. These hot-wire anemometers were mounted in the base of the sonic anemometers at levels of 5.7, 11.3, and 22.6 m.

Temperatures and temperature differences were measured with a slow-response system using shielded and aspirated platinum resistance thermometers in a Wheatstone bridge arrangement. The system was capable of obtaining mean values of temperature and temperature difference with resolutions of 0.05°C and 0.01°C, respectively. Sensors to measure temperatures were mounted at 2 and 22.63 m, and those to measure temperature differences, at 2, 4, 8, 16, 22.63, and 32 m. The temperatures and the temperature differences were sampled at the rate of one per second.

The sonic anemometers and fine platinum-wire resistance thermometers were sampled 20 times per second. To reduce aliasing effects, the sonic anemometer and platinum-wire thermometer outputs were prefiltered with a 10-Hz cutoff low-pass filter.

The Kansas experiment provided very detailed data which for many years have been and still are a source of information on the surface layer. The Kansas data allowed determining the form of the universal functions and universal constants in the Monin–Obukhov similarity theory, as well as the form of the spectral and cospectral densities. The results also pointed out the limitations in obtaining accurate measurements of turbulence in the surface layer.

8.6.4 The 1973 Minnesota experiment

The 1973 Minnesota experiment was an extension of the Kansas experiment (Izumi and Caughey, 1976; Kaimal et al., 1976). The purpose of this experiment was to study the processes in the entire boundary layer. The site of the experiment was at the middle of the southern edge of extremely flat farmland in northwestern Minnesota.

Measurements of the vertical profiles of wind velocity and temperature and of the vertical transport of momentum and heat were made from a 32-m tower. Mean wind speeds and directions were measured with two-axis sonic anemometers. The sensors were mounted at levels of 0.5, 1, 2, 4, 8, 16, 24, and 32 m on the northeast corner of the tower. The two-axis sonic anemometers and the quartz thermometers were each sampled once per second.

Three-axis sonic anemometers were used to measure fluctuating wind components. A 25-μm-diameter platinum-wire resistance thermometer was used to measure fluctuating temperatures. These probes were mounted at levels of 4 and 32 m on either the east or west side of the tower, depending on the prevailing wind, for the experimental period. A third set of probes was mounted at 4 m on top of a mast located 40 m northeast of the tower. Each fast-response sensor's output was sampled 10 times a second.

A tethered balloon (1330 m^3) was used to measure a vertical structure of the ABL. On the tethering cable of the balloon, five probes were mounted at altitudes ranging from 61 to 1220 m. The temperature sensor consisted of 180 cm of 25-μm-diameter platinum wire wound noninductively on a plastic form. A double V-shaped hot wire inclinometer mounted on a damped pendulum measured the inclination of the wind to the horizon, while a second inclinometer, oriented 90° from the first, measured the instantaneous horizontal wind direction relative to the vane direction. A magnetic flux value was used to obtain the vane direction relative to the earth's magnetic field. Wind speed was measured with a multislot photoelectric anemometer fitted with an eight-cup polystyrene rotor.

To supplement the tower and balloon observations, slow-ascent rawinsonde flights were made every 2 hours during the day and early evening, and every 3 to 4 hours during the night. Pressure, temperature, humidity, and wind velocity data were obtained to heights of about 3000 m.

The Minnesota data have provided a major source of information about the vertical distribution of the statistical moments, spectra, and cospectra in the convective and nocturnal boundary layer.

PROBLEMS

8.1. Prove that the averages of the forcing function and the response function (8.7) are the same.

8.2. Assuming that in the surface layer the integral scale $\mathscr{T}$ is proportional to height z and the wind velocity U, $\mathscr{T} \approx z/U$ and moreover that the distribution of θ is Gaussian with $\langle \theta^4 \rangle \approx 3\langle \theta^2 \rangle^2$ (Wyngaard, 1973), find the averaging time for temperature variance $\langle \theta'^2 \rangle$.

EPILOGUE

My aim in writing this text was to make the theory of the atmospheric boundary layer (ABL) easily accessible to students and easily teachable for teachers, as well as to provide a useful reference for scientists. The book discusses the current problems in ABL meteorology with a focus on the vertical structure of the ABL. The reader is introduced to the major scientific tools involved in ABL research: *similarity theory* (an elegant method of presenting empirical data), *modeling techniques* (testing our understanding of the atmosphere and also employed for its prediction), *spectral analysis* (describing properties of the ABL flow in a wide range of scales), and finally, *instrumentation and measurement methods*. The text is augmented by seventy-nine problems of various levels of difficulty to test the reader's understanding and also to extend the scope of the text. An Instructor's Manual, with solutions to these problems, figures and computer programs, is available through the publisher.

Since the importance and applications of atmospheric boundary layer meteorology were not discussed in the text, a brief comment will be worthwhile. The significance of processes in the atmospheric boundary layer is dictated by the fact that all human and biological activities take part in this layer. Its importance is also due to the strong interaction between this layer and the free atmosphere above it. The boundary layer is not an independent entity, isolated from the rest of the atmosphere. Mass and energy transfer within the ABL regulates a broad variety of processes in the entire atmosphere. Some obvious examples are the slowing of cyclonic storms and hurricanes entering inland from the ocean, and the rapid modification of cold continental air masses moving over a warm ocean or a large lake. Although the ABL contains only about 3% of the total kinetic energy of the atmosphere, it contributes as much as 30% to its total generation and 35% to its total dissipation.

The broad range of applications of boundary layer meteorology includes weather prediction, diffusion modeling for regulatory purposes and prediction, civil engineering designing, as well as various problems related to hydrology, agriculture, and astronomical observations. Perhaps the two most important are those which can be applied to engineering and weather prediction.

All meteorological elements impact the structures within the ABL and must be taken into account by engineers in their development of new designs. Strong winds and accompanying gusts have a variety of effects on skyscrapers, aircrafts, towers, bridges, chimneys, power transmission lines, etc. For example, buildings move in the wind; strong shears are hazardous to the operation of aircraft and launch vehicles near the ground. Waste gases produced by industrial processes are released to the atmosphere for disposal and their dispersal and effects are dependent on the state of the ABL.

The atmospheric boundary layer provides a critical link in the climate system. Therefore, parametrization of the ABL in short-, medium-, and large-range forecasting and climate models has an important scientific and practical meaning. Forecast and general circulation models have shown appreciable sensitivity of the thermal structure and circulation of the atmosphere on regional scales to formulation of surface fluxes, especially in long-term integration.

Many problems in the atmospheric boundary layer still remain unsolved. The stably stratified ABL is neither as well understood nor as well investigated as the convective case. Surface layer measurements are still much easier to make and interpret than those from higher levels. The interaction of radiation with clouds at the top of the ABL and shear generated turbulence is not completely comprehended. Study of the transfer of water vapor, dust and other matter from the Earth's surface through the boundary layer to the free atmosphere is still lacking.

Past and present research has concentrated on studies of the simplest, most homogenous, and undisturbed cases. As a result, our knowledge of the ABL in extreme and disturbed conditions has been scarce. Anticipated research in this direction will, no doubt, include studies on interaction between the ABL and cumulus convection associated with fronts, storms and cloud clusters. It will involve remote sensors which are particularly well suited for detection of ABL structures such as plumes, rolls, convection cells, and gravity waves.

The comparatively young science of boundary layer meteorology is growing along with other atmospheric sciences. Its development was triggered by the construction of the first computers and by the introduction of various measurement techniques. Future research on the ABL will be based on the interactions among theoretical, observational, and numerical modeling studies. A balance among these three components is a primary objective.

APPENDIX

A.1 FOURIER ANALYSIS

Any arbitrary periodic function $f(x)$ on the interval $-\pi \leq x \leq \pi$ can be expressed by the sum of the elementary trigonometric terms

$$f(x) = A_0 + \sum_{k=1}^{\infty} A_k \cos kx + \sum_{k=1}^{\infty} B_k \sin kx \tag{A.1}$$

where the coefficients A_0, A_k, and B_k are to be determined. It can be easily shown that the basic functions $\sin kx$ and $\cos kx$ are orthogonal:

$$\begin{aligned}
\int_{-\pi}^{\pi} \cos lx \cos kx \, dx &= \begin{cases} 0 & \text{if } k \neq l \\ \pi & \text{if } k = l > 0 \end{cases} \\
\int_{-\pi}^{\pi} \sin lx \sin kx \, dx &= \begin{cases} 0 & \text{if } k \neq l \\ \pi & \text{if } k = l > 0 \end{cases} \\
\int_{-\pi}^{\pi} \sin lx \cos kx \, dx &= \begin{cases} 0 & \text{if } k \neq l \\ \pi & \text{if } k = l > 0 \end{cases}
\end{aligned} \tag{A.2}$$

This yields the following set of equations for the coefficients:

$$\begin{aligned}
A_0 &= \frac{1}{2\pi} \int_{-\pi}^{\pi} f(x) \, dx \\
A_k &= \frac{1}{\pi} \int_{-\pi}^{\pi} f(x) \cos kx \, dx \\
B_k &= \frac{1}{\pi} \int_{-\pi}^{\pi} f(x) \sin kx \, dx
\end{aligned} \tag{A.3}$$

Invoking the trigonometric identity $\cos k(x - \xi) = \cos k\xi \cos kx + \sin k\xi \sin kx$ and using Euler's formula $\cos x = (e^{ix} + e^{-ix})/2$, it can be obtained from (A.1):

$$f(x) = \frac{1}{2\pi}\left\{\int_{-\pi}^{\pi} f(\xi)\,d\xi + \sum_{k=1}^{\infty}\int_{-\pi}^{\pi} f(\xi)e^{ik(x-\xi)}\,d\xi + \sum_{k=1}^{\infty}\int_{-\pi}^{\pi} f(\xi)e^{-ik(x-\xi)}\,d\xi\right\} \tag{A.4}$$

If k is allowed to be both positive and negative, the last term in (A.4) is equivalent to a summation from $-\infty$ to -1 (if $-k$ is replaced by $+k$). Recognizing also that, in the first term of (A.4), $k = 0$, we obtain:

$$f(x) = \frac{1}{2\pi}\sum_{k=-\infty}^{\infty}\int_{-\pi}^{\pi} f(\xi)e^{ik(x-\xi)}\,d\xi \tag{A.5}$$

The above equation can be rewritten as:

$$f(x) = \sum_{k=-\infty}^{\infty} c_k e^{ikx} \tag{A.6}$$

where:

$$c_k = \frac{1}{2\pi}\int_{-\pi}^{\pi} f(\xi)e^{-ik\xi}\,d\xi \tag{A.7}$$

Here c_k, the complex Fourier coefficient, is introduced in place of coefficients A_k and B_k. The relationship between these coefficients is

$$c_0 = A_0 \tag{A.8}$$

and

$$c_k = \begin{cases} \dfrac{A_k - iB_k}{2} & \text{for } k > 0 \\[2ex] \dfrac{A_{|k|} + iB_{|k|}}{2} & \text{for } k < 0 \end{cases} \tag{A.9}$$

If function $f(x)$ is periodical for $-L \le x \le L$, instead of (A.6) and (A.7) we have

$$f(x) = \sum_{k=-\infty}^{\infty} c_k e^{ik\omega x} \tag{A.10}$$

and

$$c_k = \frac{1}{2L}\int_{-L}^{L} f(\xi)e^{-ik\omega\xi}\,d\xi \tag{A.11}$$

where $\omega = \pi/L$. Substituting (A.11) into (A.10), we obtain for $L \to \infty$, $k\omega = \omega_n$ and $\Delta\omega_n = \omega_{n+1} - \omega_n = \omega = \pi/L$:

$$f(x) = \lim_{L\to\infty}\sum_{k=-\infty}^{\infty}\frac{\Delta\omega_k}{2\pi}\left[\int_{-L}^{L} f(\xi)e^{-i\omega_k\xi}\,d\xi\right]e^{i\omega_k x} \tag{A.12}$$

$$= \frac{1}{2\pi}\int_{-\infty}^{\infty}\left[\int_{-\infty}^{\infty} f(\xi)e^{-i\omega\xi}\,d\xi\right]e^{i\omega x}\,d\omega$$

The results above allow defining the Fourier transforms:

$$f(x) = \frac{1}{2}\int_{-\infty}^{\infty} F(\omega)e^{i\omega x}\,d\omega \tag{A.13}$$

where

$$F(\omega) = \frac{1}{\pi}\int_{-\infty}^{\infty} f(\xi)e^{-i\omega\xi}\,d\xi \tag{A.14}$$

A.2 THE SWEEPING METHOD

Consider the equation

$$\frac{\partial W}{\partial t} = \frac{\partial}{\partial z}k_m\frac{\partial W}{\partial z} + S \tag{A.15}$$

where W is an unknown function. Function W can be the pollutant concentration (the term S is a source/sink function) or the wind function obtained from equation (6.74) in the form

$$\begin{aligned} W &= \left[(U + iV) - (U_{g0} + iV_{g0})\right]e^{ift} \\ S &= if\Lambda z e^{ift} \\ \Lambda &= \frac{\partial U_g}{\partial z} + i\frac{\partial V_g}{\partial z} \end{aligned} \tag{A.16}$$

where $i = \sqrt{-1}$, U, and V are components of the wind velocity, U_g and V_g are the components of the geostrophic wind, f is the Coriolis parameter, and k_m is the eddy viscosity. It is assumed that W is given for $t = 0$ and additionally, that for $z = 0$ and $z = h$, the function W or its first derivative is also known. To solve this equation numerically, all derivatives are approximated by finite differences. This leads to the following system of equations (Roache, 1976):

$$\begin{aligned} \frac{W_j^{k+1} - W_j^k}{\Delta t} &= (k_m)_{j+1/2}\frac{W_{j+1}^{k+1} - W_j^{k+1}}{\Delta z^2} \\ &\quad -(k_m)_{j-1/2}\frac{W_j^{k+1} - W_{j-1}^{k+1}}{\Delta z^2} + S_j^{k+1} \end{aligned} \tag{A.17}$$

where $z = z_j = j\,\Delta z$, $t = t^k = k\,\Delta t$, $j = 0, 1, \ldots, n$, $k = 0, 1, \ldots,$ and $W(z, t) = W(z_j, t^k) = W_j^k$.

To solve the system of linear equations (A.17) with a tridiagonal matrix, the "sweeping method" can be used. To show this technique, the above set will be rewritten in the following form:

$$A_j X_{j+1} + B_j X_j + C_j X_{j-1} = F_j \qquad \text{for } j = 1, 2, \ldots, n-1 \tag{A.18}$$

where $X_j = W_j^{k+1}$ and coefficients A, B, C, and F are combinations of Δt, Δz, k_m, and Λ. The boundary conditions can be expressed in the general form

$$\begin{aligned} A_0 X_1 + B_0 X_0 &= F_0 \qquad \text{for } i = 0 \\ B_n X_n + C_n X_{n-1} &= F_n \qquad \text{for } i = n \end{aligned} \tag{A.19}$$

The solution of the system of $n + 1$ equations (A.18) and (A.19) is assumed to have the form

$$X_{j-1} = a_j X_j + b_j \qquad \text{for } i = n, n-1, \ldots, 1 \tag{A.20}$$

Substituting this equation in (A.18) yields

$$X_j = -\frac{A_j}{B_j + C_j a_j} X_{j+1} + \frac{F_j - C_j b_j}{B_j + C_j a_j} \qquad \text{for } j = 1, 2, \ldots, n-1 \tag{A.21}$$

which leads to

$$\begin{aligned} a_{j+1} &= -\frac{A_j}{B_j + C_j a_j} \\ b_{j+1} &= \frac{F_j - C_j b_j}{B_j + C_j a_j} \end{aligned} \tag{A.22}$$

From the boundary conditions it can be obtained that

$$a_1 = -\frac{A_0}{B_0} \quad \text{and} \quad b_1 = \frac{F_0}{B_0} \tag{A.23}$$

and also that

$$X_n = \frac{F_n - C_n b_n}{B_n + C_n a_n} \tag{A.24}$$

To find the solution, first all coefficients a_j and b_j are found from (A.22) and (A.23). Next, X_n is calculated from (A.24). Finally, the solution X_j is obtained from (A.21). The algorithm of the sweeping method is illustrated in Figure A.1.

A.3 VECTOR OPERATORS

There are three operations of multiplication defined for a vector $\mathbf{u} = (u_1, u_2, u_3)$: *multiplication by a scalar a*,

$$a\mathbf{u} = (au_1, au_2, au_3) \tag{A.25}$$

inner or *scalar product*,

$$\mathbf{u} \cdot \mathbf{v} = u_1 v_1 + u_2 v_2 + u_3 v_3 \tag{A.26}$$

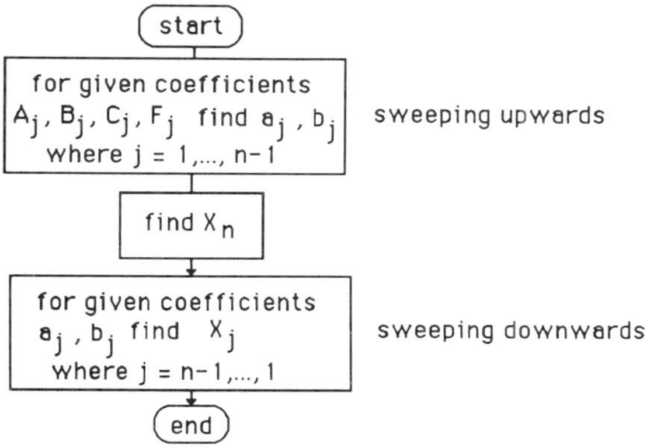

Figure A.1

and a *vector product*,

$$\mathbf{u} \times \mathbf{v} = \begin{bmatrix} i & j & k \\ u_1 & u_2 & u_3 \\ v_1 & v_2 & v_3 \end{bmatrix} \tag{A.27}$$

where $\mathbf{v}$ is a vector, $\mathbf{v} = (v_1, v_2, v_3)$. The first product (A.25) is a vector, the second (A.26) is a scalar, and the third product (A.27) is also a vector.

Differential operators of *divergence*, *gradient*, and *curl* are defined in the following way:

$$\operatorname{div} \mathbf{v} = \lim_{\mathbb{V} \to 0} \frac{\iint_{\Sigma} \mathbf{v} \cdot \mathbf{d\Sigma}}{\mathbb{V}} \tag{A.28}$$

$$\operatorname{grad} a = \lim_{\mathbb{V} \to 0} \frac{\iint_{\Sigma} a \, \mathbf{d\Sigma}}{\mathbb{V}} \tag{A.29}$$

where symbols are as explained in Figure 2.1. *Curl* is defined as

$$\mathbf{n} \cdot \operatorname{curl} \mathbf{v} = \lim_{\Sigma \to 0} \frac{\int_S \mathbf{v} \cdot \mathbf{ds}}{\Sigma} \tag{A.30}$$

where S is a closed curve, Σ is an area within this curve, $\mathbf{ds}$ is a vector tangent to the curve S, and $\mathbf{n}$ is a vector normal to Σ.

From the definition of divergence (A.28), it follows that

$$\iint_{\Sigma} \mathbf{v} \cdot \mathbf{d\Sigma} = \iiint_{\mathbb{V}} \operatorname{div} \mathbf{v} \, d\mathbb{V} \tag{A.31}$$

Equation (A.31) is called the *Gauss theorem*. Similar theorems can be obtained from (A.29) and (A.30).

Using the definition of the *nabla operator*,

$$\nabla = \left(\frac{\partial}{\partial x}, \frac{\partial}{\partial y}, \frac{\partial}{\partial z} \right) \tag{A.32}$$

the divergence, gradient, and curl in the Cartesian coordinate system can be expressed as

$$\begin{aligned} \operatorname{div} \mathbf{v} &= \nabla \cdot \mathbf{v} \\ \operatorname{grad} a &= \nabla \mathbf{a} \\ \operatorname{curl} \mathbf{v} &= \nabla \times \mathbf{v} \end{aligned} \tag{A.33}$$

From the definition, it follows that div **v** is a scalar, while grad a and curl **v** are vectors.

REFERENCES

ADRIAN, R. J., R. T. D. S. FERREIRA, AND T. BOBERG, 1986. Turbulent thermal convection in wide horizontal fluid layers. *Exp. Fluids,* 4, 121–141.

ANDRÉ, J. C., G. DE MOOR, P. LACARRÈRE, G. THERRY, AND R. DU CACHAT, 1978. Modelling the 24-hour evolution of the mean and turbulent structures of the planetary boundary layer. *J. Atmos. Sci.,* 35, 1861–1883.

ARTAZ, M. A., AND J. C. ANDRÉ, 1980. Similarity studies of entrainment in convective mixed layers. *Boundary-Layer Meteorol.,* 19, 51–66.

ARYA, S. P. S., 1975a. Geostrophic drag and heat transfer relations for the atmospheric boundary layer. *Q. J. R. Meteorol. Soc.,* 101, 147–161.

ARYA, S. P. S., 1975b. Comments on similarity theory for the planetary boundary layer of time-dependent heights. *J. Atmos. Sci.,* 32, 839–840.

ARYA, S. P. S., AND J. C. WYNGAARD, 1975. Effects of baroclinicity of wind profiles and the geostrophic drag law for the convective planetary boundary layer. *J. Atmos. Sci.,* 32, 767–778.

BALL, F. K., 1960. Control of the inversion height by surface heating. *Q. J. R. Meteorol. Soc.,* 86, 483–494.

BATCHELOR, G. K., 1960. *The Theory of Homogeneous Turbulence.* Cambridge University Press, London.

BATCHELOR, G. K., 1967. *An Introduction to Fluid Dynamics.* Cambridge University Press, London.

BÉNARD, M., 1900. Les tourbillons cellulaires dans une nappe liquide. *Rev. Gen. Sci. Pures Appl.,* II, 1261–1271.

BERKOWICZ, R., 1984. Spectral methods for atmospheric diffusion. *Boundary-Layer Meteorol.,* 30, 201–220.

BERKOWICZ, R., AND L. P. PRAHM, 1980. On spectral turbulent theory for homogeneous turbulence. *J. Fluid Mech.,* 100, 433–448.

BERKOWICZ, R., H. R. OLESEN, AND U. TORP, 1985. The Danish Gaussian air pollution model (OML): description, test and sensitivity analysis in view of regulatory application. *Proceedings NATO/CCMS 15th International Technical Meeting on Air Pollution Modeling and Its Applications,* St. Louis.

BLACKADAR, A. K., 1962. The vertical distribution of wind and turbulent exchange in neutral atmosphere. *J. Geoph. Res.,* 67, 3095–3103.

BLACKADAR, A. K., 1967. External parameters of the wind flow in the barotropic boundary layer of the atmosphere. *Proceedings GARP Study Conference,* Stockholm, Appendix 1 CSU/IUGG, Committee on Atmospheric Science, COSPAR, WMO.

BRADLEY, E. F., AND R. A. ANTONIA, 1979. Structure parameters in the atmospheric surface layer. *Q. J. R. Meteorol. Soc.,* 105, 695–705.

BRADLEY, E. F., R. A. ANTONIA, AND A. J. CHAMBERS, 1981. Temperature structure in the atmospheric surface layer. Parts 1 and 2. *Boundary-Layer Meteorol.,* 20, 275–307.

BRIGGS, G. A., 1973. Diffusion estimates for small emissions. *ATDL Contributions No. 79,* Atmospheric Turbulence and Diffusion Laboratory, Oak Ridge, Tenn.

BRIGGS, G. A., 1985. Diffusion modeling with convective scaling and effects of surface inhomogeneities. Paper presented at Specialty Conference on Air Quality Modeling of the Urban Boundary Layer, Baltimore, American Meteorological Society.

BROCK, F. V., AND C. E. NICOLAIDIS (EDS.), 1984. *Instructor's Handbook on Meteorological Instrumentation,* NCAR Tech. Note 237 + 1A.

BROST, R. A., AND J. C. WYNGAARD, 1978. A model study of the stably stratified planetary boundary layer. *J. Atmos. Sci.,* 35, 1427–1440.

BROST, R. A., J. C. WYNGAARD, AND D. H. LENSCHOW, 1982. Marine stratocumulus layers. Part 2. Turbulence budgets. *J. Atmos. Sci.,* 39(4), 818–836.

BROWN, R. A., 1970. A secondary flow model for the planetary boundary layer. *J. Atmos. Sci.,* 27, 742–757.

BROWN, R. A., 1972. In the inflection point instability of a stratified Ekman boundary layer. *J. Atmos. Sci.*, 29, 850–859.

BROWN, R. A., 1974. *Analytic Methods in Planetary Boundary Layer Modelling.* Wiley, New York.

BROWN, R. A., 1978. Similarity parameters from first-order closure and data. *Boundary-Layer Meteorol.,* 14, 381–396.

BROWN, R. A., 1980. Longitudinal instabilities and secondary flows in the planetary boundary layer. A review. *Rev. Geophys. Space Phys.,* 18, 683–697.

BROWN, R. A., 1982. On two-layer models and the similarity functions for the PBL. *Boundary-Layer Meteorol.,* 24, 451–463.

BUSINGER, J. A., J. C. WYNGAARD, Y. IZUMI, AND E. F. BRADLEY, 1971. Flux-profile relationship in the atmospheric surface layer. *J. Atmos. Sci.,* 28, 181–189.

CARL, D. M., T. C. TARBELL, AND H. A. PANOFSKY, 1973. Profiles of wind and temperature from towers over homogeneous terrain. *J. Atmos. Sci.,* 30, 788–794.

CARSON, D. L., 1973. The development of a dry-inversion-capped convectively unstable boundary layer. *Q. J. R. Meteorol. Soc.,* 99, 451–467.

CATTLE, H., AND K. J. WESTON, 1975. Budget studies of the heat flux profiles in the convective boundary layer over land. *Q. J. R. Meteorol. Soc.,* 101, 353–363.

CAUGHEY, S. J., 1977. Boundary-layer turbulence spectra in stable conditions. *Boundary-Layer Meteorol.,* 11, 3–14.

CAUGHEY, S. J., AND S. G. PALMER, 1979. Some aspects of turbulence structure through the depth of the convective boundary layer. *Q. J. R. Meteorol. Soc.,* 105, 811–827.

Caughey, S. J., J. C. Wyngaard, and J. C. Kaimal, 1979. Turbulence in the evolving stable boundary layer. *J. Atmos. Sci.,* 36, 1041–1052.

Caughey, S. J., M. Kitchen, and J. R. Leighton, 1983. Turbulence structure in convective boundary layers and implications for diffusion. *Boundary-Layer Meteorol.,* 25, 345–352.

Chalikov, D. V., 1968. Estimation of surface turbulent fluxes from the synoptic information. *Meteorol. Hydrol.,* 8, 10–19 (in Russian).

Chandrasekhar, S., 1943. Stochastic problems in physics and astronomy. *Rev. Mod. Phys.,* 15(1), 2–87. (Also in: *Selected Papers on Noise and Stochastic Processes,* N. Wax (ed.). Dover, New York, 1954.)

Clarke, R. H., 1970. Observational studies in the atmospheric boundary layer. *Q. J. R. Meteorol. Soc.,* 96, 91–114.

Clarke, R. H., and G. H. Hess, 1974. Geostrophic departure and the functions A and B of Rossby-number similarity theory. *Boundary-Layer Meteorol.,* 7, 267–287.

Clarke, R. H., A. J. Dyer, R. R. Brooke, D. G. Reid, and A. J. Troup, 1971. The Wangara experiment. Boundary-layer data. Paper no. 19. Division of Meteorol. Phys. CSIRO, Australia.

Cooney, J. A., 1972. Measurement of atmospheric temperature profiles by Raman backscatter. *J. Appl. Meteorol.,* 10, 301–308.

Corrsin, S., 1951. On the spectrum of isotropic temperature fluctuation in an isotropic turbulence. *J. Appl. Phys.,* 22, 469–473.

Cotton, W. R., and G. J. Tripoli, 1978. Cumulus convection in the shear flow-three dimensional numerical experiments. *J. Atmos. Sci.,* 35, 1503–1521.

Cvitanović, P., 1984. *Universality in Chaos.* Adam Hilger, Bristol, England.

Davydov, B. I., 1959. On statistical dynamics of noncompressible turbulent fluid. *Dokl. Akad. Nauk SSSR,* 127(4), 768–771.

Davydov, B. I., 1961. On statistical dynamics of noncompressible turbulent fluid. *Dokl. Akad. Nauk SSSR,* 136(1), 47–50.

Deardorff, J. W., 1966. The contragradient heat flux in the lower atmosphere and in the laboratory. *J. Atmos. Sci.,* 23, 503–506.

Deardorff, J. W., 1970. Convective velocity and temperature scales for the unstable planetary boundary layer and for Rayleigh convection. *J. Atmos. Sci.,* 27, 1211–1213.

Deardorff, J. W., 1972a. Parametrization of the planetary boundary layer for use in general circulation models. *Mon. Weather Rev.,* 100, 93–106.

Deardorff, J. W., 1972b. Numerical investigation of neutral and unstable planetary boundary layer. *J. Atmos. Sci.,* 29, 91–115.

Deardorff, J. W., 1973. An explanation of anomalously large Reynolds stress within the convective planetary boundary layer. *J. Atmos. Sci.,* 30, 1070–1076.

Deardorff, J. W., 1974a. Three-dimensional study of turbulence in an entraining mixed layer. *Boundary-Layer Meteorol.,* 7, 199–226.

Deardorff, J. W., 1974b. Three-dimensional study of the height and mean structure of a heated planetary boundary layer. *Boundary-Layer Meteorol.,* 7, 81–106.

Deardorff, J. W., 1978a. Progress in understanding entrainment at the top of a mixed layer. *Workshop on Planetary Boundary Layer,* Boulder, Colo., J. C. Wyngaard, ed.

DEARDORFF, J. W., 1978b. Observed characteristics of the outer layer. AMS course on the planetary boundary layer. Boulder, Colo. (unpublished manuscript).

DEARDORFF, J. W., 1980. Stratocumulus-capped mixed layers derived from three-dimensional model. *Boundary-Layer Meteorol.,* 18, 495–527.

DEARDORFF, J. W., AND G. E. WILLIS, 1975. A parametrization of diffusion into the mixed layer. *J. Appl. Meteorol.,* 14, 1451–1458.

DEARDORFF, J. W., AND G. E. WILLIS, 1985. Further results from a laboratory model of the convective planetary boundary layer. *Boundary-Layer Meteorol.,* 32, 205–236.

DEARDORFF, J. W., G. E. WILLIS, AND D. K. LILLY, 1969. Laboratory investigation of non-steady penetrative convection. *J. Fluid Mech.,* 35, 7–31.

DE BAAS, A. F., H. VAN DOP, AND F. T. M. NIEUWSTADT, 1986. An application of the Langevin equation for inhomogeneous conditions to dispersion in a convective boundary layer. *Q. J. R. Meteorol. Soc.,* 112, 165–180.

DOBSON, F., L. HASSE, AND R. DAVIS (EDS.), 1980. *Air–Sea Intersection. Instruments and Methods.* Plenum Press, New York.

DOVIAK, R. J., AND C. T. JOBSON, 1979. Dual-Doppler radar observations of clear air wind perturbations in the planetary boundary layer. *J. Geophys. Res.,* 84, 697–702.

DRAZIN, P. G., AND W. H. REID, 1980. *Hydrodynamic Stability.* Cambridge University Press.

DYER, A. J., 1977. A review of flux-profile relationships. *Boundary-Layer Meteorol.,* 7, 363–372.

DYER, A. J., AND E. F. BRADLEY, 1982. An alternative analysis of flux-gradient relationships at the 1976 ITCE. *Boundary-Layer Meteorol.,* 22, 3–19.

DYER, A. J., AND B. B. HICKS, 1970. Flux gradient transport of heat and water vapor in an unstable atmosphere. *Q. J. R. Meteorol. Soc.,* 93, 501–508.

EKMAN, V. W., 1905. On influence of the earth's rotation on ocean currents. *Ark. Mat. Astron. Fys.,* 2, 11.

ENGER, L., 1983. Numerical boundary layer model with application to diffusion. Part 2. A higher-order closure dispersion model. *Rep. No. 71.,* Department of Meteorology, University of Uppsala.

FACKRELL, J. E., AND A. G. ROBINS, 1982. Concentration fluctuations and fluxes in plumes from point sources in a turbulent boundary layer. *J. Fluid Mech.,* 117, 1–26.

FIEDLER, B. H., 1984. An integrated closure model for the vertical turbulent flux of a scalar in a mixed layer. *J. Atmos. Sci.,* 41, 674–680.

FLEAGLE, R. G., AND J. A. BUSINGER. 1980. *An Introduction to Atmospheric Physics.* Academic Press, Orlando, Fla.

GARRAT, J. R., 1972. Studies of turbulence in the surface layer over water (Lough Neagh). Part 2. Production and dissipation of velocity and temperature fluctuation. *Q. J. R. Meteorol. Soc.,* 98, 642–657.

GIFFORD, F. A., 1955. Simultaneous Lagrangian–Eulerian turbulence experiment. *Mon. Weather Rev.,* 83, 293–301.

GIFFORD, F. A., 1961. Use of routine meteorological observations for estimating atmospheric dispersion. *Nucl. Saf.,* 2(4), 47–51.

GOSSARD, E. E., AND J. H. RICHTER, 1970. The shape of internal waves of finite amplitude from high-resolution radar sounding of the lower atmosphere. *J. Atmos. Sci.*, 27, 971–973.

GOSSARD, E. E., D. R. JENSEN, AND J. H. RICHTER, 1971. Analytical study of tropospheric structure as seen by high-resolution radar. *J. Atmos. Sci.,* 28, 794–807.

GOSSARD, E. E., J. H. RICHTER, AND D. R. JENSEN, 1973. Effect of wind shear on atmospheric wave instabilities revealed by FM/CW radar observations. *Boundary-Layer Meteorol.*, 4, 113–130.

GUILLEMENT, B., H. ISAKA, AND P. MASCART, 1983. Molecular dissipation of turbulent fluctuations in the convective mixed layer. *Boundary-Layer Meteorol.,* 27, 141–162.

HANJALIC, K., AND B. E. LAUNDER, 1972. A Reynolds stress model of turbulence and its application to thin shear flows. *J. Fluid Mech.,* 52(4), 609–638.

HARDY, K. R., AND H. OTTERSEN, 1979. Radar investigation of convective patterns in the clear atmosphere. *J. Atmos. Sci.,* 26, 666–672.

HAUGEN, D. A. (ED.), 1973. *Workshop on Micrometeorology.* American Meteorological Society, Boston.

HAUGEN, D. A., J. C. KAIMAL, AND E. F. BRADLEY, 1971. An experimental study of Reynolds stress and heat flux in the atmospheric surface layer. *Q. J. R. Meteorol. Soc.,* 97, 168–180.

HAY, J. S., AND F. PASQUILL, 1959. Diffusion from a continuous source in relation to the spectrum and scale of turbulence. *Adv. Geophys.*, 6, 345–365.

HICKS, B. B., 1976. Wind profile relationships from the Wangara experiment *Q. J. R. Meteorol. Soc.,* 102, 535–551.

HINZE, I., 1959. *Turbulence.* McGraw-Hill, New York.

HÖGSTRÖM, U., 1985. Von Karman's constant in atmospheric boundary layer flow: reevaluated. *J. Atmos. Sci.*, 42, 263–270.

HÖGSTRÖM, U., 1986. Reply. *J. Atmos. Sci.,* 43, 2131–2134.

HØJSTRUP, J., 1982. Velocity spectra in the unstable planetary boundary layer. *J. Atmos. Sci.,* 39, 2239–2249.

HOOKE, W. H., F. F. HALL, AND E. E. GOSSARD, 1973. Observed generation of an atmospheric gravity wave by shear instability in the mean flow of the planetary boundary layer. *Boundary-Layer Meteorol.,* 5, 29–41.

ISAKA, H., AND B. GUILLEMENT, 1983. Molecular dissipation of turbulent fluctuations in the convective mixed layer. Part 2. *Boundary-Layer Meteorol.,* 27, 259–279.

IZUMI, Y., 1971. Kansas 1968 Field Program Data Report. *Environmental Research Papers No. 369,* AFC RL-72 0041, Air Force Cambridge Research Lab., Bedford, Mass.

IZUMI, Y., AND J. S. CAUGHEY, 1976. Minnesota 1973 Atmospheric Boundary Layer Experiment Data Report. *Air Force Cambridge Research Papers No. 547.*

JONES, W. P., AND B. E. LAUNDER, 1972. The prediction of laminarization with a two-equation model of turbulence. *Int. J. Heat Mass Transfer,* 15, 301–314.

KAI, K., 1982. Statistical characteristics of turbulence and the budget of turbulent energy in the surface boundary layer. *Environmental Research Center Paper No. 1,* University of Tsukuba.

KAIMAL, J. C., 1978. Horizontal velocity spectra in an unstable surface layer. *J. Atmos. Sci.,* 35(11), 18–23.

KAIMAL, J. C., J. C. WYNGAARD, Y. IZUMI, AND A. R. COTÉ, 1972. Spectral characteristics of surface layer turbulence. *Q. J. R. Meteorol. Soc.,* 98, 563–689.

KAIMAL, J. C., J. C. WYNGAARD, D. A. HAUGEN, O. R. COTÉ, Y. IZUMI, S. J. CAUGHEY, AND C. J. READINGS, 1976. Turbulence structure in the convective boundary layer. *J. Atmos. Sci.,* 33(11), 2152–2168.

KAIMAL, J. C., R. A. EVERSOLE, D. H. LENSCHOW, B. B. STANKOV, P. H. KAHN, AND J. A. BUSINGER, 1982. Spectral characteristics of the convective boundary layer over uneven terrain. *J. Atmos. Sci.,* 39(5), 1098–1114.

KAMPÉ DE FÉRRIET, J., 1939. Les fonctions aléatories stationnaires et la théorie statistique de la turbulence homogène. *Ann. Soc. Sci. Bruxelles,* 59, 145–194.

KAZANSKII, A. B., AND A. S. MONIN, 1961. On the dynamical interaction between the atmosphere and earth's surface. *Izv. Akad. Nauk SSSR Ser. Geofiz.,* 5, 514–515.

KLEMP, J. B., AND R. B. WILHELMSON, 1978. The simulation of three-dimensional convective storm dynamics. *J. Atmos. Sci.*, 35, 1070–1096.

KOLMOGOROV, A. N., 1941. The local structure of turbulence in compressible turbulence for very large Reynolds numbers. *C. R. Akad. Nauk SSSR,* 30, 301–305.

KORRELL, A., H. A. PANOFSKY, AND R. J. ROSSI, 1982. Wind profiles at the Boulder Tower. *Boundary-Layer Meteorol.,* 22, 297–312.

KUETTNER, J., 1959. The band structure of the atmosphere. *Tellus,* 11, 267–296.

KUETTNER, J. P., 1971. Cloud bands in the earth's atmosphere. *Tellus,* 23, 4040–4426.

LAICHTMAN, D. L., 1970. *Physics of the Atmospheric Boundary Layer.* Gidrometeorizdat, Leningrad (in Russian).

LAIKHTMAN, D. C., AND D. L. YORDANOV, 1979. Vertical velocity at the upper boundary of the unsteady-state planetary boundary layer. *Izv. Akad. Nauk SSSR, Ser. Fiz. Atmos. Okeana (Bull. Acad. Sci. USSR Atmos. Oceanic Phys.),* 15, 309–310.

LAMB, R. G., 1979. The effect of release height on material dispersion in the convective planetary boundary layer. *Preprints 4th Symposium on Turbulence, Diffusion, and Air Pollution,* Reno, Nev., American Meteorological Society.

LAMB, R. G., 1982. Diffusion in the convective boundary layer. In: *Atmospheric Turbulence and Air Pollution Modelling,* F. T. M. Nieuwstadt and H. van Dop (eds.). D. Reidel, Dordtrecht, The Netherlands.

LANDAU, L. D., AND E. M. LIFSHITZ, 1944. *Fluid Mechanics.* English ed. Pergamon Press, London, 1959.

LAUNDER, B. E., 1976. Heat and mass transport. In: *Turbulence,* P. Bradshaw (ed.). Springer-Verlag, Berlin.

LAUNDER, B. E., G. J. REECE, AND W. RODI, 1975. Progress in the development of Reynolds-stress turbulence closure. *J. Fluid Mech.,* 68, 537–565.

LEMONE, M. A., 1973. The structure and dynamics of horizontal roll vortices in the planetary boundary layer. *J. Atmos. Sci.,* 30, 1077–1091.

LENSCHOW, D. H., 1970. Airplane measurements of planetary boundary layer structure. *J. Appl. Meteorol.,* 9, 1339–1343.

LENSCHOW, D. H., 1972. The measurement of air velocity and temperature using the NCAR Buffalo Aircraft Measuring system. *Technical Note NCAR / EDD-74,* NCAR, Boulder, Colo.

LENSCHOW, D. H. (ED.), 1986. *Probing the Atmospheric Boundary Layer.* American Meteorological Society, Boston.

LENSCHOW, D. H., AND P. L. STEVENS, 1980. The role of thermals in the convective boundary layer. *Boundary-Layer Meteorol.,* 19, 509–532.

LENSCHOW, D. H., J. C. WYNGAARD, AND W. T. PENNEL, 1980. Mean-field and second-moment budgets in a baroclinic convective boundary layer. *J. Atmos. Sci.,* 37, 1313–1326.

LENSCHOW, D. H., L. XINGSHEN, Z. CUIJUAN, AND B. B. STANKOV, 1988. The stably-stratified boundary layer over Great Plains. Part 1. Mean and turbulence structure. *Boundary-Layer Meteorol.,* 42, 95–121.

LENSCHOW, D. H., 1974. Model of the height variation of the turbulent kinetic energy budget in the unstable planetary boundary layer. *J. Atmos. Sci.,* 31, 465–474.

LESLIE, D. C., 1973. *Developments in the Theory of Turbulence.* Clarendon Press, Oxford.

LETTAU, H. H., AND B. DAVIDSON, 1957. *Exploring the Atmosphere's First Mile.* Pergamon Press, Oxford.

LORENZ, E. N., 1963. Deterministic non-periodic flow. *J. Atmos. Sci.,* 20, 130–141.

LEWELLEN, W. S., AND M. TESKE, 1973. Prediction of the Monin–Obukhov similarity functions from an invariant mode of turbulence. *J. Atmos. Sci.,* 30, 1340–1345.

LILLY, D. K., 1968. Models of cloud-topped mixed layers under a strong inversion. *Q. J. R. Meteorol. Soc.,* 94, 292–309.

LILLY, D. K., AND D. H. LENSCHOW, 1971. Aircraft measurement of the atmospheric mesoscales using an inertial reference system. Facilities for Atmospheric Research, NCAR, *Rep. No. 19,* 2–8.

LIN, C. C., AND W. H. REID, 1962. Turbulent flow. In: *Handbuch der Physik,* Vol. 8, Pt. 2. Springer-Verlag, Berlin, pp. 438–523.

LUMLEY, J. L., 1979. Computational modelling of turbulent flows. *Adv. Appl. Mech.,* 18, 123–176.

LUMLEY, J. L., AND H. A. PANOFSKY, 1964. *The Structure of Atmospheric Turbulence.* Interscience, New York.

MAHRT, L. J., AND D. H. LENSCHOW, 1976. Growth dynamics of the convective mixed layer in mesoscale models with coarse resolution. *J. Atmos. Sci.,* 32, 2288–2308.

MAHRT, L. J., J. C. ANDRÉ, AND R. C. HEALD, 1982. On the depth of the nocturnal boundary layer. *J. Appl. Meteorol.,* 21, 90–92.

MANDELBROT, B. B., 1982. *Fractal Geometry of Nature.* W. H. Freeman, San Francisco.

MANINS, P. C., 1979. Partial penetration of an elevated inversion layer by chimney plumes. *Atmos. Environ.,* 13, 733–741.

MCALISTER, L. G., J. R. POLLARD, A. R. MAHONEY, AND P. J. R. SHAW, 1969. Acoustic sounding: a new approach to the study of atmospheric structure. *Proc. IEEE,* 57, 579–587.

McBean, G. A., and J. A. Eliot, 1975. The vertical transports of kinetic energy by turbulence and pressure in the boundary layer. *J. Atmos. Sci.,* 32, 753–766.

McNider, R. T., and R. A. Pielke, 1981. Diurnal boundary-layer development over sloping terrain. *J. Atmos. Sci.,* 38, 2198–2212.

Melgarejo, J. W., and J. W. Deardorff, 1974. Stability functions for the boundary-layer resistance laws based upon observed boundary-layer heights. *J. Atmos. Sci.,* 31, 1324–1333.

Mellor, G. L., 1973. Analytic properties of stratified planetary surface layers. *J. Atmos. Sci.,* 30, 1061–1069.

Mellor, G. L., and T. Yamada, 1974. A hierarchy of turbulence closure models for planetary boundary layers. *J. Atmos. Sci.,* 31, 1791–1806.

Merry, M., and H. A. Panofsky, 1976. Statistics of vertical motion over land. *Q. J. R. Meteorol. Soc.,* 102, 255–260.

Miyake, M., M. Donelan, G. McBean, C. A. Paulson, and E. Leavitt, 1970. Comparison of turbulent fluxes over water determined by profile and eddy correlation techniques. *Q. J. R. Meteorol. Soc.,* 96, 132–137.

Moeng, C.-H., 1984. A large-eddy-simulation model for the study of planetary boundary-layer turbulence. *J. Atmos. Sci.,* 41, 2052–2062.

Moeng, C.-H., and J. C. Wyngaard, 1984. Statistics of conservative scalars in the convective boundary layer. *J. Atmos. Sci.,* 41, 3161–3169.

Monin, A. S., 1959. On the similarity of turbulence in the presence of a mean vertical temperature gradient. *J. Geophys. Res.,* 64(12), 2196–2197.

Monin, A. S., and A. M. Obukhov, 1954. Basic laws of turbulent mixing in the atmosphere near the ground. *Tr. Akad. Nauk SSSR Geoph. Inst.,* No. 24(151), 1963–1987.

Monin, A. S., and A. S. Yaglom, 1965. *Statistical Fluid Mechanics.* Nauka, Moscow (English translation, 1971).

Monin, A. S., and S. S. Zilitinkevich, 1967. Planetary boundary layer and large scale atmospheric dynamics. *Proceedings GARP Study Conference,* Stockholm, Appendix VI ICSU/IUGG, Committee on Atmospheric Science, COSPAR, WMO.

Monji, N., 1973. Budgets of turbulent energy and temperature variance in the transition zone from forced to free convection. *J. Meteorol. Soc. Jpn.,* 2, 133–145.

Moore, D. S., and J. W. Yackel, 1974. *Applicable Finite Mathematics.* Houghton Mifflin, Boston.

Nieuwstadt, F. T. M., 1984. The turbulent structure of the stable, nocturnal boundary layer. *J. Atmos. Sci.,* 41, 2202–2216.

Nieuwstadt, F. T. M., and H. van Dop (eds.), 1984. *Atmospheric Turbulence and Air Pollution Modelling.* D. Reidel, Dordrecht, The Netherlands.

Nieuwstadt, F. T. M., and A. P. van Ulden, 1978. A numerical study on vertical dispersion of passive contaminants from a continuous source in the atmospheric surface layer. *Atmos. Environ.,* 12, 2119–2124.

O'Brien, J. J., 1970. A note on the vertical structure of eddy-exchange coefficient in the planetary boundary layer. *J. Atmos. Sci.,* 27, 1213–1215.

OBUKHOV, A. M., 1946. Turbulence in thermally nonhomogeneous atmosphere. *Tr. Inst. Teor. Geofiz. Akad. Nauk SSSR,* 1, 95–115 (in Russian).

OBUKHOV, A. M., 1959. On influence of Archimedian forces on a structure of temperature field in turbulent flow. *Dokl. Akad. Nauk SSSR Ser. Geograph. and Geoph.,* 125(6), 1246–1248 (in Russian).

OKAMOTO, M., AND E. K. WEBB, 1970. The temperature fluctuations in a stable stratification. *Q. J. R. Meteorol. Soc.,* 96, 591–600.

PANOFSKY, H. A., AND J. A. DUTTON, 1984. *Atmospheric Turbulence.* Wiley, New York.

PANOFSKY, H. A., H. TENNEKES, D. H. LENSCHOW, AND J. C. WYNGAARD, 1977. The characteristics of turbulent components in the surface layer under convective conditions. *Boundary-Layer Meteorol.,* 11, 355–361.

PAQUIN, J. E., AND S. POND, 1971. Spectra of temperature and humidity fluctuations and of the fluxes of moisture and sensible heat in the marine boundary layer. *J. Atmos. Sci.,* 28, 918–928.

PASQUILL, F., 1961. The estimation of the dispersion of windborne material. *Meteorol. Mag.,* 90, 33–49.

PASQUILL, F., 1974. *Atmospheric Diffusion.* Wiley, New York.

PAULSON, C. A., 1970. The mathematical representation of wind speed and temperature profiles in the unstable atmospheric surface layer. *J. Appl. Meteorol.,* 9, 857–861.

PIELKE, R. A., 1984. *Mesoscale Meteorological Modeling.* Academic Press, Orlando, Fla.

PLATE, E. J. (ED.), 1982. *Engineering Meteorology.* Elsevier, New York.

PRANDTL, L., 1932. Meteorologische Anwendungen der Strömungslehre. *Beitr. Phys. Atmos.,* 19, 188–202.

PRIESTLEY, C. B. B., 1955. Free and forced convection in the atmosphere near the ground. *Q. J. R. Meteorol. Soc.,* 81, 139–143.

PRIESTLEY, C. B. B., 1959. *Turbulent Transport in the Lower Atmosphere.* University of Chicago Press, Chicago.

RAO, K. S., AND H. F. SNODGRASS, 1978. The structure of the nocturnal planetary boundary layer. Environmental Research Laboratories, Air Resources Atmospheric Turbulence and Diffusion Laboratory, Oak Ridge, Tenn.

RAYLEIGH, LORD (J. W. STRUTT), 1916. On convection currents in a horizontal layer of fluid when the higher temperature is on the other side. *Philos. Mag.,* Ser. 6, 32, 529–546. (Reprinted in *Theory of Thermal Convection,* B. Saltzman (ed.), Dover, New York, 1962.)

RICHARDSON, L. F., 1926. Atmospheric diffusion shown a distance-neighbour graph. *Proc. R. Soc.,* A110(765), 709–737.

RICHTER, J. H., D. R. JENSEN, R. A. PAPPERT, AND V. R. NONKESTER, 1973. New development in FM-CW radar sounding. *Boundary-Layer Meteorol.,* 4, 179–199.

ROACHE, P. J., 1976. *Computational Fluid Dynamics.* Hermosa Publishers, Albuquerque, N.Mex.

ROTTA, J. C., 1951. Statistische Theorie nighthamogener Turbulenz. *Z. Phys.,* 129, 547.

SCHMIDT, K. F., C. A. FRIEHE, AND C. H. GIBSON, 1979. Structure of marine surface layer turbulence. *J. Atmos. Sci.,* 36, 602–618.

SCHOTLAND, R. M., 1985. Remote ground-based observing systems. *Handbook of Applied Meteorology*, 361–377. J. Killey & Sons.

SCHUMANN, U., 1977. Realizability of Reynolds stress turbulence models. *Phys. Fluids,* 20, 721–725.

SCHUSTER, H. B., 1984. *Deterministic Chaos.* Physik-Verlag, Weinheim, West Germany.

SEGAL, M., R. T. MCNIDER, R. A. PIELKE, AND D. S. MCDOUGHAL, 1982. A numerical model simulation of the regional air pollution meteorology of the greater Chesapeake Bay area—summer day case study. *Atmos. Environ.,* 16, 1–17.

SETHURAMAN, S., R. E. MEYERS, AND R. M. BROWN, 1978. A comparison of an Eulerian and a Lagrangian time scale for over water atmospheric flows during stable conditions. *Boundary-Layer Meteorol.,* 14, 557–565.

SMITH, F. B., 1968. Conditioned particle motion in a homogeneous turbulent field. *Atmos. Environ.,* 2, 491–508.

SOMMERIA, G., AND M. A. LEMONE, 1978. Direct testing of the three-dimensional model of the planetary boundary layer against experimental data. *J. Atmos. Sci.,* 35, 25–39.

SORBJAN, Z., 1979. Urban boundary layer. In: *WMO Tech. Note No. 165.*

SORBJAN, Z., 1983. Rossby-number similarity in the atmospheric boundary layer over a slightly inclined terrain. *J. Atmos. Sci.,* 40, 718–728.

SORBJAN, Z., 1984. A model study of the stably stratified steady-state atmospheric boundary layer over as lightly inclined terrain. *J. Atmos. Sci.,* 41, 1863–1894.

SORBJAN, Z., 1986a. On similarity in the atmospheric boundary layer. *Boundary-Layer Meteorol.,* 34, 377–397.

SORBJAN, Z., 1986b. On vertical distribution of passive species in the atmospheric boundary layer. *Boundary-Layer Meteorol.,* 35, 73–81.

SORBJAN, Z., 1986c. Local similarity of spectral and cospectral characteristics in the stable-continuous boundary layer. *Boundary-Layer Meteorol.,* 35, 257–275.

SORBJAN, Z., 1988. Local similarity in the atmospheric boundary layer. *Preprint of the 8th Symp. on Turbulence and Diffusion*, April 1988, San Diego, Calif., American Meteorological Society, 353–356.

STAGE, S. A., AND J. A. BUSINGER, 1981. A model for entrainment into a cloud-topped marine boundary layer. Part II. Discussion of model behavior and comparison with other models. *J. Atmos. Sci.,* 38, 2230–2242.

STULL, R. B., 1973. Inversion-rise model based on penetrative convection. *J. Atmos. Sci.,* 30, 1092–1099.

STULL, R. B., 1984. Transilient turbulence theory. Part 1. The concept of eddy-mixing across finite distances. *J. Atmos. Sci.,* 41, 3351–3367.

STULL, R. B., AND E. B. KRAUS, 1987. Transilient model of the upper ocean. *J. Geoph. Res.,* 92, 10745–10755.

STULL, R. B., AND A. G. M. DRIEDONKS, 1987. Application of the transilient turbulence parametrization to atmospheric boundary-layer simulations. *Boundary-Layer Meteorol.,* 40, 209–239.

STULL, R. B., 1988. *An Introduction to Boundary Layer Meteorology*. Kluwer Academic Publishers, Norwell, Mass.

SUN, W. Y., AND C. Z. CHANG, 1986a. Diffusion model for a convective layer. Part 1. Numerical simulation of convective boundary layer. *J. Climate Appl. Meteorol.,* 25, 1445–1453.

SUN, W. Y., AND C. Z. CHANG, 1986b. Diffusion model for a convective layer. Part 2. Plume released from a continuous point source. *J. Climate Appl. Meteorol.,* 25, 1454–1463.

SUNDARARAJAN, A., 1975. Significance of the neutral height scale for the convective barotropic planetary boundary layer. *J. Atmos. Sci.,* 32, 2285–2287.

SUTTON, O. G., 1953. *Micrometeorology.* McGraw-Hill, New York.

SYKES, R. I., W. S. LEWELLEN, AND S. F. PARKER, 1986. A Gaussian plume model of atmospheric dispersion based on second-order closure. *J. Climate Appl. Meteorol.,* 25, 322–331.

TAKEUCHI, K., 1961. On the structure of the turbulent field in the surface boundary layer. *J. Meteorol. Soc. Jpn.,* Set II, 39(6), 346–364.

TATARSKII, V. I., 1961. *Wave Propagation in a Turbulent Medium.* McGraw-Hill, New York.

TAYLOR, G. I., 1921. Diffusion by continuous movements. *Proc. London Math. Soc.,* 20, 196–211.

TAYLOR, G. I., 1938. The spectrum of turbulence. *Proc. R. Soc. London.* Ser. A164, 476–490.

TELFORD, J. W., AND J. A. BUSINGER, 1986. Comments on "Von Karman's constant in atmospheric boundary layer flow: reevaluated." *J. Atmos. Sci.,* 43, 2127–2130.

TELFORD, J. W., AND J. WARNER, 1964. Fluxes of heat and vapor in the lower atmosphere derived from aircraft observations. *J. Atmos. Sci.,* 21, 539–548.

TENNEKES, H., 1973. A model for the dynamics of the inversion above a convective boundary layer. *J. Atmos. Sci.,* 30, 558–567.

TENNEKES, H., 1975. Reply to comments by S. S. Zilitinkevich. *J. Atmos. Sci.,* 32, 992–995.

TENNEKES, H., AND J. L. LUMLEY, 1972. *A First Course in Turbulence.* MIT Press, Cambridge, Mass.

THERRY, G., AND P. LACARRÉRE, 1983. Improving the eddy kinetic energy model for planetary boundary layer description. *Boundary-Layer Meteorol.,* 25, 63–88.

THORPE, S. A., 1979. Instability and waves. *Weather,* 34, 102–105.

TILLMAN, J. E., 1972. The indirect determination of stability, heat and momentum fluxes in the atmospheric boundary layer from simple scalar variables during dry unstable conditions. *J. Appl. Meteorol.,* 11(5), 783–792.

WEBB, E. K., 1970. Profile relationships: the log-linear range and extension to strong stability. *Q. J. R. Meteorol. Soc.,* 96, 67–90.

WEIL, J. C., 1983. Application of advances in planetary boundary layer understanding to diffusion modeling. *Proceedings of 6th Symposium on Turbulence and Diffusion,* Boston, American Meteorological Society, pp. 42–46.

WEIL, J. C., 1985. Updating applied diffusion models. *J. Climate Appl. Meteorol.,* 24, 1111–1130.

WENDLER, G., AND Y. KODAMA, 1984. On climate of Dome C, Antarctica in relation geographical setting. *J. Climatol.,* 4, 495–508.

WESELY, M. H., 1976. Magnitudes of turbulent fluctuations in the atmospheric surface layer. *Proceedings 1974 Symposium on Atmospheric Diffusion and Air Pollution,* Santa Barbara, American Meteorological Society, Boston, 15–18.

WETZEL, P., 1982. Toward parametrization of the stable boundary layer. *J. Appl. Meteorol.,* 21, 7–13.

WEXLER, A., 1965. *Humidity and Moisture,* Vol. 1. Reinhold, New York.

WILCZAK, J. M., AND J. A. BUSINGER, 1983. Thermally indirect motions in the convective atmospheric boundary layer. *J. Atmos. Sci.,* 40, 343–358.

WILLIS, G. E., AND J. W. DEARDORFF, 1974. A laboratory model of the unstable planetary boundary layer. *J. Atmos. Sci.,* 31, 1297–1307.

WILLIS, G. E., AND J. W. DEARDORFF, 1976. A laboratory study of dispersion from an elevated source in a convective mixed layer. *Atmos. Environ.,* 12, 1305–1313.

WILLIS, G. E., AND J. W. DEARDORFF, 1978. A laboratory study of diffusion into a convective planetary boundary layer. *Q. J. R. Meteorol. Soc.,* 105, 109–117.

WILLIS, G. E., AND J. W. DEARDORFF, 1981. A laboratory study of dispersion from a source in the middle of the convective mixed layer. *Atmos. Environ.,* 12, 1305–1313.

WIPPERMANN, F., 1972a. Empirical formulae for the universal functions M and N in the resistance law for a barotropic and diabatic planetary boundary layer. *Beitr. Phys. Atmos.,* 45, 305–311.

WIPPERMANN, F., AND D. YORDANOV, 1972. A note on the Rossby similarity for flows of barotropic planetary boundary layers. *Beitr. Phys. Atmos.,* 45, 66–71.

WIPPERMANN, F., D. ELTING, AND H. LEYKAUF, 1973. Effects of non-stationarity on the planetary boundary layer. *Beitr. Phys. Atmos.,* 46, 34–56.

WYATT, L. R., 1978. The entrainment interface in a stratified fluid. *J. Fluid Mech.,* 86, 293–311.

WYNGAARD, J. C., 1973. On surface layer turbulence. In: *Workshop on Micrometeorology,* D. A. Haugen (ed.), American Meteorological Society, Boston, pp. 101–148.

WYNGAARD, J. C., 1975. Modeling the planetary boundary layer-extension to the stable case. *Boundary-Layer Meteorol.,* 9, 441–460.

WYNGAARD, J. C., 1981. Cup, propeller, vane and sonic anemometer in turbulence research. *Ann. Rev. Fluid Mech.,* 13, 399–423.

WYNGAARD, J. C., 1982. *Boundary-Layer Modelling. Lecture Notes on Atmospheric Turbulence and Air Pollution Modelling.* D. Reidel, Dordrecht, The Netherlands, Chap. 3.

WYNGAARD, J. C., 1983. Lectures on the planetary boundary layer. In: *Mesoscale Meteorology—Theories, Observations and Models,* D. K. Lilly and T. Gal-Chen (eds.), D. Reidel, Dordrecht, The Netherlands.

WYNGAARD, J. C. (ED.), 1984. Large-eddy simulation: guidelines for its application to planetary boundary layer research. Final report from The Working Group on Large Eddy Simulation. The U.S. Army Research Office. *Rep. No. 84101-GS-Z.*

WYNGAARD, J. C., AND R. A. BROST. 1984. Top-down and bottom-up diffusion in the convective boundary layer. *J. Atmos. Sci.,* 41, 102–112.

WYNGAARD, J. C., AND O. R. COTÉ, 1971. The budgets of turbulent kinetic energy and temperature variance in the atmospheric surface layer. *J. Atmos. Sci.,* 28, 190–201.

WYNGAARD, J. C., AND O. R. COTÉ, 1972. Cospectral similarity in the atmospheric surface layer. *Q. J. R. Meteorol. Soc.,* 98, 590–603.

WYNGAARD, J. C., AND O. R. COTÉ, 1974. The evolution of the convective planetary boundary layer—a higher-order-closure model study. *Boundary-Layer Meteorol.,* 7, 284–308.

WYNGAARD, J. C., AND M. A. LEMONE, 1980. Behavior of the refractive index structure parameter in the entraining convective boundary layer. *J. Atmos. Sci.,* 37, 1573–1585.

WYNGAARD, J. C., O. R. COTÉ, AND Y. IZUMI, 1971. Local free convection, similarity and budgets of shear stress and heat flux. *J. Atmos. Sci.,* 28, 1171–1182.

WYNGAARD, J. C., O. R. COTÉ, AND K. S. RAO, 1974a. Modelling the atmospheric boundary layer. *Adv. Geophys.,* 18A, 193–211.

WYNGAARD, J. C., S. P. S. ARYA, AND O. R. COTÉ, 1974b. Some aspects of the structure of convective planetary boundary layers. *J. Atmos. Sci.,* 31, 747–754.

WYNGAARD, J. C., W. T. PENNELL, D. H. LENSCHOW, AND M. A. LEMONE, 1978. The temperature-humidity covariance budget in the convective boundary layer. *J. Atmos. Sci.,* 35(1), 47–58.

YAMADA, T., 1976. On the similarity functions *A*, *B* and *C* of the planetary boundary layer. *J. Atmos. Sci.,* 33, 781–793.

YAMADA, T., AND G. L. MELLOR, 1975. A simulation of the Wangara atmospheric boundary layer data. *J. Atmos. Sci.,* 32, 2309–2329.

YAMAMOTO, S., M. GAMO, AND O. YOKOYAMA, 1977. Airborne measurements of turbulent heat flux. *J. Meteorol. Soc. Jpn.,* 55, 533–545.

YOKOYAMA, O., M. GAMO, AND S. YAMAMOTO, 1979. The vertical profiles of the turbulence quantities in the atmospheric boundary layer. *J. Meteorol. Soc. Jpn.,* 57(3), 264–272.

YORDANOV, D., 1980. A note on the Rossby similarity of a nonstationary atmospheric boundary layer. *Beitr. Phys. Atmos.,* 53, 167–171.

YORDANOV, D., AND F. WIPPERMANN, 1972. The parametrization of turbulent fluxes of momentum, heat and moisture at the ground in a baroclinic planetary boundary layer. *Beitr. Phys. Atmos.,* 45, 58–65.

ZEMAN, O., 1975. The dynamics of entrainment in planetary boundary layers. A study in turbulence modeling and parametrization. Ph.D. thesis, Pennsylvania State University.

ZEMAN, O., 1981. Progress in the modelling of planetary boundary layers. *Ann. Rev. Fluid Mech.,* 13, 253–272.

ZEMAN, D., AND J. L. LUMLEY, 1976. Modeling buoyancy-driven mixed layers. *J. Atmos. Sci.,* 33, 1974–1983.

ZEMAN, D., AND H. TENNEKES, 1977. Parametrization of the turbulent energy budget at the top of the daytime atmospheric boundary layer. *J. Atmos. Sci.,* 34, 111–123.

ZHANG, S. F., S. P. ONCLEY, AND J. A. BUSINGER, 1988. A critical evaluation of the von Karman constant from a new atmospheric surface layer experiment. *Proceedings 8th Symposium on Turbulence and Diffusion*, San Diego, 148–150.

ZHOU, M. Y., D. H. LENSCHOW, AND B. B. STANKOV, 1985. Wave and turbulence structure in a shallow baroclinic convective boundary layer and overlying inversion. *J. Atmos. Sci.,* 42, 47–57.

ZILITINKEVICH, S. S., 1970. *Dynamics of the Atmospheric Boundary Layer.* Gidrometeorizdat, Leningrad (in Russian).

ZILITINKEVICH, S. S., 1972. On the determination of the height of the Ekman boundary layer. *Boundary-Layer Meteorol.,* 3, 141–145.

ZILITINKEVICH, S. S., 1975a. Resistance laws and prediction equations for the depth of the planetary boundary layer. *J. Atmos. Sci.,* 32, 741–753.

ZILITINKEVICH, S. S., 1975b. Comments on "A model for the dynamics of the inversion above a convective boundary layer." *J. Atmos. Sci.,* 32, 991–992.

ZILITINKEVICH, S. S., AND D. V. CHALIKOV, 1968a. The laws of resistance and of heat and moisture exchange in the interaction between the atmosphere and an underlying surface. *Izv. Acad. Nauk SSSR Fiz. Atmos. Okeana (Bull. Acad. Sci. USSR Atmos. Oceanic Phys.),* 4, 438–441.

ZILITINKEVICH, S. S., AND D. V. CHALIKOV, 1968b. On the determination of the universal wind and temperature profiles in the surface layer of the atmosphere. *Izv. Akad. Nauk SSSR Fiz. Atmos. Okeana (Bull. Acad. Sci. USSR Atmos. Oceanic Phys.),* 4, 294–302.

ZILITINKEVICH, S. S., AND D. L. LAIKHTMAN, 1965. On closure of a system of equations for turbulent flow in the atmospheric boundary layer. *Tr. GGO,* 167, 44–48 (in Russian).

INDEX